Snow Ecology

Snow Ecology is the first book to integrate the study of snow and ice in the physical, chemical, and biological sciences into a multidisciplinary overview of life in, on, and under snow.

The book opens up a new perspective on snow cover as a habitat for organisms under extreme environmental conditions and as a key factor in the ecology of much of the Earth's surface. Acknowledged experts in the disciplines that constitute snow science provide an understanding of the interrelationships between snow structure and life. The authors describe the fundamental physical and chemical processes that characterize the evolution of snow and the feedback mechanisms between snow cover and the life cycles of true snow organisms and biota of cold regions. The book also provides a study of the relationship between snow and climate and the paleoecological evidence for the influence of past snow regimes on plant communities.

Snow Ecology will be used as a main textbook on advanced courses in biology, ecology, geography, environmental science, and earth science where an important component is devoted to the study of the cryosphere. It will also be useful as a reference text for graduate students, researchers, and professionals at academic institutions and in government and nongovernmental agencies with environmental concerns.

H. G. Jones is a professor at the Institut national de la recherche scientifique, Université du Québec, Ste-Foy, Québec, Canada.

J. W. Pomeroy is a senior lecturer at the Centre for Glaciology, Institute of Geography and Earth Sciences, University of Wales, Aberystwyth, Wales, UK.

D. A. Walker is a professor at the Institute of Arctic Biology and Department of Biology and Wildlife, University of Alaska, Fairbanks, Alaska, USA.

R. W. Hoham is a professor in the Department of Biology, Colgate University, Hamilton, New York, USA.

Snow Ecology

An Interdisciplinary Examination of Snow-Covered Ecosystems

Edited by

H. G. Jones
J. W. Pomeroy
D. A. Walker
R. W. Hoham

CAMBRIDGE
UNIVERSITY PRESS

CAMBRIDGE UNIVERSITY PRESS
Cambridge, New York, Melbourne, Madrid, Cape Town,
Singapore, São Paulo, Delhi, Tokyo, Mexico City

Cambridge University Press
The Edinburgh Building, Cambridge CB2 8RU, UK

Published in the United States of America by Cambridge University Press, New York

www.cambridge.org
Information on this title: www.cambridge.org/9780521188890

First published 2001
First paperback edition 2011

A catalogue record for this publication is available from the British Library

Library of Congress Cataloguing in Publication data

Snow ecology : an interdisciplinary examination of snow-covered ecosystems / edited by
  H.G.Jones . . . [et al.].
      p.   cm.
    Based on papers from a conference held in Quèbec City, Canada in 1993.
    Includes bibliographical references.
    ISBN 0-521-58483-3
    1. Snow ecology – Congresses.   1. Jones, H. G. (H. Gerald), 1936-

  QH541.5.S57 S66   2000
  577.5′86 – dc21   00-023665

ISBN 978-0-521-58483-8 Hardback
ISBN 978-0-521-18889-0 Paperback

Additional resources for this publication at www.cambridge.org/9780521188890

The editors and authors of this work would like to acknowledge and thank all the researchers in snow ecosystems who have gone before us to pave the way for this book. One who will be particularly missed is Dr. Dwight Billings, who passed away after coauthoring a chapter for this book. He will be fondly remembered for his pioneering contributions to alpine snow ecology in the United States and his contributions to this book.

Contents

List of Authors and Affiliations

Dr C. W. Aitchison[1]
Department of Zoology
University of Manitoba
Winnipeg, Manitoba
R3T 2N2 Canada

Dr Yves Bégin
Centre d'études nordiques
Département de Géographie
Université Laval
Ste-Foy, Québec
G1K 7P4 Canada

Dr W. D. Billings[2]
Department of Botany
Duke University
Durham, North Corolina
27706 USA

Dr Simon Boivin
Centre d'études nordiques
Département de Géographie
Université Laval
Ste-Foy, Québec
G1K 7P4 Canada

Dr Eric Brun
Centre d'études da la neige
Météo France
1441 rue de la Piscine
Domaine Universitaire
St-Martin d'Hères
38406 France

Dr Trevor D. Davies
School of Environmental Sciences
University of East Anglia
Norwich
NR4 7TJ UK

Brian Duval
Department of Microbiology
University of Massachusetts
Amherst, Massachusetts
01003 USA

Dr Pavel Ya. Groisman[3]
State Hydrological Institute
St Petersburg
Russia

Dr R. W. Hoham
Department of Biology
Colgate University
Hamilton, New York
13346 USA

Dr H. Gerald Jones
INRS-Eau, Université du Québec
2700 Einstein, CP7500
Ste-Foy, Québec
G1V 4C7 Canada

Dr J. D. de Molenaar
Dienst Landbouwkundig
Onderzoek
Instituut voor Bosbouw en
Broenbeheer (IBG-LDO)
Post bus 23, 6700 AA
Wageningen
The Netherlands

Dr John W. Pomeroy[4]
National Water Research Institute
11 Innovation Boulevard
Saskatoon, Saskatchewan
SN7 3H5 Canada

Dr Martyn Tranter
School of Geographical Sciences
University of Bristol
Bristol
BS8 1SS UK

Dr D. A. Walker[5]
Institute of Arctic and Alpine
Research
University of Colorado
Boulder, Colorado
80309 USA

Dr Roman I. Zlotin[6]
Department of Biogeography
Institute of Geography
Russian Academy of Sciences
Moscow
Russia

[1]Current address:
Dr C. W. Williams
4664 Spurraway Road
Kamloops, British Columbia
V2H 1M7 Canada

[2]deceased.

[3]Current address:
National Environmental Satellite
Data and Information Service
National Climatic Center
Federal Building
151 Patton Avenue
Ashville, North Corolina
28801 USA

[4] Current address:
Centre for Glaciology
Institute of Geography and Earth
Sciences
University of Wales, Aberystwyth
Aberystwyth, Wales
SY23 3DB UK

[5] Current address:
Institute of Arctic Biology
University of Alaska,
Fairbanks
P.O. Box 757000
Fairbanks, Alaska
99775-7000 USA

[6] Current address:
Indiana School of Public
and Environmental
Affairs
Indiana University
Bloomington, Indiana
47405 USA

Preface

Snow ecology is the science of the relationships between organisms and their environment, whether it be in snow cover or as part of a snow ecosystem. A snow ecosystem may be envisioned as a set of interacting, interdependent biotic and abiotic subsystems, in which one of the abiotic subsystems is a snow cover. Although snow has long been the subject of scientific investigation, the vast majority of the studies have been restricted to the scope of various disciplines in the physical and biological sciences. As a result, very few works permit a full appreciation of snow cover and/or snow-covered regions as functional snow ecosystems. However, as snow scientists became more involved in research that required an interdisciplinary approach, and more familiar with the results of research on snow in disciplines other than their own, the need to facilitate the exchange of information and understanding of basic snow sciences across a wide range of snow-related research became apparent. In 1991 the International Commission of Snow and Ice of the International Association of Hydrological Sciences formed a Snow Ecology Working Group (Jones et al., 1994), which drew together snow scientists from many disciplines. The goal of the working group was to compile basic information that could lead to the study of snow from an ecosystem perspective. At a meeting in Quebec City, Canada, in 1993, the working group reviewed the state of snow ecology and embarked on the publication of a volume that would achieve this objective. The result is this book.

The purpose of the book is to introduce some of the basic scientific principles that enable us to understand the functioning of snow ecosystems. The approach is to examine the physical, chemical, and biological processes that influence the evolution of snow and/or depend on the dynamics of the snow cover itself. The subject matter is mainly directed to the study of seasonal snow cover, although much of the material is applicable to permanent snow and ice fields. Seasonal snow cover is a dynamic habitat of limited duration. Its dynamics and duration are controlled by physical metamorphism, phase changes, and chemical transformations, which themselves are driven by interactions with atmospheric and soil systems and plant and animal communities. In particular, its influence in moderating winter meteorological conditions within snow cover permits the activity of particular communities to continue throughout the winter season. These subnivean communities themselves sustain populations that live above snow or visit the snow, and so, as winter inevitably must give way to spring and summer, snow cover provides an essential link in the interannual cycles of biological and hydrogeochemical production of high latitude and altitude areas of the Earth.

The book consists of chapters that outline the basic physical, chemical, and biological principles and processes that are critical to snow ecology. The chapter sequence and the bridging between the individual chapters allow the reader to read the work as a whole or select certain chapters of interest. In the opening chapter of the book, Groisman and Davies describe the relationship between snow cover and climate and illustrate some of the global-scale feedback mechanisms that cause far-ranging weather patterns to change from year to year. Examples such as the influence of the snow cover in western Eurasia and the Tibetan Plateau on the Indian summer monsoon and the relationship between the El Niño southern oscillation and the extent of snow covers in North America and Eurasia lead Groisman and Davies to argue that seasonal snow covers have a bearing on the dynamics of *all* the Earth's ecosystems. The understanding of snow cover and climate is thus an important foundation in foreseeing the implications of climate change for the biosphere.

But how will changes in the accumulation, duration, extent, and internal structure of seasonal snow cover affect the physical ability of snow to support and influence life? These questions are pertinent, as the second chapter, by Pomeroy and Brun, on snow physics demonstrates. Life can continue in and under snow because of the unique physical texture of the milieu. In the subnivean world, organisms and soil rely on the insulating capacity of snow for heat retention so that extreme thermal fluctuations in the atmosphere are dampened at the soil surface. The insulating capacity itself is controlled by snow depth and density, which can be enhanced by the ability of exposed plants to reduce near-ground wind speeds and hence trap blowing and falling snow. Conversely, evergreen forests may significantly reduce snow accumulation through interception of snow in their canopies. However, snow mediates not only heat but also light between the atmosphere and the ground and many microorganisms and plants have adapted to light levels in snow that are optimal for photosynthesis. The chapter also considers the role of snow in the energy and water balance of forests, tundra, and steppes. The structure and heat exchange of snow with the atmosphere, above-snow vegetation, and the soil are key elements in the timing of runoff and soil moisture replenishment in spring. This is the time when many ecosystems receive most of their annual water input and when the melting snowpack itself possesses a flourishing microbiological and invertebrate community.

xvi

Snow cover, however, is not only a hydrological reservoir but also a source of nutrients and chemicals, on which the true nival and subnivean organisms rely or are otherwise influenced during growth and reproduction. In the chapter on the chemistry of snow, Tranter and Jones present the processes of chemical acquisition of snow during formation in the atmosphere. The chapter also outlines the transformation of nutrients within the snow cover, particularly during melt and discharge of meltwater to soil and streams. Particular attention is payed to the dynamics of nitrogen species in snow and during snowmelt, as nitrogen is the limiting nutrient in many ecosystems, but along with other snow chemicals has reached toxic levels in certain developed regions of the Earth. In snow ecosystems, the snow cover, as recipient of snowfall, winter rain, and dry deposition, is an important medium of transfer for nitrogen species to the soil from the atmosphere. Although in the short term most nutrient transfer in soil and vegetation is the result of internal cycling, the atmosphere is ultimately the source for nitrogen. In snow-covered soils, nutrient cycling continues throughout the winter, and, in particular, gives rise to gaseous emissions (e.g., CO_2, N_2O) at the snow–soil interface. Gaseous emissions under snow may represent a significant part of the annual flux of carbon fixed by photosynthesis.

The rest of the book chapters are devoted to the biology of snow and its environment. In the chapter on the microbial ecology of ice and snow, Hoham and Duval present an overview of microbial populations in snow cover and discuss some of the origins, distribution, physiology, and evolutionary aspects of these organisms. Particular attention is paid to the true snow algal populations that grow and reproduce wholly within the water retained by melting snow. The algae possess structural and reproductive adaptations that permit them to successfully complete these essential phases of their life cycle during the relatively short melt season. The authors discuss the relationship between snow chemistry and algal distribution and the depletion of snow-nutrient content during growth. In the case of forest snow covers, canopy fallout onto the snow provides growth factors and soluble phosphorus species for stimulation of algal growth and increased assimilation of nutrients. Other organisms such as bacteria, yeasts, and snow fungi coexist with the algae and make up an important component of the true snow microbial population.

The microbial populations are preyed upon by some invertebrates, which can be considered the basis of an episodic nival food web during the snowmelt period. Small mammals that are active beneath and on the snow can feed on the invertebrates, thus extending the food chain into colder periods. In the chapter on the effect of snow on small animals, Aitchison reviews the techniques to determine the activity and populations of the principal winter-active invertebrate fauna, including spiders, mites,

xvii

springtails, beetles, flies, and other insects. The chapter presents the global distribution of these small animals and the physiological mechanisms and morphological adaptations that allow them to tolerate low temperatures and find food. The section on small vertebrates concentrates on the study of small mammals, most of which weigh less than 250 grams. Particular emphasis is placed on the ecology of shrews that feed on invertebrates, and on voles, which are herbivorous. Physiological mechanisms to combat heat loss in these small animals include inactivation of certain glands and specialized mechanisms for heat production. Small mammals can also undergo seasonal morphological change to lessen heat loss or to increase their mobility in snow. The section also references techniques to trap these small animals for population estimates.

Vegetation in cold regions is closely related to the snow cover regimes. In the chapter on snow-vegetation interactions in arctic and alpine environments, Walker, Billings, and de Molenaar review the snow-related ecological factors that influence the physiology and distribution of plants in arctic and alpine regions. The snow-related patterns of plant species and plant communities are examined at different scales, with discussion on ecosystem characteristics such as snow patch vegetation, soil properties, and decomposition rates. Physiological adaptation of plants to snow cover includes the ability to synthesize chlorophyll at low levels of light under snow, storage of photosynthate below ground or in the leaves, and very rapid flowering during or immediately after snowmelt. The changes in vegetation due to altered snow accumulation regimes are described by a review of snow-fence experiments in which the depth of snow is increased by trapping blowing snow. The possibility of extrapolating the changes in snow regimes to changes in landscape patterns and the linking of ecosystem models to maps of vegetation through the use of remote sensing images and geographic information systems are also discussed.

The final chapter in the book, on tree-ring dating by Bégin and Boivin, introduces the reader to dendochronology and its use in determining past snow regimes. In the Subarctic, trees and snow are intimately related. The distribution of trees is patchy and considerably influenced by the accumulation of wind-blown snow and duration of snow cover, which, in turn, is subjected to control by the trees themselves. By an analysis of tree morphology and the dating of tree rings one can reconstruct snow depths of the past. The precise method is to focus on events such as the development of new stems, stem scars, and abrupt changes in growth, which can be dated from the tree rings. Trees that are completely protected by snow cover do not show any of the characteristic snow-related physical injuries due to desiccation and abrasion by wind-blown snow that they are subjected to once they are exposed above the snow cover. The effect of heavy snow accumulation on trees can also lead to deformation as a result of branch bending, tearing, and breakage. The chapter contains examples that show these types of injury and the date of the injury from the tree-ring analysis.

This brief introduction to the book describes some of the chapter content. Snow ecology is a broad and evolving field, and not all developing facets have been treated in this work. The book focuses on the snow environment, ecosystem interactions, and subnivean organisms. Those who seek additional material on the life of large mammals in the cold can find many references in the current literature (e.g., wolves [Huggard, 1993], muskoxen [Nellemenn, 1997], caribou [Ouellet, Heard, and Boutin, 1993], goshawks [Tornberg, 1997]). We also recommend the seminal work on the ecology of mammals and birds by Formazov (1946).

Finally, we acknowledge that humanity is now playing a major role in changing the atmosphere, the climate, and the ecology of cold regions; the role and direct impacts of mankind are, however, not within the scope of this book. The relationship between mankind and snow and ice throughout history is ambivalent and has often been epitomized as a struggle for survival in the adverse circumstances of cold, storms, isolation, starvation, and deep snow. On the other hand, snow and ice have fascinated mankind, intrigued the curious mind, and lent substance to artistic expression and sporting achievements. Through the method of science, we have reconstructed the past and found that the Earth has endured long ages of ice and snow when vast areas of the continents were covered by glaciers and ice caps for hundreds of thousands of years. In understanding the sparse energy environment of snow ecosystems, we may be able to anticipate conditions for life on other worlds. Technology now facilitates human existence, agriculture, and industry in cold regions, but the impact of such activity on snow ecosystems is considerable. We need improved knowledge about how snow ecosystems function so that the technological and natural communities can coexist while maintaining the diversity of life in such harsh environments. This book sets out to take one on a journey into that familiar yet exotic world of snow and ice and life in the cold; we wish the reader well.

H.G. Jones, J.W. Pomeroy, D.A. Walker, R.W. Hoham, and R. Zlotin.

References

Formazov, A.N., 1946. Snow cover as an integral factor of the environment and its importance in the ecology of mammals and birds. *Moscow Society of Naturalists, Materials for Fauna and Flora U.S.S.R. Zoology. Section, New Series.* 5, 1–152. [English translation:

Prychodko, W., and Pruitt, W.O., 1963. Occasional Paper 1, Boreal Inst. Univ. Alberta, Edmonton.]

Huggard, D.J., 1993. The effect of snow depth on predation and scavenging by Gray Wolves. *J. Wildl. Manage.*, 57, 382–88.

Jones, H.G., Pomeroy, J.W., Walker, D.A., and Hoham, R.W., 1994. Interdisciplinary group investigates snow ecology. *Eos Trans. Amer. Geophys. Union*, 75(14), 162–63.

Nellemenn, C., 1997. Grazing strategies of muskoxen (*Ovibus moschatus*) during winter in Angujaartorfiup Nunaa in western Greenland. *Can. J. Zool.*, 75, 1129–34.

Ouellet, J.P., Heard, D.C., and Boutin, S., 1993. Range impacts following the introduction of Caribou on Southampton Island, Northwest Territories, Canada. *Arctic Alpine Res.*, 25, 136–41.

Tornberg, R., 1997. Prey selection of the goshawk *Accipter gentilis* during the breeding season: the role of prey profitability and vulnerability. *Ornis Fennica*, 74, 15–28.

Acknowledgments

The editors express their gratitude to the reviewers of the book chapters; their comments and suggestions were extremely helpful and very much appreciated.

We would also like to thank the institutions that contributed financially to the original workshop, which gave rise to the book, and/or to the subsequent meetings of the editorial board. They are:

Canadian Polar Commission (Ottawa, Canada).

Natural Sciences and Engineering Research Council of Canada (Government of Canada, Ottawa, Canada).

Hydro-Québec (Montréal, Québec, Canada).

l'Institut national de la recherche scientifique (Université du Québec, Québec, Canada).

Ministère de l'Environnement et Faune (Gouvernement du Québec, Québec, Canada).

Environment Canada (National Water Research Institute, National Hydrology Research Centre, Saskatoon, Canada).

Department of Canadian Heritage (Prince Albert National Park, Waskesiu Lake, Saskatchewan, Canada).

1 Snow Cover and the Climate System

PAVEL YA. GROISMAN AND TREVOR D. DAVIES

1.1 Introduction

At the end of each winter, large areas in high latitudes and altitudes of both hemispheres are covered by snow. In a few months most of it disappears, only to start building up again in the late autumn (Gutzler and Rosen, 1992; Robinson, Dewey, and Heim, 1993; Groisman et al., 1994a). In many regions, particularly in the Southern Hemisphere, this seasonal expansion of snow cover is restricted by the limits of land masses. In the polar regions, however, the seasonal expansion of sea ice over the ocean allows snow cover to expand, significantly changing the properties of the sea ice (Ledley, 1991). The depth of snow cover varies widely, and in some mountainous regions (e.g., Alps, Rocky Mountains) it can exceed 3 m. The first snow is light and white with a density of around 100 kg m^{-3}. At the end of winter, most of the snow on the ground is much denser (up to 500 kg m^{-3}) and dirtier, and so it loses its whiteness (see Pomeroy and Brun, Chapter 2). The first autumn snowfall, when the daytime temperature remains below freezing, can start the formation of seasonal snow cover, but later advection of warm air and/or solar radiation may melt it. A single weather event can extend continental snowlines equatorward by up to 1,000 km (Lamb, 1955; Cohen and Rind, 1991), although it usually takes a few weeks between the first snowfall and formation of the stable snow cover in high latitudes (e.g., Russia). In lower latitudes (e.g., China, United States) snow cover is more ephemeral and can melt and regrow several times during the winter. After being established, however, feedback processes tend to support the existence of the snow cover, mostly because of the high reflectivity (albedo) that reduces the surface radiation balance and thus the energy available for snowmelt.

Because it covers more than half the Northern Hemispheric land area each year, and it has such varying properties, seasonal snow cover is an important ecological factor in many extratropical regions. Snow cover is a radiative sink. The high shortwave albedo (reflectivity), which, combined with its high thermal emissivity, increases the amount of infrared radiation lost near the earth's surface (Male and Gray, 1981).

1

The radiative losses are not quickly replaced by heat fluxes from below because of the insulating properties of the snow. These insulating properties are particularly effective at night. For example, the temperature of the upper surface of a 10-cm snow cover may drop by more than 10°C overnight but the temperature of the underlying soil surface may drop by less than 1°C. Snow on the ground, therefore, keeps the near-surface soil relatively warm, preventing it from deep freezing and protecting the root systems of plants from damage. The presence of a sufficiently deep snow cover is vital in many areas to maintain sufficiently warm subnivean temperatures for the winter grain crop. In a similar fashion, snow cover provides a relatively favorable winter habitat for small mammals (e.g., field mice) that are unable to migrate far from their summer homes.

For some species, snowmelt represents the most important environmental perturbation, or stimulus, in the whole year (Hoham and Duval, Chapter 4). Changes in the timing and character of snowmelt (Davies and Vavrus, 1991) will therefore affect biospheric systems within the sphere of influence of the snowmelt regime.

Seasonal snow cover has important effects on ecosystems that are remote from the snow cover itself. For example, the Indian summer monsoon is influenced by the previous season's snow cover in western Eurasia and the Tibetan Plateau (Hahn and Shukla, 1976; Verma, Subramaniam, and Dugam, 1985; Barnett et al., 1989). Indeed, seasonal snow cover is such an important part of the global climate system that it can be argued that *all* ecosystems – around the globe – are indirectly affected by snow cover because of its role as a component of the climate system.

Whether the focus of ecological interest is directed toward populations that are intimately associated with snow cover – as is the case with this volume – or is widened to those ecosystems with a more remote dependence, an understanding of snow cover climatology and its links with the atmosphere are important foundations. This importance is heightened by the prospect of a large-scale climate change. Satellite observations of Northern Hemisphere snow cover are available from 1966 to present (Dewey and Heim, 1981, 1982; Robinson et al., 1993). These records show a strong inverse link between Northern Hemisphere snow cover and temperature (Figure 1.1). The same conclusions can be drawn from the analysis of paleoclimatic data and sensitivity experiments with global climate models (Budyko and Izrael, 1991; Mokhov, 1984; Cohen, 1994). The prospect of a continuing enhanced greenhouse gas effect (Intergovernmental Panel on Climate Change [IPCC], 1996) has clear implications for snow cover ecosystems. The timing, depth, internal structure, and extent of seasonal snow cover will change, affecting these ecosystems. Because of the spatial scale of the two-way interactions between snow cover and the atmosphere, a changing climate, coupled with a

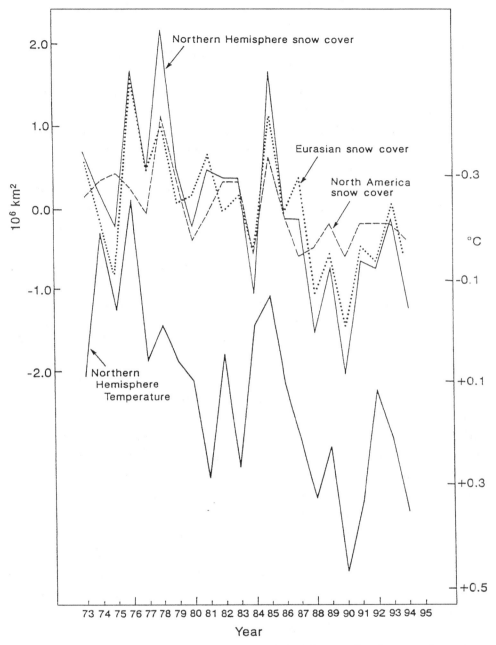

Figure 1.1. Annual snow cover variations derived from satellite shortwave observations (anomalies from the 1973–1994 means) for Northern Hemisphere, North America, Eurasia (left-hand scale), together with Northern Hemisphere combined land and sea temperature (anomalies from 1951–1981 mean) (right-hand scale; inverted). (Matson, 1986; Ropelewski, 1993, updated; courtesy of D. Garrett, National Oceanic and Atmospheric Administration, National Weather Service, Washington, DC; Jones, 1994.)

3

changing snow cover, has ecological implications beyond the geographical confines of snow cover.

In addition to being an important forcing factor in atmospheric change, the two-way interaction means that snow cover is a useful detector and monitor of this change (Barry, 1985; Schlesinger, 1986). Changes in snow cover extent, amount, and density, when reliably detected, will also indicate a change in the global hydrological cycle and ecosystems over the entire extratropical land area (Watson, Zinyowera, and Moss, 1996). Consequently, in this chapter, in addition to considering the nature of snow cover–atmosphere interactions, we also describe snow cover observations.

1.2   Snow Cover Observations and Data Sources

The importance of snow cover information for various human activities was a reason for early establishment of snow observation systems. In the first millennium, snow cages installed upstream in the mountains were used for flood forecasting in China (Biswas, 1970). Currently, several northern countries have accumulated a century-long time series of in situ snow cover observations (Jackson, 1978; Pfister, 1985; Mestcherskaya, Belyankina, and Golod, 1995; Brown and Goodison, 1996; Hughes and Robinson, 1996; Easterling et al., 1997). Initially, in the first generation of observations, snow on the ground was characterised only by snow depth measurements. These measurements were made by snow sticks. In some countries (e.g., Russia) snow depth measurements were made separately at field and forest locations to account for significant differences in snow accumulation. The need to know the water equivalent of snowpack, and a requirement for a better spatial representativeness of snow monitoring, led to the establishment of snow course observations. This is when an observer surveys a transect, collecting many samples of snowpack and recording its amount and microphysical properties (type, density, presence of crust, etc.). This arduous procedure proved to be reliable for many practical reasons in agro–meteorology and hydrology and stays mostly intact in many countries. The first digital data sets of snow courses recently became available from the World Data Center for Glaciology (National Snow and Ice Data Center [NSIDC], 1996). Nineteen years ago, in the western United States and Alaska, snow courses were gradually replaced by an automated system (SNOTEL) (Rallison, 1981) that currently delivers the point snowpack water equivalent information from more than 600 remote sites in the Rockies and Cordilleras (Aguado et al., 1992; Cayan, 1996; Redmond and Schaefer, 1997).

4

The major uncertainties relating to ground-based (in situ) observations concern the initial accuracy of point or snow-course measurements and the consequent up-scaling to area averages (Sevruk, 1992). A recent discussion of the requirements for and techniques of ground-based snow observations is provided by Pomeroy and Gray (1995). In windblown areas, snow tends to accumulate in forested areas and ravines, leaving open plain areas with less snow than in the forest. Snow melts on exposed sites more quickly than in protected locations and can be blown away or sublimate during snow flurries (Vershinina and Volchenko, 1974; Pomeroy and Jones, 1996; Mestcherskaya et al., 1995). In mountainous regions, snow accumulation is extremely heterogeneous because of variable wind exposure, orographic effects, and adiabatic effects, which influence snowmelt (Pomeroy and Brun, Chapter 2). This makes the spatial generalization of snow cover information a difficult task. Moreover, it is impossible to derive regional information in relatively remote areas (e.g., northern Canada, Siberia), where in situ measurements are rare. Consequently, remote techniques provide the only feasible method of acquiring regional snow cover distributions (Karl et al., 1989). Remote techniques can be divided into airborne and satellite-borne methods. Advantages and disadvantages of the different techniques are shown in Figure 1.2.

Airborne passive observations are based on the fact that soundings of natural γ-radiation from the land surface are decreased by water (liquid or frozen) overlying the land and in the upper 20 cm of the soil. Hence, comparative soundings made before snowfall and before snowmelt will give estimates of the total amount of water equivalent of snow (SWE) on the ground, combined with upper soil water, along a flight route (Kogan, Nazarov, and Fridman, 1969; Peck, Carrol, and Vandmark, 1980). A limitation is the separation of the above- and below-ground water proportions. In spite of this and other shortcomings (cf. Jones and Carroll, 1983; Glynn et al., 1988; Carroll and Carroll, 1989), the method has proved useful for a wide range of users. Estimates of snow cover extent and SWE with a mean areal accuracy of approximately 10 mm are routinely provided in parts of North America (Carroll, 1995).

Satellite-borne observations of snow cover are made in the shortwave (visible and near-infrared) and microwave bands. The data in Figure 1.1 were derived from satellite shortwave measurements. The original data consist of digitised maps of weekly snow cover extent prepared from National Oceanic and Atmospheric Administration (NOAA) polar orbiting satellites, supplemented by GOES and METEOSAT observations (Matson and Wiesnet, 1981; Wiesnet et al., 1987; Dewey and Heim, 1981, 1982; Matson, 1986; Ropelewski, 1993; Robinson, Keimig, and Dewey, 1991; Robinson et al., 1993). Although the observations started in 1966, the series is not regarded as being

5

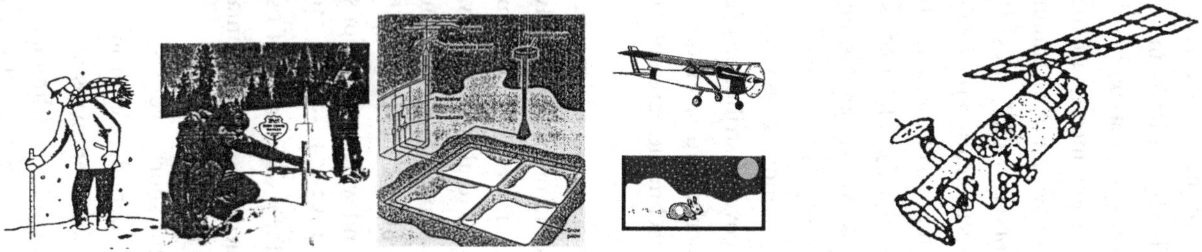

	IN SITU MEASUREMENTS			AIRBORNE	SATELLITE-BORNE MEASUREMENTS		
	Snow rulers	Snow courses	Automated stations	Gamma-radiation sounding	Visible imagery	Microwave passive sounding	Microwave active sounding
Measured snow characteristic	Snow depth at a given point	Snow depth and snow water equivalent; SWE for each landscape and average over the region	SWE at a given point	SWE along the flight track	Snow cover extent over the grid cell	Snow cover extent and SWE over the grid cell	SWE over the grid cell
Advantages	Long time series (up to 100 years)	Most reliable source of the regional SWE; relatively long time series	Can be installed in remote areas; "labor-free"	Provides reliable regional estimates of SWE	Global coverage; relatively long time series (since 1966)	Global coverage; superior spatial (25 km x 25 km) and temporal (daily) resolution	Superior spatial resolution (up to tens of meters)
Disadvantages	Often are not representative of the surrounding terrain; do not provide SWE estimates	Extremely laborious technique	Problems with spatial representativeness of the site can be developed with time; short time series	Application of the technique requires high investments; short time series	Time increment (week) does not satisfy most of the users; does not provide SWE estimates	Absence of reliable universal algorithm of the SWE evaluation; short time series (since 1987)	Absence of reliable universal algorithm of the SWE evaluation; there are no products with global coverage
Key reference	Gray & Male (1981)	Gray & Male (1981)	Rallison (1981)	Carroll (1995)	Robinson et al. (1993)	Armstrong (1993)	Matzler and Schanda (1984)

Figure 1.2. Advantages and disadvantages of different observation techniques (SWE = snow water equivalent).

homogeneous until the end of 1972 (Robinson et al., 1993). The spatial resolution of the data varies by latitude, ranging from a grid cell resolution of 20,643 km^2 at high latitudes to 42,394 km^2 at lower latitudes. The proportion of the Northern Hemisphere land area that has experienced snow cover during at least one week in the last 20 Januarys amounts to 55 percent (54.2 × 10^6 km^2). On average, 23 percent of the entire global surface (50 percent of the land) is permanently or temporarily covered by snow during the year (Sevruk, 1992).

Microwave measurements of snow cover extent and water equivalent have been made by sensors aboard geostationary satellites since 1987 (Weaver, Morris, and Barry, 1987; Barry, 1991). The scattering of microwave energy emitted by the ground is related to the amount of dry snow (number of grains). Although melting snow is indistinguishable from wet bare soil in these observations, passive microwave remote sensing does have a considerable advantage of being operable in nearly all weather conditions and in darkness. Active microwave remote sensing has the potential for very high resolution (tens of meters) snow cover observation, but the delivery of acceptable SWE observations is problematic (Matzler and Schanda, 1984; Leconte, 1993). Although snow cover extent observations are now regarded as reliable (Ferraro et al., 1994; Basist and Grody, 1994), sadly, there are continuing problems with reliable indications of SWE with these measurements.

The old "surface-based" observational network remains a viable and important component of the snow observing system that should not be replaced by the satellite network. They are, in fact, complementary. Notwithstanding current problems, satellite-based snow cover/depth observations probably will cover the future needs of many users of snow data (Armstrong, 1993; Ferraro et al., 1994). Economics and scientific objectives now require a merging of all available snow information in one superior enhanced data set. It is necessary to ensure homogeneity in the transfer from the old surface-based network to the new observational network during this merging. This transfer has not gone smoothly (Robinson et al., 1991; Armstrong, 1993; Barry et al., 1994), but the effort is still under way.

The situation with snow data availability is changing quickly and, no doubt, after this book is published new sources will have emerged. At the moment, processing of some very large data sets (e.g., raw satellite images) requires too many computer resources for many users. Nevertheless, current snow data availability is itemised as follows:

- A large global archive (more than 8,000 stations worldwide) of mean monthly precipitation (including frozen precipitation) has been accumulated in

7

the Global Historical Climatology Network data set (Vose et al., 1992, Peterson and Vose, 1997). This is available free of charge via the Carbon Dioxide Information Analysis Center, Oak Ridge, Tennessee, USA. In the former Soviet Union (FSU), each station (10,000 of them) has an adjustment set to produce unbiased long-term values of precipitation. Data from 622 Soviet stations (adjustments and monthly values) through 1999 are now available through World Data Center-A, Asheville, North Carolina, USA.

- Recently (1991–1992) a data set containing daily snow data for a selected set of first-order stations, together with specially selected snow data for the heavily wooded areas totally 284 stations of the FSU, has been received by the U.S. National Climatic Data Center (NCDC) from Russia in the framework of the USA-FSU bilateral exchange of data and was quality controlled at the NSIDC in Boulder, Colorado. These data are currently available on CD-ROM (NSIDC, 2000). A large data set of snow depth and SWE has been collected over the entire FSU at the State Hydrological Institute (St. Petersburg, Russia), Institute of Geography of the Russian Academy of Science (Moscow), and World Data Center-B (Obninsk, Russia). More than half these data (more than 1,300 stations) are currently available from NSIDC.

- A data set of daily snow depth measurements from 195 stations in the United States is available free of charge from the U.S. Carbon Dioxide Information Data Center (Easterling et al., 1997).

- A database comprising parallel observations of snowfall, its water equivalent, and snow on the ground for 335 Canadian stations has been collected at World Data Center-A (up to 1992) and is available from the Meteorological Service of Canada, Downsview, Ontario.

- A snow course data set for the western United States for the past 50 years is currently computerised (Shafer and Huddleston, 1986).

- The Historical Daily Climatic Data Set for the United States, for more than 1,000 stations with snowdepth and snowfall data before 1988, has been assembled and quality controlled (Robinson, 1988, 1993). These data are available at the NCDC and may be augmented for the last years from the "Summary of the Day," the U.S. national archive that is currently being updated and quality controlled at the Data Operations Branch of the NCDC (Groisman, Sun, and Heim, 2000).

- Alaska is virtually a snow laboratory because it contains maritime, continental, and Arctic climates in close proximity. The data from intensive snow-related studies (snow courses, passive microwave sounding, studies of microphysics of snow on the ground) performed during the past 30 years by the Geophysical Institute of the University of Alaska at Fairbanks (Benson, 1992; Friedman, Benson, and Gleason, 1991) are now in computer form and available to the scientific community.
- European and Asian countries maintain their snow-related data separately. An exception is the North Atlantic Climatological Data Set Project (Frich, Brodsgaard, and Cappelen, 1991; Frich, 1994), an international venture between the meteorological services of eleven countries with a goal of compiling a joint representative archive of reference meteorological data. This project has been completed in 1996.

Each summer the U.S. National Operational Hydrologic Remote Sensing Center/ NOAA compiles and distributes a CD-ROM that contains the previous season's airborne and satellite snow data and products for most of North America (Carroll, 1995).

Global snow cover data (visible imagery) are available from NOAA/National Environmental Satellite and Data Service and from the World Data Center for Glaciology. This center provides a wide range of snow-related products from conventional observations and remote sensing, including the unique Global Climatology of Mean Monthly Snow Depth (Schutz and Bregman, 1988).

Special Sensor Microwave/Image products are being prepared for distribution at the World Data Center for Glaciology on CD-ROM.

The U.S. NOAA-NASA Pathfinder Program (NOAA-NASA, 1994) was established to play an integral role in the development of Earth-related data and information systems of these two agencies. All initial projects of this program involve remotely sensed, space-based observations. Two types of these observations (advanced very high-resolution radiometer, and SSM/I) provide snow cover information in visible and near infrared and in the microwave wavelengths, respectively. The progress of this program is associated with the creation of several data services (NASA Distributive Active Archive Centers and NOAA Data Centers) that can be conveniently accessed on the Internet and the World Wide Web. Currently, a significant part of snow cover information is available from these centers. It is expected that, after 1997, most of this information will be readily available to users worldwide.

9

1.3   Snow Cover Climatology

As indicated previously, relatively reliable global-scale observations have been available only since the early 1970s. From a climatological point of view, the period since then has not been stable. Besides the warming (possibly induced partly by greenhouse gas increases), there have been other climate forcings. The eruptions of El Chichón and Pinatubo injected large amounts of aerosol into the stratosphere (Minnis et al., 1993), and some pronounced El Niño and La Niña events affected the global heat balance (Quinn and Neil, 1992). Some of the warmest years of the century (and possibly for several centuries) have occurred in the period (Bradley and Jones, 1993). This instability needs to be borne in mind when considering global snow cover climatology. It prevents us from using satellite data for precise evaluation of an average climatic state of seasonal snow cover extent. But it helps us use these data for studying the snow cover sensitivity to various external factors as well as for evaluation of internal relationships between snow cover and the other climate variables.

1.3.1   Spatial and Temporal Distributions of Snow Cover Extent

Figure 1.3 (see plate section) shows the seasonal distributions of snow cover extent in the Northern Hemisphere (Groisman et al., 1994a). The snow cover seasons are different from the usual climatological seasons because of the extensive snow cover in March and the relative paucity of snow cover in September. Some regions experience large interannual variability in seasonal snow cover, reflecting large fluctuations and/or systematic changes during the past 20 years (Barnett et al., 1989; Gutzler and Rosen, 1992; Karl et al., 1993; Barry et al., 1994; Groisman et al., 1994a; Groisman, Karl, and Knight, 1994b; Brown, Hughes, and Robinson, 1995). The regions of high variability (Figure 1.4 [see plate section]) were termed snow transient regions by Groisman et al. (1994a). Climatic changes most probably will occur in the snow transient regions and/or in their vicinity; thus, these regions are of special interest because of the strong potential for climate-induced ecosystem impact.

Caution must be exercised when referring to trends over relatively short periods, but the negative trends in Northern Hemispheric, North American, and Eurasian snow cover shown in Figure 1.1 for the period 1973–1994 are all statistically significant at the 0.05 level (Groisman et al., 1994a). Overall, the extent of annual Northern Hemisphere snow cover has declined by around 7 percent over the past 26 years (1973–1998). The correlations with Northern Hemispheric temperatures are –0.67 (Northern Hemisphere), –0.57 (North America), and –0.59 (Eurasia) (all significant at the 0.01 level). The decrease in snow cover has been most pronounced in the April–September

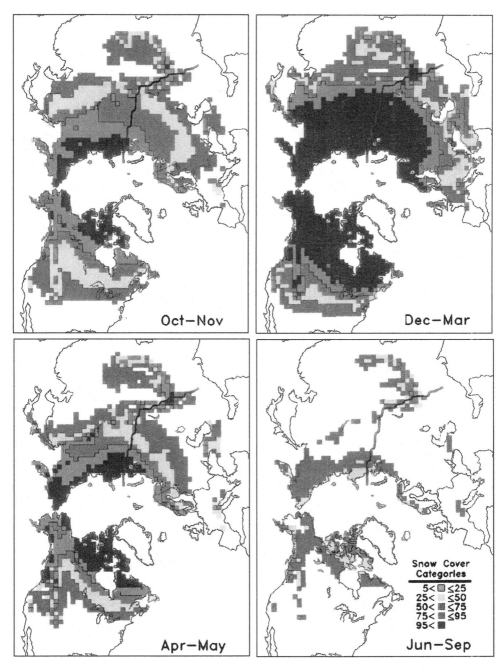

Figure 1.3. Northern Hemisphere seasonal snow cover distribution. White is the least frequent snow cover; blue is the most frequent. Snow cover categories assigned to each grid cell are based on the fraction of time with snow cover over the whole period (1973–1992). Category 0, <0.05 of the time with snow cover; 1, 0.05–0.25; 2, 0.25–0.50; 3, 0.50–0.75; 4, 0.75–0.95; 5, >0.95. See Groisman et al. (1994a) for full details.

A colour version of these plates is available for download from
www.cambridge.org/9780521188890

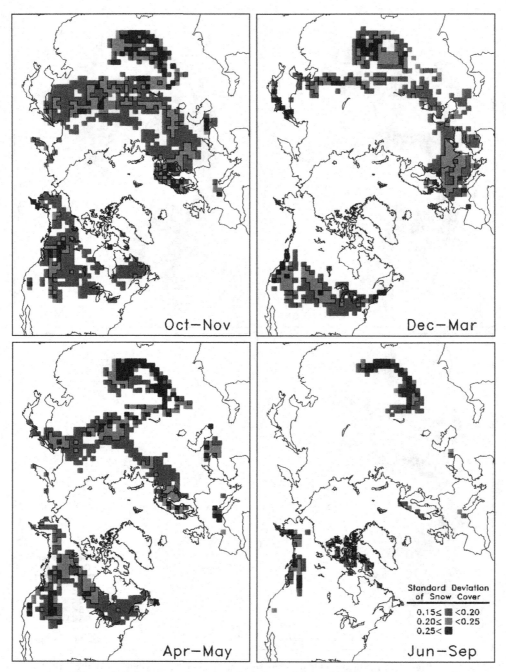

Figure 1.4. Interannual standard deviations of the probability of the presence of snow cover for each season. Regions marked in colors are considered "snow transient regions," representing areas where most variability has occurred in the past 20 years (Groisman et al., 1994a).

A colour version of these plates is available for download from
www.cambridge.org/9780521188890

12

period with the maximum (winter) snow cover extent showing insignificant changes (Groisman et al., 1994a). Century-long trends in near-surface temperature have been particularly strong in the spring season (Jones and Briffa, 1992), and it is possible that this warming has been reinforced by the corresponding changes in snow cover (Groisman et al., 1994b). Snow cover and temperature links are further considered later in this chapter.

Several recent studies describe a longer time scale of snow cover variations for various regions and paint a more complex picture. Brown and Goodison (1996) analysed the snow cover extent over southern Canada since 1915 and combined efforts with American colleagues (Brown, Hughes, and Robinson, 1995; Hughes and Robinson, 1996) to assess the long-term variability of snow cover extent over the interior of North America. Their analyses show that, on a century time scale, a gradual increase of 11×10^3 km^2 year^{-1} was documented in winter (December–February) snow cover mostly due to a snowfall increase, whereas in spring (March–May) snow cover gradually decreased by an average -6×10^3 km^2 year^{-1}. They also show a large multidecadal variability in regional snow cover extent time series and indicate that current (1980s to 1990s) low snow cover values over North America are still within the expected range of natural variability. In particular, this means that a century-long trend in North American spring snow cover is statistically insignificant (Brown, 1997). Of particular importance in these North American studies was a proper linkage between the continental-scale satellite-based snow cover extent time series and those derived from the in situ observations for regions along the snow line in southern Canada. For the period of common observations (after 1972) both time series correlated closely, thus giving a new dimension to the results that otherwise would be more regionally confined (Brown and Goodison, 1996). It is important to note that in these studies an increase in snow cover over North America was reported for winter season (the last 25 years of satellite observations show significant variations in winter snow cover extent but no clear tendency). A century-long decrease in spring snow cover reported by Brown and Goodison (1996) matches the conclusions of Groisman et al. (1994a) for the past two decades and shows that, at least for North America, this trend has a longer origin in the past (see also, Groisman et al., 2000). There are indications that the same is true for the FSU and, thus, for a significant part of Eurasia (Mestcherskaya et al., 1995; Georgievsky et al., 1996; Brown, 1997). Brown (1997) shows a statistically significant decrease of -19×10^3 km^2 year^{-1} in Eurasian spring (March–May) snow cover extent during the period 1915–1985 but indicates that the trend in North American spring snow cover for this period

13

was statistically insignificant. The decade after 1985 had exceptionally low hemispheric snow cover extent, but again the most prominent decrease occurred in Eurasia (Figure 1.1).

1.3.2   Spatial and Temporal Distributions of Snow Depth

The mean winter and spring (conventional climatological seasons) Northern Hemisphere snow depth distributions, based on in situ observations, are shown in Figure 1.5. The maps are based mainly on sources in the Western Hemisphere; they are, therefore, less reliable in the Eastern Hemisphere (Marshall, Roads, and Glatzmaier, 1994) where, for example, they do not show snow over central Asia and western China. Up to 80 cm of snow is reported in winters in Canada and central Russia, with the snow line going south of 40°N. In spring the map presents significant snow depth (above 20 cm) in northern Canada and the Labrador Peninsula, but in Eurasia reports of such deep snow are found only north of 60°N. This last statement shows how misleading three-month averaging in intermediate seasons can be for seasonal snow cover. In fact, March is the month with maximum values of snow depth over most of Russia (Kopanev, 1982). Later in the spring, snow cover becomes more dense. Sublimation and then snowmelt quickly reduce its depth, but there is quite enough water in the spring snowpack to generate regular floods of snowmelt origin over the Great Russian Plain in May and further to the east and north in June and July (Koren', 1991; Kuchment, Demidov, and Motovilov, 1983).

Snow depth data are available for the area of the FSU (Kopanev, 1982; NSIDC, 2000; Mestcherskaya et al., 1995), the United States (Easterling et al., 1997), and for the People's Republic of China (Li, 1987). Mestcherskaya et al. (1995) analysed the end-of-February snow depth time series over major agricultural regions of the FSU for the period 1891 through 1992. They reported a systematic decrease over time in regionally averaged snow depth. Over the western (European) regions of the FSU this century-long decrease was monotonic and more pronounced (15 percent per 100 years). In the regions east of the Ural Mountains (Kazakhstan, western Siberia) a strong decrease of snow accumulation in the first half of the twentieth century was replaced by an increase during the past 30 years (1963 to 1992). Ye, Cho, and Gustafson (1997) concluded that, during the period 1936 through 1983 winter (January through March), snow depth over the FSU increased in northern parts but decreased in southern regions. Both these analyses used a large amount of snow depth information (Mestcherskaya et al. used data from 287 stations and Ye et al. used data from 119 stations), but it is difficult to compare their results because of the differences in regions and time intervals considered. For

14

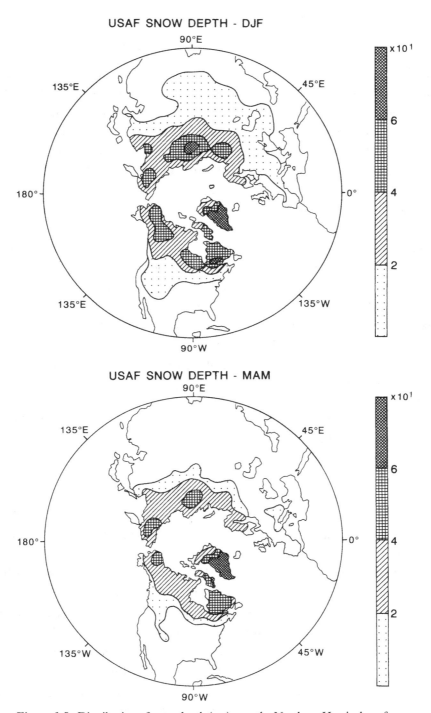

Figure 1.5. Distribution of snow depth (cm) over the Northern Hemisphere for December–January–February (DJF) and March–April–May (MAM). The contour interval is 10 cm, with the lowest contour indicating the location of the snowline (from Marshall et al., 1994; based on Foster and Davy, 1988).

15

example, the regions of northern Russia, where Ye et al. found the most significant increase in snow depth, are not included in Mestcherskaya et al.'s analysis; the period 1936–1983 has no pronounced trend in global surface air temperature, whereas during the period 1891–1992 global temperature increased significantly (Jones, 1994). Li (1987, 1989) presents the map of average annual snow depth over China and describes the countrywide changes in snowfall and snow depth for 1951–1980. He reports a significant increase in snowfall and large-scale fluctuations of snow depth during this period.

1.3.3   Information about SWE

This very important element of climatic systems is not yet well documented globally because of the unresolved problems in existing retrieval algorithms of satellite SWE observations (Armstrong, 1993). The first regional SWE climatologies recently became available for the United States and Russia. They are based on in situ observations (snow courses, SNOTEL network) and on airborne γ-radiation soundings (Cayan, 1996; Carroll, 1995; Haggerty and Armstrong, 1996). Early analyses of these data show that the snowpack is a sensitive component of regional climate: its significant and systematic changes (retreat) were documented in the western part of the FSU for 1966–1990 (Haggerty and Armstrong, 1996), and its correlations with regional circulation patterns have been published for the western United States (Changnon, McKee, and Doesken, 1993; Cayan, 1996).

1.3.4   Seasonal Snow Cover Classifications

Several classifications of seasonal snow cover have been developed during the past 70 years (Formosov, 1946; Richter, 1954; Benson, 1969; McKay and Gray, 1981; Pruitt, 1984; Sturm, Holmgren, and Liston, 1995). Here we present the latest classification by Sturm et al. (1995) because, using the criteria shown in Table 1.1, they created a concise hemispheric map of their classes for the time of maximum snow extent (Figure 1.6). Discriminant analysis of snow cover and meteorological observations in Alaska allowed the authors to define clusters of meteorological conditions most closely related to each type of snow cover. The approach was tested with Russian snow observations and classifications and then applied to worldwide meteorological information (wind, temperature, and precipitation) to construct the map shown. A further discussion of this classification is provided by Pomeroy and Brun (Chapter 2). Consideration of both Table 1.1 and Figure 1.6 gives the reader an initial perception of the snow properties that can be expected in the regional ecosystems of the Northern Hemisphere.

16

Table 1.1. *Description of snow classes (according to Sturm et al., 1995). Geographical distribution of these classes is shown in Figure 1.6.*

Snow cover class	Description	Depth range (cm)	Bulk density (g cm^{-3})	Number of layers
Tundra	A thin, cold, windblown snow. Maximum depth, ~75 cm. Usually found above or north of tree line. Consists of a basal layer of depth hoar overlain by multiple wind slabs. Surface zastrugi common. Melt features rare.	10–75	0.38	0–6
Taiga	A thin to moderately deep low-density cold snow cover. Maximum depth, 120 cm. Found in cold climates in forests where wind, initial snow density, and average winter air temperatures are all low. By late winter consists of 50–80% depth hoar covered by low-density new snow.	30–120	0.26	>15
Alpine	An intermediate to cold deep snow cover. Maximum depth, ~250 cm. Often alternate thick and thin layers, some wind affected. Basal depth hoar common as well as occasional wind crusts. Most new snowfalls are low density. Melt features occur but are generally insignificant.	75–250	no data	>15
Maritime	A warm deep snow cover. Maximum depth can be in excess of 300 cm. Melt features (ice layers, percolation columns) very common. Coarse-grained snow due to wetting ubiquitous. Basal melting common.	75–500	0.35	>15
Ephemeral	A thin, extremely warm snow cover. Ranges from 0 to 50 cm. Shortly after it is deposited, it begins melting, with basal melting common. Melt features common. Often consist of a single snowfall, which melts away; then a new snow cover re-forms at the next snowfall.	0–50	no data	1–3

(*cont.*)

17

Table 1.1. (*cont.*)

Snow cover class	Description	Depth range (cm)	Bulk density (g cm^{-3})	Number of layers
Prairie	A thin (except in drifts) moderately cold snow cover with substantial wind drifting. Maximum depth, ~100 cm. Wind slabs and drifts common.	0–50	no data	<5
Mountain, special class	A highly variable snow cover, depending on solar radiation effects and local wind patterns. Usually deeper than associated type of snow cover from adjacent lowlands.		no data	variable

1.4   Snow–Climate Interactions and Sensitivity

Snow–climate interactions and sensitivity can be assessed empirically with observational data and numerical experiments with global climate models (GCMs). In this section we first describe the results of observational studies. Initially, relationships between snow cover and local climatic variables are considered. Then, links between snow cover and the atmosphere on larger scales are discussed, followed by an assessment of global-scale snow–climate interactions. At the end of this section, we briefly review modeling studies.

1.4.1   Interactions Between Snow Cover and Other Climatic Variables

Snow cover is associated with lower air temperatures near the surface (Voeikov, 1889). Lamb (1955) points out that this is a two-way relationship, with low temperatures needed for snow cover development; once the snow cover has formed, the greater surface albedo produces lower near-surface air temperatures. Foster, Owe, and Rango (1983) indicated geographical differences; the statistical link between winter snow cover and winter atmospheric temperature was much stronger in North America than in Eurasia, whereas the link between autumn snow cover and winter temperature is stronger in Eurasia. Other studies that have linked cold season temperatures with regional snow cover variations include those of Namias (1960, 1962, 1985), Wagner (1973), and Dewey (1987). Typical mean monthly depressions of near-surface air temperature due to the presence of snow cover are around 5°C (Voeikov, 1889;

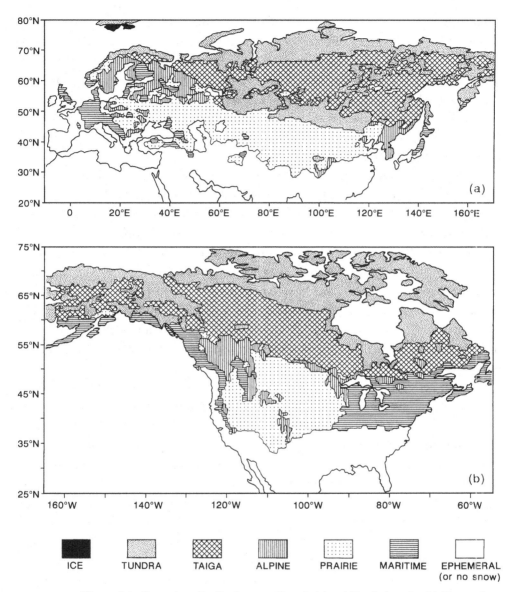

Figure 1.6. Snow class distribution over Eurasia (a) and North America (b) (figure after Sturm et al., 1995). Classes are described in Table 1.1. Special case: mountainous snow cover is not shown because of coarse map resolution. Map is based on meteorological information that was preliminary linked (clustered) with snow classes in Alaska and independently tested on Russian snow data.

Pokrovskaya and Spirina, 1965; Namias, 1985; Leathers, Ellis, and Robinson, 1995). Table 1.2 shows the mean differences in temperature, and a range of other climatic variables, associated with the presence of snow cover. The data are classified by major biome zones, and the presentation of data for clear sky conditions reduces the dependence of the differences on synoptic situations. There has been some modeling confirmation of these empirical associations; for example, Yeh, Wetherald, and Manabe (1983, 1984) showed that early removal of snow cover led to increased temperatures in spring and early summer.

In Table 1.2 we compare the situations with and without snow on the ground at each station for a given period of the Julian year (in 10-day intervals) and with a given type of synoptic situation (for clear sky conditions only). The relationship between snow cover and temperatures at the surface-atmosphere boundary (i.e., significantly lower surface temperatures with snow on the ground) predefines many relationships between snow cover and the other variables shown in this table. With lower temperatures, the atmosphere cannot store much water vapour and the vertical turbulent heat exchange (under the same synoptic conditions) is suppressed. These predefine the negative signs of P_w and F in Table 1.2, but the higher relative humidity with snow on the ground is consistent with the documented increase of cloud cover when the land is snow covered.

The temperature effect can be factored out to an extent by comparing the values of the other variables over snow, and over bare ground, under the same surface temperatures (Table 1.3). When compared with the information in Table 1.2, it is shown that removing the temperature effect produces a strong association between snow and cloud cover and strong effects of snow cover on the establishment of near-surface temperature inversions and sensible heat fluxes.

Snow cover-induced temperature anomalies can change the static stability of the atmosphere through a considerable depth and so may also be expected to lead to effects on precipitation (Namias, 1985; Walsh, 1987; Johnson et al., 1984). Considering fluctuations in North American snow cover extent, Karl et al. (1993) found a strong inverse relationship with regional temperatures. They identified marked subcontinental differences in behaviour over time. Some areas within North America are responsible for most of the interannual variability of snow cover extent, tending to be along the southern border of the snow cover limits. The amount of snowfall in these areas south of 55°N decreased in the 1980s, whereas the mean maximum temperature increased. This strongly contributed to the snow cover retreat in that decade.

Table 1.2. *Mean differences in climatic variables associated with the presence of snow on the ground, from synoptic observations of 223 FSU stations within the same 10-day intervals when both "snow on the ground" and "no snow on the ground" conditions were reported. Data are classified by biome zones. Period of observations used is 1936–1985. Climatic variables are atmospheric pressure at the elevation of the station (P_a), surface air and ground temperatures (T_a and T_g), water vapour pressure (P_w), relative humidity (H), wind speed at the anemometer height (W), vertical turbulent sensible heat flux (Q_H) near the surface, and total and low cloudiness (for cloudiness mean values for snow covered [S] and bare [NS] soil conditions are also shown; from data in Groisman et al. 1996, 1997).*

Biome zone	P_a (hPa)	T_a (K)	T_g (K)	P_w (hPa)	H (%)	W (%)	Q_H (W m^{-2})
Average skies							
Tundra	−2	−4	−4	−1	3	9	−6
Taiga	−2	−5	−6	−1	3	4	−8
Forest	−1	−5	−5	−1	3	3	−4
Forest steppe	−0	−5	−6	−1	3.5	3	−1
Steppe	1	−6	−6	−2	3.5	7	1
Semidesert and desert	5	−8	−8	−2	9	−5	3
Average	0.15	−5.6	−6.0	−1.4	4.3	3	−4.1

Biome zone	P_a (hPa)	T_a (K)	T_g (K)	P_w (hPa)	H (%)	W (%)	Q_H (W m^{-2})
Clear skies							
Tundra	−2	−5	−6	−1	2	−6	−10
Taiga	−1	−6	−8	−1	2	−8	−14
Forest	−0	−7	−8	−1.5	1	−8	−7
Forest steppe	1	−7	−8	−2	2.5	−11	−3
Steppe	2	−8	−8	−2	2	−3	0
Semidesert and desert	4.5	−9	−9	−2	8	−8	1
Average	0.97	−7.1	−8.2	−1.6	2.8	−7	−6.3

Biome zone	Mean cloudiness				Difference in cloudiness (S-NS)	
	Total		Low			
	S	NS	S	NS	Total	Low
Tundra	0.75	0.69	0.49	0.46	0.05	0.04
Taiga	0.76	0.71	0.47	0.42	0.05	0.05
Forest	0.71	0.66	0.47	0.41	0.05	0.05
Forest steppe	0.71	0.68	0.49	0.46	0.03	0.04
Steppe	0.70	0.68	0.48	0.45	0.02	0.03
Semidesert and desert	0.63	0.59	0.41	0.30	0.04	0.11
Average	0.70	0.66	0.45	0.40	0.04	0.05

Table 1.3. *Differences in climate variable values over snow and bare ground with ground surface temperature effects factored out. Based on the data set from synoptic observations of 223 FSU stations (same as in Table 1.1) within the same small ground surface temperature intervals (0.5 K) when both "snow on the ground" and "no snow on the ground" conditions were reported. Data are classified by day (08:00–17:00 LST), night (21:00–05:00 LST), and intermediate hours (06:00– 07:00 and 18:00–20:00 LST), and by clear sky and average climate conditions. Period of observations used is 1936-1985. Climatic variables are atmospheric pressure at the elevation of the station (P_a), surface air temperatures (T_a), percent of near-the-surface temperature inversions (I), water vapour pressure (P_w), relative humidity (H), wind speed at the anemometer height (W), total and low cloudiness, and vertical turbulent sensible heat flux (Q_H), near the surface (from data in Groisman et al., 1996, 1997).*

Cloud conditions	P_a (hPa)	T_a (K)	I (%)	P_w (hPa)	H (%)	W (%)	Cloudiness		Q_H (W m^{-2})
Average skies							Total	Low	
Daytime hours only	−2	−1	−14	−0.2	0	12	0.09	0.09	17
Intermediate hours only	−2	−0	−5	0.0	3	17	0.13	0.12	1
Nighttime hours only	−2	−0	−3	0.0	2	18	0.16	0.10	−1
Clear skies									
Daytime hours only	0	−2	−21	−0.6	−4	11			38
Intermediate hours only	−1	−0	−1	−0.0	−0	11			−0
Nighttime hours only	−1	0	0	0.0	0	15			−1

1.4.2 Relationships with Synoptic Circulations

The influence of snow cover on radiation processes should not hide the fact that, during the cold season at high latitudes, radiation factors do not represent the only, or even the major, heat source/sink. Heat advection by the synoptic-scale atmospheric circulation is important (Lamb, 1955) and significantly affects the association between snow cover and temperature (Pokrovskaya and Spirina, 1965). Kopanev (1982) showed that 75 to 80 percent of winter thaws in eastern Europe were initiated by heat advection by synoptic circulations (i.e., by extratropical cyclones).

Snow cover, atmospheric circulation, and temperature are interdependent, and frequently it is very difficult to unravel cause and effect (Lamb, 1955; Cohen and

Rind, 1991). It is much better to characterise their relationships in terms of "feedbacks" (Cess et al., 1991; Randall et al., 1994). Feedback processes can often be discerned in a less ambiguous way in modeling studies than in empirical studies. Nevertheless, many such studies do produce "suggestive" indications (Walsh, 1987) and so are worthy of review.

Work by Lamb (1955, 1972) and by Walsh, Tucek, and Peterson (1982) indicated that atmospheric thickness is smaller over snow-covered surfaces. Lamb (1972) argued that extensive snow cover may reinforce synoptic-scale troughs in following seasons. A slow retreat of seasonal regional snow cover means that the ground remains cold and saturated. This encourages persistence of a trough and delays the springtime increase of temperature in western Europe, for example, by several weeks (Lamb, 1972). Brinkmann and Barry (1972) and Williams (1978) indicated that similar effects may occur in northeast Canada, as did Toomig (1981) in northern Russia. Confirmation came from the work of Williams (1975), who modeled a stronger cold trough over the eastern part of the continent with an expanded North American snow cover.

The early work of Namias (1962) uncovered the positive feedbacks in the interactions between snow cover and synoptic circulations in North America. An extensive and/or long-lived snow cover enhances the land–sea temperature contrast in northeastern North America. This greater baroclinicity leads to a greater frequency of depressions along the Atlantic coast (Dickson and Namias, 1976), followed by cold polar anticyclones' producing cold air advection over northeastern North America, which represents a positive feedback loop. An effective way of illustrating the effect of snow cover on synoptic systems was adopted by Walsh (1987) by comparing the forecast with observed sea-level pressures for cases with extensive snow cover. He demonstrated that the forecast pressures were too high over the eastern seaboard, indicating that extensive snow cover is associated with more intense depressions in that area (Figure 1.7). Walsh and Ross (1988) also found that more extensive snow cover leads to lower sea-level pressure over the east coast of Asia.

Lack of snow cover can also initiate particular circulation patterns. Namias (1964) pointed out that a restricted snow cover over a land area adjacent to a cold ocean surface (such as Scandinavia) can produce a large horizontal temperature gradient upon a relatively rapid increase in land surface temperature. In turn, this may produce a strong high-pressure ridge that could lead to a persistent blocking pattern affecting the regional climate for weeks or months.

Recent studies by Aguado et al. (1992), Changnon et al. (1993), Dettinger and Cayan (1995), and Cayan (1996) contain detailed analyses of the interaction of atmospheric

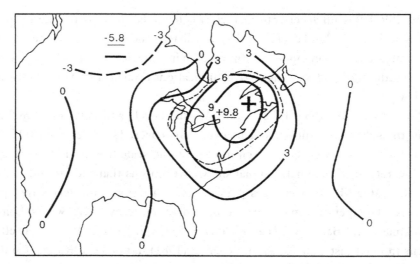

Figure 1.7. Composite difference of 24 hour in persistence forecasts of sea-level pressure (millibar); values represent means for cases with heavy snow cover minus means for light snow cover. Values inside thin dashed line are statistically significant. Diagram shows that the forecasts for the heavy snow cover cases underpredict pressure over the northeastern seaboard of North America (after Walsh, 1987).

circulation and snowpack and its hydrological consequences over the western United States. However, they stress the direct effects – large-scale shifts in atmospheric circulation that result in snowpack changes. Cayan (1996) links the exceptionally low SWE over the western United States at the beginning of snowmelt with a particular circulation pattern, having a strong low in the central North Pacific and high pressure over the Pacific Northwest region of the United States, and concludes that the exceptionally high SWE in various regions of the country are associated with separate regional circulation patterns.

1.4.3   Large-Scale Interactions

A number of empirical studies have linked above-normal snow cover in Eurasia and/or the Tibetan plateau with a delayed or weakened Indian monsoon (lower-than-normal precipitation) (e.g., Hahn and Shukla, 1976; Chen and Yan, 1978). Barnett et al. (1989) argued that this interaction might be part of a much larger effect of Eurasian snow cover in global climate dynamics (Barnett, 1985; Yasunari, 1987). Barnett et al. (1989) conducted GCM experiments in which, when snow amount over Eurasia was increased, the Southeast Asian monsoon was weakened, but there were also effects on atmospheric fields over the tropical Pacific and North America.

24

Barnett et al. argued that the greater snow cover (and its eventual melting and evap-oration of the meltwater) weakened the summer heating of the land mass and, hence, the land/sea temperature contrast that drives the monsoon. The links to more remote regions of the earth's atmosphere were affected through the very large-scale waves in the atmosphere.

El Niño Southern Oscillation (ENSO) is the major large-scale low-frequency varia-tion (with a period of 5 to 7 years) of atmospheric pressure, wind, and surface temper-ature fields in the tropical Pacific (Diaz and Markgraf, 1992). It has global intercon-nections, including those with large-scale snow cover over different parts of the globe (e.g., Li, 1989; Barnett et al., 1989; Yang, 1993; Groisman et al., 1994a). Barnett et al. (1989) point out the general similarity between some of the large-scale atmospheric responses to Eurasian snow cover and some features associated with ENSO. Generally, there is greater snow cover over Eurasia during an ENSO event, although the rela-tionship is reversed for North America. However, there are geographical and seasonal variations (Groisman et al., 1994a). For example, in winter (December through March) ENSO events are associated with an expansion of snow cover over the United States and a reduction of cover in Canada. This pattern is related to the increased strength of the subtropical jetstream over the southern United States and increased zonal flow over higher latitudes (Karl et al., 1993). For China, winter snowfall and snowpack accumulation are positively correlated with ENSO (Li, 1989). ENSO events influence the large-scale temperature fields, but partial correlation analysis (based on data from Groisman et al., 1994a) indicates that there are ENSO influences on snow cover in addition to that of temperature.

There are documented century-long trends in cold season precipitation over north-ern Eurasia and Canada (Groisman et al., 1991, Karl et al., 1993; Groisman and East-erling, 1994). Snowfall over northern Canada (north of 55°N) has been increasing at around 5 percent per decade over the period 1950–1990 (Figure 1.8). It is possible that this increase, in the more northern latitudes, is related to more available moisture with higher global temperatures or results from a change in synoptic patterns because of the diminishing snow cover. Karl et al. (1993) described areas where snow cover frequency was strongly dependent on continental or hemispheric mean annual tem-perature as temperature-sensitive regions (TSRs). Groisman et al. (1994a) extended this approach to the whole of the Northern Hemisphere; the TSRs closely coincided with the snow transient regions shown in Figure 1.4.

The inverse relationship between snow cover extent and temperature (Figure 1.1) may be due, in part, to the snow–albedo–temperature feedback (Budyko, 1974; Robinson and Dewey, 1990; Cess et al., 1991; Leathers and Robinson, 1993;

25

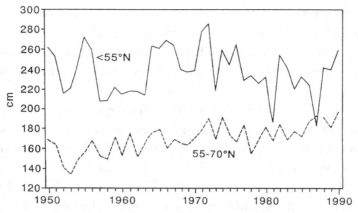

Figure 1.8. Annual Canadian area average snowfall (from Karl et al., 1993).

Groisman et al., 1994a, 1994b). Groisman et al. (1994b) examined snow cover–temperature feedback from an empirical standpoint. They determined that the positive feedback was most prominent in spring (April and May) despite the maximum snow cover occurring in winter. Furthermore, Groisman et al. (1994b) suggested that the feedback of the spring snow cover variations over the past decades may have made a substantial contribution to the observed increases in spring temperatures over the extratropical Northern Hemisphere land area. Brown (1997) shows that this hypothesis is supported by the data for Eurasia, which experienced a statistically significant reduction in spring snow cover and a statistically significant increase in spring air temperature over the twentieth century (specifically, he considered the period 1915–1985). However, he indicated that there were no systematic changes in spring snow cover or air temperature over North America during the same period, which decreased the statistical significance of spring snow cover reduction at the hemispheric scale. This hemispheric reduction of spring snow cover still is statistically significant at the 0.05 level if the snow cover extent data are analysed for the period from 1915 to the present (Brown, 1997).

It should be noted that the regions where the most prominent changes in snow cover extent were detected in the past decades, and then linked to the global temperature changes (spring and summer TSRs), are mostly located in subpolar and polar regions of the Northern Hemisphere. These are the areas where snow cover is an integral part of the ecosystem and where the ecosystems are highly sensitive to change. As a word of caution, let us also note that regional analyses (e.g., Li, 1989; Gutzler and Rosen, 1992; Groisman et al., 1994a; Brown et al., 1995) show that there are seasons (e.g., winter) and regions (e.g., China) where snow cover changes are not

26

correlated with global temperature or where the sign of these correlations is not even positive.

The links with ENSO and global temperature demonstrate that annually varying large-scale snow cover plays a significant role in important global-scale atmospheric circulations that act over time scales of seasons to years. Such research confirms the importance of understanding some of the feedback processes involved in snow cover–atmosphere interactions. It is possible that, with an enhanced greenhouse effect operating, these feedbacks will be strengthened. It is necessary to consider model simulations and feedback processes in more detail. General circulation model simulations help promote this understanding.

1.4.4 Modeling Snow Cover–Atmosphere Interactions and Feedback Processes

The first generation of GCMs (e.g., Manabe, 1969) dealt with the snow cover in a simple way that distorted the seasonal and spatial variations of the hydrological cycle and surface heat balance (Yeh et al., 1983). This generation of models reproduced ground surface temperature (under the snow cover) with up to a $10°C$ bias (Verseghy, 1991, 1992) and rendered them of limited applicability for assessing effects on terrestrial ecosystems.

The next generation of GCMs incorporated much better parameterizations of land–atmosphere processes. Several layers in the soil were introduced that significantly improved the representation of the seasonal cycle of soil temperature and humidity (Philips, 1994). Moreover, the inclusion of albedo–temperature relationships for snow-covered land was a significant step forward (Marshall et al., 1994; Randall et al., 1994).

Although some GCMs exhibit substantial errors in snow cover area or timing of spring retreat on a regional scale, most of the current generation of GCMs capture the gross features of seasonal snow cover (Davies, 1994; Frei and Robinson, 1995). They have been used to provide insight into feedback processes, although the complexities of models and processes still can make it difficult to resolve cause and effect (e.g., the sign of snow cover–temperature feedback). Cess et al. (1991) reviewed the results from a range of GCM experiments. The complexity of feedbacks was highlighted by the fact that, although most models produced the expected positive feedback in the snow–albedo–temperature system, some produced a weak negative feedback. In some cases the negative feedback was attributed to a redistribution of clouds, whereas in others it appeared to be related to cloud-induced long-wave radiation interactions. There was also significant long-wave feedback associated with snow retreat in some models.

27

Many aspects of the negative feedback findings in some of the GCM studies reviewed by Cess et al. (1991) were complementary to GCM results reported by Cohen and Rind (1991). They found that part of the cycle forced on the energy balance by the presence of snow cover appeared to have a negative feedback built into it. The initial cooling caused by the greater snow cover increased atmospheric stability, which suppressed the flux of sensible and latent heat away from the surface. Moreover, the lower temperatures reduced the amount of long-wave radiation emitted by the surface. These effects were large enough to reverse the negative heating trend, so, in part of the cycle, the snow cover anomaly (i.e., an extra snow cover artificially introduced in a model run) worked to destroy itself.

Even with the existing benefit of GCM experiments, much remains to be done to elucidate snow–temperature feedbacks. However – and guarding against overconfidence in complex GCMs – the model approach does have advantages over empirical methods with respect to the difficulty of separating out the thermodynamic effects of snow cover from the dynamical influence of the regime that produced it in the first place.

Interactions of snow cover with vegetated surfaces represent a special interest for ecological modelers. Thomas and Rowntree (1992) developed a scheme that allows the maximum albedo of a snow cover to be prescribed as a function of vegetation type. They modeled the responses of the Northern Hemisphere climate to boreal deforestation (a possible transitory consequence of global warming, in the eventuality of the warming's being too rapid for contemporaneous species migration). The deforestation led to an increased albedo with a related reduction in net radiation in April and May. There were associated reductions in the fluxes of sensible and latent heat from the surface and a decrease in precipitation in the deforested regions because of the decreased evaporation. Removal of the boreal forest in the model also delayed snowmelt, while increasing its peak magnitudes. The problem of using simplified land surface representations to examine surface interactions with climate change is illustrated by this result, as surface observations show that, in fact, a quickening of snowmelt occurs upon boreal deforestation (Pomeroy and Granger, 1997).

Much remains to be done on the incorporation of ecosystem–atmosphere interaction into GCMs. Several soil–vegetation–atmosphere transfer schemes have been developed (e.g., Sellers et al., 1986; Dickinson et al., 1986; Wood, 1991; Xue et al., 1991). A diversity of landscapes and canopies can be introduced in the model grid cells and the effect of each landscape type interaction with snow cover can then be determined separately (Sellers et al., 1986: Dickinson et al., 1993). Approaches have also been developed that take into account the effects of complex terrain (Wood, Lettenmaier, and Zartarian, 1992; Wigmosta, Vail, and Lettenmaier, 1994) and the way snow cover develops different radiative properties as it ages (Dickinson, Henderson-Sellers, and Kennedy, 1993). However, these schemes are complicated and further development is needed for GCMs to run efficiently with their inclusion.

The more advanced snow parameterization schemes consider snow cover as a separate thermal layer from the underlying soil (Verseghy, 1991) and simulate a multilayer snow cover (Loth, Graf, and Obehuber, 1993; Lynch-Stieglitz, 1994). So far, only one land-surface scheme routinely applied in GCM simulations (used in the Canadian Climate Centre) considers the heat transfer processes within the snowpack and treats the underlying surface separately (Vershegy, 1991). It is anticipated that snow cover processes and snow cover interactions with other systems will be better identified and modeled in the future (as, indeed, it is anticipated that GCM simulations of the climate will improve). Notwithstanding present shortcomings, it still is instructive to consider the consequences of possible future climate.

GCM projections of future climate suggest enhanced warming in regions of seasonal snow cover (Mitchell et al., 1990; Gates et al., 1992). Because of snow cover–atmosphere feedbacks, it may be argued that terrestrial systems affected by spring snow cover variations may be under particular "pressure" with continuing climate change (Davies and Vavrus, 1991; Groisman et al., 1994a). Besides stronger warming and reduced seasonal snow cover, $2 \times CO_2$ GCM (models perturbed by a doubling of the CO_2-equivalent concentration) projections indicate a general increase in soil moisture content in northern latitudes because of more precipitation, greater snowmelt, and a small rate of increase in potential evaporation due to relatively low winter temperatures (IPCC, 1990). The projected earlier snowmelt is predicted to lead to greater summer drying. The hydrological regime of high latitude lands may be strongly affected for such reasons.

Figure 1.9 is an example of a $2 \times CO_2$ GCM projection. A significant decrease in snow cover and snow mass is projected throughout the year. Another experiment, using a Canadian Climate Centre $2 \times CO_2$ model, projected a 20 percent reduction in winter seasonal snow cover in the Northern Hemisphere and as much as 50 percent in the summer (Cohen, 1994). Some scenarios suggest quite dramatic effects on a regional scale; for example, Brown et al. (1995) suggested that a plausible continental warming of $4°C$ may result in a 70 percent decrease in snow cover duration over the Great Plains and a 40 percent decrease over the Canadian prairies.

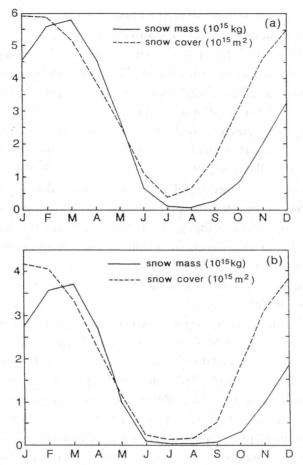

Figure 1.9. Monthly average of snow cover (m^2; dashed line) and snow mass (kg; solid line) for the Northern Hemisphere in the Goddard Institute for Space Studies GCM for current climate (a) and $2 \times CO_2$ climate (b) (after Cohen, 1994).

1.5   The Impact of Possible Future Climatic Changes on Snow Cover and its Ecology

Changes in the climate system (resulting from enhanced greenhouse gas warming) are likely to have major effects on seasonal snow cover, thus initiating complex feedback processes. The impacts range from the local scale (e.g., an increase in diurnal temperature range in areas with reduced cover [Cao, Mitchell, and Lavery, 1992]) to the hemispheric and global repercussions. Hydrological regimes may be substantially altered.

Current state of knowledge about the ongoing climatic change is expressed in recently completed reports of the IPCC (IPCC, 1996; Watson et al., 1996). According to these reports we may expect global warming that will be especially pronounced in high latitudes. This warming will be accompanied by an enhanced hydrologic cycle. This means there will be competing tendencies: warming produces earlier snowmelt, but the enhanced hydrologic cycle initially leads to increased snowfall. Thus far, we can assert that in spring the first factor has been more effective globally (at least in the past 25 years). For the winter season (December through March) we have no clear-cut answer about the global snow cover behaviour during the past several decades. However, this suggests a possible increase of interannual variability in snow cover conditions with global warming. It is possible that two competing factors, X and Y, that emerge with climate change can contribute to the snow cover variations and have a mean combined effect that is close to zero but, because they are independent of each other, induce a significant variance:

$$E(X + Y) \approx 0; \quad \text{var}(X + Y) \gg 0$$

Below we sketch possible climatic changes in snow-covered regions that we can expect with warmer global and regional temperatures:

- Changes in snowfall regimes: more frequent mixed precipitation; more frequent rain on snow; redistribution of frozen and liquid precipitation totals; general increase in cold season precipitation totals. The last two features actually have been documented in Canada (Karl et al., 1993; Groisman and Easterling, 1994; Brown and Goodison, 1996). Cold season precipitation over Russia has also increased during the past century (Groisman, 1991).

- Changes in snowpack properties: shorter duration of stable and unstable snow cover; decrease in snow depth and decrease in albedo due to more frequent snow metamorphism with thaws and rain-on-snow events – consequently, snowpacks may become more icy and dense, affecting the insulation properties.

- Changes in amount and timing of snowmelt will change the entire seasonal runoff pattern, making it less seasonal: more snowmelt water will go to

31

the soil to be released later during the low-flow period of the year and less snowpack will be available for spring snowmelt flooding. This hydrologic pattern already has been observed over the European part of the FSU during the post World War II period (Georgievsky et al., 1995, 1996; Shiklomanov et al., 1995).

- Changes in evaporation and soil moisture regimes: a longer summer (period without snow on the ground) will facilitate evaporation and lead to summer dryness but the larger amount of precipitation will compete with this tendency. The above consideration of two competing factors leading to an increase in variability also may be pertinent to soil moisture conditions in high latitudes with global warming.
- It is certain that the rate of change will be more disruptive to some components of the ecosystems that are related to snow cover than the direction of the change itself. The climate has changed in the past, sometime very dramatically, and biomes had to adjust themselves to these changes. But these changes rarely occurred at the speed with which they do now (Stouffer, Manabe, and Vinnikov, 1994). If we assume that an important reason for the recent increase of global near-surface air temperature by 0.5 °C per century has been anthropogenic emissions, then these rates will only increase in the next several decades (IPCC, 1996).

Ecological communities that inhabit the snowpack, or are currently in the sub-snow cover environment, will be affected by climate change. At the very least, there will be a change in their spatial distributions and, possibly, modifications to their seasonal cycles if they cope with perturbations to the annual snow cover regime. Beyond the environs of the seasonal snow cover, possibly at very remote locations, other ecological communities will be changing in response to the changing climate due in part to climate–snow cover feedbacks. In turn, changes to some of these bio-genic systems (e.g., the vegetation zones) will have feedbacks to the climate–snow cover system. The interfaces between physical and ecological change, across the spectrum of space and time scales, represent an enormous challenge to scientific research. The challenge is much greater for an emerging research area, such as snow ecology, where so much remains to be done on descriptive characterisation, on identifying and modeling the processes that sustain the living communities, and on elucidating the effect of these communities on the physical and geochemical properties of the snow.

32

1.6   Conclusions

Snow cover is an important part of the climate system. It interacts with other components of the climate system on hemispheric, regional, local, and microscales. It is therefore an important influence not only on ecosystems that are within or beneath the snow cover or are part of those regions directly affected by its seasonally shifting pattern but also on ecosystems that are remote from the direct effects of the lying snow. The behaviour of snow cover is therefore of interest to ecologists not only in terms of establishing the dynamic relationships between it and components of ecosystems but also because of the character of long-term change. Although there are observational data sets of snow cover that extend back for a century, there are concerns about the representativeness and homogeneity of the observations. Satellite observations started in the mid-1960s, but they are not generally regarded as being reliable until the early 1970s; thus, homogeneous analysis of regional snow cover behaviour over time is possible only for a relatively short period. Over this recent period, Northern Hemispheric snow cover has decreased by about 10 percent, with the most pronounced changes occurring in the summer months. Given the caveats about data quality already expressed, it appears that such a decline has been occurring in North America and Eurasia for most of the twentieth century. This, of course, has implications for the ecosystems of the seasonally transient snow character of cover areas, because the character of snowmelt seasons will have changed. Such changes immediately draw attention to the prospect of continued change in the future. Future enhanced greenhouse gas warming will lead, naturally, to higher surface temperatures but also to changes in atmospheric circulation patterns and to a strengthened hydrological cycle. It is possible that some areas will experience greater snow cover; indeed, there are such indications in the observations. Whatever the future changes, and plausible projections are described in the previous section, it is very clear that changes in snow cover will affect ecosystems via the feedback links with the rest of the climate system or – and more pronounced – by directly changing the immediate physical environment of organisms that live within, or in close proximity to, lying snow.

1.7   Future Research Needs

The nature of the feedback processes between snow cover and the rest of the climate system is poorly known. As part of improving knowledge in this area, snow cover processes need to be better understood and incorporated into predictive models. Included in this quest will have to be more sophisticated snow–ecosystem–atmosphere interaction studies and acceptable ways to include these processes in models. Because we have much to learn about the ecology of, and the ecological dynamics in, regions affected by seasonal snow cover, there is much to do.

1.8   References

Aguado, E., Cayan, D.R., Riddle, L., and Roos, M. 1992. Climatic fluctuations and the timing of West Coast streamflow. *J. Climate*, **5**, 1468–1483.

Armstrong, R.L. 1993. Detection of fluctuations in global snow cover using passive microwave remote sensing. *Proc. Snow Watch '92 Conference*, Niagara-on the-Lake, Ontario, Canada, March 30 to April 1, 1992, 52–56.

Barnett, T.P. 1985. Variations in near-global sea level pressure. *J. Atmos. Sci.*, **42**, 478.

Barnett, T.P., Dumenil, L., Schlese, U., Roeckner, E., and Latif, M. 1989. The effect of Eurasian snow cover on regional and global climate variations. *J. Atmos. Sci.*, **46**, 661–685.

Barry, R.G. 1985. The cryosphere and climate change. In *Detecting the Climatic Effects of Increasing Carbon Dioxide*, M.C. MacCracken and F.M. Luther (Eds.). U.S. Department of Energy, Washington D.C. DOE/ER 0235, pp. 109–148.

Barry, R.G. 1991. Cryospheric products from DMSP-SSM/I: status and research applications. *Palaeogeogr. Palaeoclimatol. Palaeoecol.*, **90**, 231–234.

Barry, R.G., Armstrong, R.L., Krenke, A.N., and Kadomtseva, T. 1994. *Cryospheric Indices of Global Change*. Final report, NSF Grant SES-91-12420, Boulder, CO.

Basist, A., and Grody, N.C. 1994. Identification of snow cover using SSM/I measurements. *Proc. 6th Conference on Climate Variations* Jan. 24–28, 1994, Nashville, TN. American Meteorological Society, Boston, 252–256.

Benson, C.S. 1969. *The Seasonal Snow Cover in Arctic Alaska*. Report 51, The Arctic Institute of North America, Calgary, Canada.

Benson, C.S. 1992. An overview of research on snow in Alaska. *Proc. Snow Watch '92 Conference*. Niagara-on the-Lake, Ontario, Canada, March 29 to April 1, 1992.

Biswas, A.K. 1970. *History of Hydrology*. Elsevier, Amsterdam.

Bradley, R.S., and Jones, P.D. 1993. 'Little Ice Age' summer temperature variations: their nature and relevance to recent global warming trends. *Holocene*, **3**, 367–376.

Brinkmann, W.A.R., and Barry, R.G. 1972. Palaeoclimatological aspects of the synoptic climatology of Keewatin, Northwest Territories, Canada. *Palaeogeogr. Palaeoclimatol. Palaeoecol*, **11**, 77–91.

Brown, R.D., and Goodison, B.E. 1996. Interannual variability in reconstructed Canadian snow cover, 1915-1992. *J. Climate*, **9**, 1299–1318.

Brown, R.D., Hughes, M.G., and Robinson, D.A. 1995. Characterizing the long-term variability of snow-cover extent over the interior of North America. *Ann. Glaciol.*, **21**, 45–50.

Brown, R.D. 1997. Historical variability in Northern Hemisphere spring snow covered area. *Ann. Glaciol.*, **25**, 340–346.

Budyko, M.I. 1974. *Climate and Life*. Academic Press. New York, NY.

Budyko, M.I., and Yu.A. Izrael (Eds.) 1991. *Anthropogenic Climatic Change*. The University of Arizona Press. Tucson, Arizona.

Cao, H.X., Mitchell, J.F.B., and Lavery, J.R. 1992. Simulated diurnal range and variability of surface temperature in a global climate model for present and doubled CO_2 climates. *J. Climate*, **5**, 920–943.

Carroll, S.S., and Carroll, T.R. 1989. Effect of forest biomass on airborne snow water equivalent estimates obtained by measuring terrestrial gamma radiation. *Remote Sens. Environ.*, **27**, 313–320.

Carroll, T.R. 1995. GIS used to derive operational hydrologic products from in situ and remotely sensed snow data. In *Geographical Information Systems in Assessing Natural Hazards*, A. Carrara and F. Guzzetti (Eds.). Kluwer Academic, Netherlands, pp. 335–342.

Cayan, D.R. 1996. Interannual climate variability and snowpack in the western United States. *J. Climate*, **9**, 928–948.

Cess, R.D., Potter, G.L., Zhang, M.-H., et al. 1991. Interpretation of snow-climate feedback as produced by 17 general circulation models. *Science*, **253**, 888–892.

Changnon, D., McKee, T.B., and Doesken, N.J. 1993. Annual snowpack patterns across the Rockies: Long-term trends and associated 500 Mb synoptic patterns. *Mon. Wea. Rev.*, **121**, 633–647.

Chen, L.T., and Yan, Z.X. 1978. A statistical analysis of the influence of anomalous snow cover over Qinghai-Tibetan Plateau during the winter-spring monsoon of early summer. *Proc. Conf. Medium and Long Term Hydrometeorological Prediction in the Basin of the Yangtze River*, Vol. 1, Hydroelectric Press, Beijing, China.

Cohen, J. 1994. Snow cover and climate. *Weather*, **49**, 151–156.

Cohen J., and Rind, D. 1991. The effect of snow cover on climate. *J. Climate*, **4**, 689–706.

Davies, T.D. 1994. Snowcover-atmosphere interactions. In *Snow and Ice Covers: Interactions with the Atmosphere and Ecosystems*, H.G. Jones, T.D. Davies, A. Ohmura, and E. Morris (Eds.). IAHS Publication No. 223, IAHS Press, Wallingford, UK, pp. 3–13.

Davies, T.D., and Vavrus, S.J. 1991. Climatic change and seasonal snowcovers: a review of factors regulating the chemical evolution of snow cover and a predictive case study for north-eastern North America. In *NATO ASI Series, G: Ecol. Sci., Vol. 28, Seasonal Snowpacks, Processes of Compositional Change*, T.D. Davies, M. Tranter, and H.G. Jones (Eds.). Spinger-Verlag, Berlin, pp. 421–456.

Dettinger, M.D., and Cayan, D.R. 1995. Large-scale atmospheric forcing of recent trends toward early snowmelt runoff in California. *J. Climate*, **8**, 606–623.

Dewey, K.F. 1987. Snow cover/atmospheric interactions. In *Large Scale Effects of Seasonal Snow Cover*, B.E. Goodison, R.G. Barry, and J. Dozier (Eds.). IAHS Publication No. 166, IAHS Press, Wallingford, UK, pp. 27–42.

Dewey, K.F., and Heim, R.R. Jr. 1981. *Satellite Observations of Variations in Northern Hemisphere Seasonal Snow Cover.* NOAA Technical Report NESS 87, National Oceanic and Atmospheric Administration, Washington, DC.

Dewey, K.F., and Heim R. Jr. 1982. A digital archive of Northern Hemisphere snow cover, November 1966 through December 1980. *Bull. Am. Meteorol. Soc.,* **63**, 1132–1141.

Diaz, H.F., and Markgraf, V. (Eds.). 1992. *El Niño: Historical and Paleoclimatic Aspects of the Southern Oscillation.* Cambridge University Press, Cambridge, England.

Dickinson, R.E., Henderson-Sellers, A., Kennedy, P.J., and Wilson, M.F. 1986. *Biosphere-Atmosphere Transfer Scheme (BATS) for the NCAR Community Climate Model.* National Center for Atmospheric Research, Boulder, CO.

Dickinson, R.E., Henderson-Sellers, A., and Kennedy, P.J. 1993. *Biosphere Atmosphere Transfer Scheme (BATS) Version 1e As Coupled to the NCAR Community Climate Model.* National Center for Atmospheric Research Technical Note, National Center for Atmospheric Research, Boulder, CO.

Dickson, R.R., and Namias, J. 1976. North American influence on the circulation and climate of the North Atlantic sector. *Mon. Wea. Rev.,* **104**, 1255–1265.

Easterling, D.R., Jamason, P.B., Bowman, D.P., Hughes, P.Y., Mason, E.H., and Allison, L.J. 1997. *Daily Snow Depth Measurements from 195 Stations in the United States.* ORNL/CDIAC-95 NDP-059, Environmental Science Division Publication No.4610, U.S. Carbon Dioxide Information Data Center, Oak Ridge National Laboratory, Oak Ridge, TN.

Ferraro, R., Grody, N., Forsyth, D., Carey, R., Basist, A., Janowiak, J., Weng, F., Marks, G.F., and Yanamandra, R. 1994. Microwave measurements produce global climatic, hydrologic data. *EOS,* **75**, 337–338, 343.

Formosov, A.N. 1946. Snow cover as an integral factor of the environment and its importance in the ecology of mammals and birds. *Materials for Fauna and Flora of the USSR. New Series, Zoology.* (English edition published by Boreal Institute, University of Alberta, Edmonton, Canada, Occasional Paper 1), **5**, 1–152.

Foster, D.J. Jr., and Davy, R.D. 1988. *Global Snow Depth Climatology.* USAFETAC/TN-88/006. USAF Environment Technical Applications Center, Scott Air Force Base, IL.

Foster, J., Owe, M., and Rango, A. 1983. Snow cover and temperature relationships in North America and Eurasia. *J. Climate Appl. Meteorol.,* **22**, 460–469.

Frei, A., and Robinson, D.A. 1995. Northern Hemisphere snow cover extent: comparison of AMIP results to observations. *Proc. First International AMIP Scientific Conference,* Monterey, CA, May 1995, pp. 499–504.

Frich, P., Brodsgaard, B., and Cappelen, J. 1991. *North Atlantic Climatological Data Set (NACD), Present Status and Future Plans.* DMI Technical Report 91-8. Danish Meteorological Institute, Copenhagen, Denmark.

Frich, P. 1994. North-Atlantic Climatological Data Set (NACD), towards a European Climatic Data Center (ECDC). In *Climate Variations in Europe,* R. Heino (Ed.). Publications of the Academy of Finland 3/94, Helsinki, 386 pp. Proc. European Workshop on Climate Variations, Kirkkonommi (Majvik), Finland, 15-18 May 1994, pp. 81–96.

36

Friedman, I., Benson, C.S., and Gleason, J. 1991. Isotopic changes during snow metamorphism. In *Stable Isotope Geochemistry: A Tribute to Samuel Epstein*, H.P. Taylor, Jr., J.R. O'Neil, and I.R. Kaplan (Eds.). The Geochemical Society Special Publication No. 3, pp. 211–221.

Gates, W.L., Mitchell, J.F.B., Boer, G.J., Cubasch, U., and Meleshko, V.P. 1992. Climate modeling, climate prediction and model validation. In *Climate Change 1992; The Supplementary Report to the IPCC Scientific Assessment*, J.T. Houghton, B.A. Callander, and S.K. Varney (Eds.). Cambridge University Press, Cambridge, England, pp. 97–134.

Glynn, J.E., Carroll, T.R., Holman, P.B., and Grasty, R.L. 1988. An airborne gamma ray snow survey of a forest covered area with a deep snowpack. *Remote Sens. Environ.*, **26**, 149–160.

Georgievsky, V.Yu., Zhuravin, S.A., and Ezhov, A.V. 1995. Assessment of trends in hydrometeorological situation on the Great Russian Plain under the effect of climate variations. *Proc. AGU 15th Annual Hydrology Days*, H.J. Morel-Seytoux (Ed.). Atherton, CA, pp. 47–58.

Georgievsky, V.Yu., Ezhov, A.V., Shalygin, A.L., Shiklomanov, I.A., and Shiklomanov, A.I. 1996. Evaluation of possible climate change impact on hydrological regime and water resources of the former USSR rivers. *Russian Meteorol. Hydrol.*, No. 11, 89–99 (in Russian).

Gray, D.M., and Male, D.H. 1981 (Eds.). *Handbook of Snow: Principles, Processes, Management and use.* Pergamon Press, Oxford, England.

Groisman, P.Ya. 1991. Data on present-day precipitation changes in the extratropical part of the Northern hemisphere. In *Greenhouse-Gas-Induced Climatic Change: A Critical Appraisal of Simulations and Observations*, M.E. Schlesinger (Ed.). Elsevier, Amsterdam, pp. 297–310.

Groisman, P.Ya., Koknaeva, V.V., Belokrylova, T.A., and Karl, T.R. 1991. Overcoming biases of precipitation measurement: a history of the USSR experience. *Bull. Am. Meteorol. Soc.*, **72**, 1725–1733.

Groisman, P.Ya., and Easterling, D.R. 1994. Variability and trends of precipitation and snowfall over the United States and Canada. *J. Climate*, 7, 184–205.

Groisman, P.Ya., Karl, T.R., Knight, R.W., and Stenchikov, G.L. 1994a. Changes of snow cover, temperature, and the radiative heat balance over the Northern Hemisphere. *J. Climate*, 7, 1633–1656.

Groisman, P.Ya., Karl, T.R., and Knight, R.W. 1994b. Observed impact of snow cover on the heat balance and the rise of continental spring temperatures. *Science*, **263**, 198–200.

Groisman, P.Ya., Genikhovich, E.L., and Zhai, P.-M. 1996. "Overall" cloud and snow cover effects on internal climate variables: The use of clear sky climatology. *Bull. Am. Meteorol. Soc.*, **77**, 2055–2065.

Groisman, P.Ya., Genikhovich, E.L., Bradley, R.S., and Ilyin, B.M. 1997. Assessing surface-atmosphere interactions from former Soviet Union standard meteorological data. Part 2. Cloud and snow cover effects. *J. Climate*, **10**, 2184–2199.

Groisman, P.Ya., Sun, B., and Heim R.R. Jr. 2000. Trends in spring snow cover retreat over the U.S. and the effects of observation time bias. *Proc. of 11th Symposium on Global*

Change Studies, 9–14 Jan. 2000, Long Beach, CA, American Meteorological Society, Boston, 54–57.

Gutzler, D.S., and Rosen, R.D. 1992. Interannual variability of wintertime snow cover across the Northern Hemisphere. *J. Climate*, **5**, 1441–1447.

Hahn, D.G., and Shukla, J. 1976. An apparent relationship between Eurasian snow cover and Indian monsoon rainfall. *J. Atmos. Sci.*, **33**, 2461–2462.

Haggerty, C.D., and Armstrong, R.L. 1996. Snow trends within the former Soviet Union. *Suppl. EOS, Trans. AGU*, **77**, [abstracts of papers presented at the fall meeting of the American Geophysical Union, December 1996, San Francisco, CA].

Hughes, M.G., and Robinson, D.A. 1996. Historical snow cover variability in the Great Plains regions of the USA: 1910 through to 1993. *Int. J. Climatol.*, **16**, 1005–1018.

Intergovernmental Panel on Climate Change (IPCC) 1990. *Climate Change. The IPCC Scientific Assessment*, J.T. Houghton, G.J. Jenkins, and J.J. Ephraums (Eds.). Cambridge University Press, New York.

Intergovernmental Panel on Climate Change (IPCC) 1996. *Climate Change 1995: The Science of Climate Change. The Second IPCC Scientific Assessment*, J.T. Houghton, L.G. Meira Filho, B.A. Callendar, N. Harris, A. Kattenberg, and K. Maskell (Eds.). Cambridge University Press, New York.

Jackson, M.C. 1978. Snow cover in Great Britain. *Weather*, **33**, 298–309.

Johnson, R.H., Young, G.S., Toth, J.J., and Zehr, R.M. 1984. Mesoscale weather effects of variable snow cover over Northeast Colorado. *Mon. Wea. Rev.*, **112**, 1141–1152.

Jones, P.D. 1994. Hemispheric surface air temperature variations: a reanalysis and an update to 1993. *J. Climate*, **7**, 1794–1802.

Jones, P.D., and Briffa, K.R. 1992. Global surface air temperature variations during the twentieth century: Part 1, spatial, temporal and seasonal details. *Holocene*, **2**, 165–179.

Jones, W.K., and Carroll, T.R. 1983. Error analysis of airborne gamma radiation soil moisture measurements. *Agric. Meteorol.*, **28**, 19–30.

Jordan, R. 1991. *A One-Dimensional Temperature Model for a Snow Cover*. Technical documentation for SNTHERM. 89, Special Report 91-16. U.S. Army Corps of Engineers, Cold Regions Research & Engineering Laboratory, Hannovee, NH.

Karl, T.R., Tarpley, D., Quayle, R.G., Diaz, H.F., Robinson, D.A., and Bradley, R.S. 1989. The recent climate record: what it can and cannot tell us. *Rev. Geophys.*, **27**, 405–430.

Karl, T.R., Groisman, P.Ya., Knight, R.W., and Heim, R.R. Jr. 1993. Recent variations of snow cover and snowfall in North America and their relation to precipitation and temperature variations. *J. Climate*, **6**, 1327–1344.

Kogan, R.M., Nazarov, I.M., and Fridman, Sh.D. 1969. *Basis of Environmental Gamma-Spectrometry*. Atomizdat, Moscow.

Kopanev, I.D. 1982. *Climatological Aspects of Snow Cover Evolution Studies*. Gidrometeoizdat, Leningrad (in Russian).

Koren', V.N. 1991. *Mathematical Models for the Forecast of Runoff*. Gidrometeoizdat, Leningrad (in Russian).

Kuchment, L.S., Demidov, V.N., and Motovilov, Yu.G. 1983. *Streamflow Formation*, Nauka, Moscow (in Russian).

Lamb, H.H. 1955. Two-way relationships between the snow cover or ice limit and 100-500 Mb thickness in the overlying atmosphere. *J.R. Meteorol. Soc.*, **81**, 172–189.

Lamb, H.H. 1972. *Climate: Present, Past and Future, Volume 1. Fundamentals and Climate Now*. Methuen, London.

Leathers, D.J., and Robinson, D.A. 1993. The associations between extremes in North American snow cover extent and United States temperatures. *J. Climate*, **6**, 1345–1355.

Leathers, D.J., Ellis, A.W., and Robinson, D.A. 1995. Characteristics of temperature depressions associated with snow cover across the Northeast United States. *J. Appl. Meteorol.*, **34**, 381–390.

Leconte, R. 1993. Modeling the interactions between microwaves and a snow/soil system. *Proc. Snow Watch '92 International Workshop*, Niagara-on the-Lake, Ontario, Canada, March 30 to April 1, 1992, pp. 188–203.

Ledley, T.S. 1991. Snow on sea ice: competing effects in shaping climate. *J. Geophys. Res.*, **96**, 17195–17208.

Li, P. 1987. Seasonal snow resources and their fluctuation in China. In *Large-Scale Effects of Seasonal Snow Cover*, B.E. Goodison, R.G. Barry, and Y. Dozier (Eds.). IAHS Publication No. **166**, IAHS Press, Wallingford, UK, pp. 93–104.

Li, P. 1989. Recent trends and regional differentiation of snow variation in China. In *Snow Cover and Glacier Variations*, S.C. Colbeck (Ed.). IAHS Publication No. **183**, IAHS Press, Wallingford, UK, pp. 3–10.

Loth, B., Graf, H.-F., and Obehuber, J.M. 1993. Snow cover model for global climate simulations. *J. Geophys. Res.*, **98**, 10451–10464.

Lynch-Stieglitz, M. 1994. The development and validation of a simple snow model for the GISS GCM. *J. Climate*, **7**, 1842–1855.

Male, D.H., and Gray, D.M. 1981. Snow ablation and runoff. In *Handbook of Snow: Principles, Processes, Management and use*, D.M. Gray and D.H. Male (Eds.). Pergamon Press, Oxford, England, pp. 360–430.

Manabe, S. 1969. Climate and the ocean circulation. The atmospheric circulation and the hydrology of the Earth's surface. *Mon. Wea. Rev.*, **97**, 739–774.

Marshall, S., Roads, J.O., and Glatzmaier, G. 1994. Snow hydrology in a general circulation model. *J. Climate*, **7**, 1251–1269.

Matson, M. 1986. The NOAA satellite-derived snow cover data base: Past, present, and future. In *Glaciological Data, Report GD-18: Snow Watch '85*, G. Kukla, A. Hecht, R.G. Barry, and D. Wiesnet (Eds.). World Data Center-A for Glaciology, Boulder, CO, pp. 115–124.

Matson, M., and Wiesnet, D.R. 1981. New data base for climate studies. *Nature*, **287**, 451–456.

Matzler, C., and Schanda, E. 1984. Snow mapping with active microwave sensors. *Int. J. Remote Sens.*, **5**, 409–422.

McKay, G.A., and Gray, D.M. 1981. The distribution of snowcover. In *Handbook of Snow:*

Principles, Processes, Management and use, D.M. Gray and D.H. Male (Eds.). Pergamon Press, Oxford, England, pp. 153–190.

Mestcherskaya, A.V., Belyankina, I.G., and Golod, M.P. 1995. Monitoring of snow depth in the main wheat-producing regions of the former USSR during the period of instrumental observations. *News Russian Acad. Sci. Seria Geograph.*, 101–110 (in Russian).

Minnis, P., Harrison, E.F., Stowe, LL, Gibson, G.G., Denn, F.M., Doelling, D.R., and Smith, W.L. Jr. 1993. Radiative climate forcing by the Mount Pinatubo eruption. *Science*, **259**, 1411–1415.

Mitchell, J.F.B., Manabe, S., Meleshko, V., and Tokioka, T. 1990. Equilibrium climate change—and its implications for the future. In *Climate Change, the IPCC Scientific Assessment*, J.T. Houghton, L.G. Meira Filho, B.A. Callander, N. Harris, A. Kattenberg, and K. Maskell (Eds.). Cambridge University Press, Cambridge, England, pp. 131–172.

Mokhov, I.I. 1984. The temperature sensitivity of cryosphere area of Northern Hemisphere. *Izv. AS USSR. Phys. Atmos. Ocean*, **20**, 136–143 (in Russian).

Namias, J. 1960. Snowfall over eastern United States: factors leading to its monthly and seasonal variations. *Weatherwise*, **13**, 238–247.

Namias, J. 1962. Influence of abnormal heat sources and sinks on atmospheric behaviour. *Proc. Int. Symp. Numerical Weather Prediction*, Tokyo 1960. Meteorological Society of Japan, Tokyo, pp. 615–627.

Namias, J. 1964. Seasonal persistence and recurrence of blocking during 1958–1960. *Tellus*, **16**, 394–407.

Namias, J. 1985. Some empirical evidence for the influence of snow cover on temperature and precipitation. *Mon. Wea. Rev.*, **113**, 1542–1553.

The NOAA-NASA Pathfinder Program. 1994. University Corporation for Atmospheric Research, Boulder, Colorado 1994. 23 pp.

National Snow and Ice Data Center (NSIDC). 1995. *NSIDC Notes*, Issue No. 13, CIRES, University of Colorado, Boulder, CO.

National Snow and Ice Data Center (NSIDC). 1996. *The Historical Soviet Snow Water Equivalent Data from 1000 Stations of the Former USSR*. (Data acquired in the framework of the U.S.-Russia Agreement in the Field of Environmental Protection, Working Group VIII [The Influence of Environmental Changes on Climate: 1995 Protocol for Working Group VIII]).

National Snow and Ice Data Center (NSIDC). 2000. *The Historical Soviet Daily Snow Depth*, Version 2 (HSDSD), CD-ROM.

Peck, E.L., Carrol, T.R., and Vandmark, S.C. 1980. Operational aerial snow surveying in the United States. *Hydrol. Sci. Bull.*, **25**, 51–65.

Peterson, T.C., and Vose, R.S. 1997. An overview of the Global Historical Climatology Network Temperature Database. *Bull. Am. Meteorol. Soc.*, **78**, 2837–2849.

Pfister, C. 1985. Snow cover, snow lines and glaciers in central Europe since the 16th century. In *The Climate Scene*, M.J. Tooley and G.M. Sheail (Eds.). George Allen and Unwin, London, pp. 154–174.

Philips, T.J. 1994. *A Summary Documentation of the AMIP Models*. Program for Climate Model Diagnosis and Intercomparison, Report No. 18, Lawrence Livermore National Laboratory Publ., Livermore, CA.

Pokrovskaya, T.V., and Spirina, L.P. 1965. Estimates of snowcover impact on spring air temperature over the European part of the USSR. *Trans. Main Geophys. Observatory*, **181**, 110–113.

Pomeroy, J.W., and Gray, D.W. 1995. Snow accumulation, relocation and management. National Hydrology Research Institute Science Report No. 7, Minister of Environment: Saskatoon, Saskatchewan. 144 pp.

Pomeroy, J.W., and Jones, H.G. 1996. Wind blown snow: Sublimation, transport and changes to polar snow. In *Chemical Exchange Between the Atmosphere and Polar Snow*, E.W. Wolff and R.C. Bales (Eds.). NATO ASI Series, 143, Springer-Verlag, Berlin, pp. 453–489.

Pomeroy, J.W., and Granger, R.Y. 1997. Sustainability of the western Canadian borial forest under changing hydrological conditions. Snow accumulation and ablation. In *Sustainability of Water Resources under Increasing Uncertainty*, D. Rosjberg, N. Boutayeb, A. Gustard, Z. Kundzewicz, and P. Rasmussen (Eds.). IAHS Publ. 240, IAHS Press, Wallingford, England, pp. 237–242.

Pruitt, W.O. Jr. 1984. Snow and living things. *Northern Ecology and Resource Management*, R. Olson (Ed.). University of Alberta Press, pp. 51–77.

Quinn, W.N., and Neil, V.T. 1992. The historical records of El Niño events. In *Climate Since A.D. 1500*, R.S. Bradley and P.D. Jones (Eds.). Routledge, London, pp. 623–648.

Randall, D.A., Cess, R.D., Blanchet, J.P. et al. 1994. Analysis of snow cover feedbacks in 14 general circulation models. *J. Geophys. Res.*, **99**, 20757–20771.

Rallison, R.E. 1981. Automated system for collecting snow and related hydrological data in mountains of the western United States. *Hydrol. Sci. Bull.*, **26**(1), pp. 83–89.

Razuvaev, V.N., Apasova, E.B., Martuganov, R.A., and Kaiser, D.P. 1995. *Six- and Three-Hourly Meteorological Observations from 223 USSR Stations*. ORNL/CDIAC-66, NDP-048, Carbon Dioxide Information Analysis Center, Oak Ridge National Laboratory, Oak Ridge, TN.

Redmond, K.T., and Schaefer, G.L. 1997. Quality control of the historical SNOTEL database. *AMS Proc. 13th Conference Hydrology*, Long Beach, CA, February 2–7. American Meteorological Society, Boston, pp. 24–26.

Richter, G.D. 1954. *Snow Cover, Its Formation and Properties*. U.S. Army Cold Regions Research and Engineering Laboratory, Transl. 6, NTIS AD 045950, Hanover, NH.

Robinson, D.A. 1988. Construction of a United States historical snow data base. *Proc. 1988 Eastern Snow Conference*, Lake Placid, NY, pp. 50–59.

Robinson, D.A. 1993. Historical daily climatic data for the United States. Preprints. *Eighth Conference on Applied Climatology*, Anaheim, CA, American Meteorological Society, pp. 264–269.

Robinson, D.A., and Dewey, K.F. 1990. Recent secular variations in the extent of Northern Hemisphere snow cover. *Geophys. Res. Lett.*, **17**(10), 1557–1560.

Robinson, D.A., Dewey, K.F., and Heim, R.R. Jr. 1993. Global snow cover monitoring: an update. *Bull. Am. Meteorol. Soc.*, **74**(9), 1689–1696.

Robinson, D.A., Keimig, F.T., and Dewey, K.F. 1991. Recent variations in Northern Hemisphere snow cover. *Proc. 15th Annual Climate Diagnostics Workshop*, Asheville, NC, National Oceanic and Atmospheric Administration, Washington, DC, pp. 219–224.

Ropelewski, C.F. 1993. Northern Hemisphere snow cover and temperature pattern in the 1980s. *Proc. Snow Watch '92 International Workshop*, Niagara-on the-Lake, Ontario, Canada, March 30 to April 1, 1992, pp. 26–31.

Schlesinger, M.E. 1986. CO_2-induced changes in seasonal snow cover simulated by the OSU coupled atmosphere-ocean general circulation model. In *Snow Watch '85, Glaciological Data*, G. Kukla, A. Hecht, R.G. Barry, and D. Wiesnet (Eds.). World Data Center for Glaciology, Boalder, CO. Report GD-18, pp. 249–270.

Sellers, P.J., Mintz, Y., Sud, Y-C., and Dalcher, A. 1986. A simple biosphere model (SiB) for use within general circulation models. *J. Atmos. Sci.*, **43**, 505–531.

Sevruk, B. (Ed.). 1992. *Snow Cover Measurements and Areal Assessment of Precipitation and Soil Moisture*. Operational Hydrology Report 35, Publication 749. World Meteorological Organanisation, Geneva.

Shafer, B.A., and Huddleston, J.M. 1986. A centralised forecasting system for the Western United States. *Proc. 54th Western Snow Conference*, Phoenix, AR, April 15–17, 1986, pp. 61–70.

Schutz, C., and Bregman, L.D. 1988. *Global Annual Snow Accumulation by Months*. Note N-2687-RC, Rand Corporation. Santa Monica, CA.

Shiklomanov, I.A., Georgievsky, V.Yu., Ezhov, A.V., Shalygin, A.L., Shereshevsky, A.I., Zheleznyak, M.I., Trofimova, I.V., and Groisman, P.Ya. 1995. *An Assessment of the Influence of Climate Uncertainty on Water Management in the Dnipro River Basin*. Final Report on the Water Management Implications of Global Warming: The Dnipro River Basin. Rusian State Hydrological Institute, St. Petersburg, Russia.

Stouffer, R.J., Manabe, S., and Vinnikov, K.Ya. 1994. Model assessment of the role of natural variability in recent global warming. *Nature*, **367**, 634–636.

Sturm, M., Holmgren, J., and Liston, G.E. 1995. A seasonal snow cover classification system for local to global applications. *J. Climate*, 8, 1261–1283.

Thomas, G., and Rowntree, P.R. 1992. The boreal forests and climate. *Qt. J.R. Meteorol. Soc.*, **505**(B), 469–497.

Toomig, K.G. 1981. Correlations of mean annual albedo and short-wave radiation balance with these parameters in early spring. *Meteorologiya Gidrologiya*, No. 5, pp. 48–52 (in Russian).

Verma, R.K., Subramaniam, K., and Dugam, S.S. 1985. Interannual and long-term variability of the summer monsoon and its possible link with northern hemispheric surface air temperature. *Proc. Indian Acad. Sci. (Earth Planet. Sci.)*, **94**(3), 137–148.

Verseghy, D.L. 1991. A Canadian land surface scheme for GCMs. I. Soil model. *J. Climatol.*, **11**, 111–133.

Verseghy, D.L. 1992. Snow modeling in general circulation models. *Proc. Snow Watch*

'92 International Workshop, Niagara-on the-Lake, Ontario, Canada, March 30 to April 1, 1992, pp. 86–92.

Vershinina, L.K., and Volchenko, V.N. 1974. Maximum water storage in snow cover in the Northeast of the European part of the USSR. *Trans. State Hydrol. Inst.*, **214**, 37–53 (in Russian).

Voeikov, A.I. 1889. Snow cover, its effects on soil, climate, and weather and methods of investigations. *Notes Russian Geograph. Soc. Gen. Geogr.*, **18**(2) (in Russian).

Vose, R.S., Schmoyer, R.L., Steurer, P.M., Peterson, T.C., Heim, R.R. Jr., Karl, T.R., and Eischeid, J.K. 1992. *The Global Historical Climatology Network: Long-Term Monthly Temperature, Precipitation, Sea Level Pressure, and Station Pressure Data*. Carbon Dioxide Information Analysis Center, Oak Ridge National Laboratory, U.S. Dept. of Energy, Environmental Sciences Division Publication No. 3912.

Wagner, A.J. 1973. The influence of average snow depth on monthly mean temperature anomaly. *Mon. Wea. Rev.*, **101**, 624–626.

Walsh, J.E. 1987. Large-scale effects of seasonal snow cover. In *Large Scale Effects of Seasonal Snow Cover*, B.E. Goodison, R.G. Barry, and J. Dozier (Eds.). IAHS Publication No. 166. IAHS Press, Wallingford, UK, pp. 3–14.

Walsh, J.E., and Ross, B. 1988. Sensitivity of 30-day dynamical forecasts to continental snow cover. *J. Climate*, **1**, 739–754.

Walsh, J.E., Tucek, D.R., and Peterson, M.R. 1982. Seasonal snow cover and short-term climatic fluctuations over the United States. *Mon. Wea. Rev.*, **110**, 1474–1485.

Watson, R.T., Zinyowera, M.C., and Moss, R.H. 1996. *Climate Change 1995-Impact, Adaptations and Mitigation of Climate Change*. Scientific-Technical Analyses, Cambridge University Press, Cambridge, England.

Weaver, R., Morris, C., and Barry R.G. 1987. Passive microwave data for snow and ice research: Planned products from the DMSP SSM/I system. *EOS*, **68**, 769–777.

Wiesnet, D.R., Ropelewski, C.F., Kukla, G.J., and Robinson, D.A. 1987. A discussion of the accuracy of NOAA satellite-derived global seasonal snow cover measurements. In *Large Scale Effects of Seasonal Snow Cover, Proc. Vancouver Symposium*, B.E. Goodison, R.G. Barry, and J. Dozier (Eds.). IAHS Publication No. **166**, Wallingford, UK, pp. 291–304.

Wigmosta, M.S., Vail, L.W., and Lettenmaier, D.P. 1994. A distributed hydrology-vegetation model for complex terrain. *Water Resources Res.*, **30**, 1665–1679.

Williams, J. 1975. The influence of snow cover on the atmospheric circulation and its role in climatic change: An analysis based on results from the NCAR global circulation model. *J. Appl. Meteorol.*, **14**, pp. 137–152.

Williams, L.D. 1978. Ice-sheet initiation and climatic influences of expanded snow cover in Arctic Canada. *Quatern. Res.*, **10**, 141–149.

Wood, E.F. 1991. Global scale hydrology: advances in land surface modeling. U.S. National Report to International Union of Geodesy and Geophysics, 1987–1900. *Rev. Geophys.*, **29** Supplement, pp. 193–201.

Wood, E.F., Lettenmaier, D.P., and Zartarian, V.G. 1992. A land-surface hydrology

43

parameterization with subgrid variability for general circulation models. *J. Geophys. Res.*, **97**(D3), 2717–2728.

Xue, Y., Sellers, P.J., Kinter, J.L., and Shukla, J. 1991. A simplified biosphere model for global climate studies. *J. Climate*, **4**, 345–364.

Yang, S. 1993. ENSO-snow-monsoon associations and seasonal prediction. *Proc. Fourth Symp. Global Change Studies*, Anaheim, CA, American Meteorological Society, Boston, pp. 418–422.

Yasunari, T. 1987. Global structure of the El Niño/Southern Oscillation. Part 2. Time evolution. *J. Meteorol. Soc. Japan Series II*, **65**, pp. 81–102.

Ye, H., Cho, H.-R., and Gustafson, P.E. 1997. The changes of Russian winter snow accumulation during 1936–1983 and its spatial patterns. *J. Climate*, **11**, 856–863.

Yeh, T.C., Wetherald, R.T., and Manabe, S. 1983. A model study of the short-term climatic and hydrologic effects of sudden snow-cover removal. *Mon. Weath. Rev.*, **111**, 1013–1024.

Yeh, T.C., Wetherald, R.T., and Manabe, S. 1984. The effect of soil moisture on the short-term climate and hydrology change – a numerical experiment. *Mon. Weath. Rev.*, **112**, 474–490.

2 Physical Properties of Snow

J.W. POMEROY AND E. BRUN

2.1   Introduction: Snow Physics and Ecology

As demonstrated in the previous chapter, snow interacts strongly with the global climate system, both influencing and forming as a result of this system. The following chapters discuss the interaction of snow with the chemical and biological systems. This chapter discusses the physical properties of snow as the habitat and regulator of the snow ecosystem. In this sense, the physical snow cover may be perceived not only as the medium but also as the mediator of the snow ecosystem that transmits and modifies interactions between microorganisms, plants, animals, nutrients, atmosphere, and soil. The role of snow as a habitat for life is discussed by Hoham and Duval (Chapter 4) and Aitchison (Chapter 5). Snow cover mediates the snow ecosystem because it functions as an:

1. Energy bank: snow stores and releases energy. It stores latent heat of fusion and sublimation and crystal bonding forces (Langham, 1981; Gubler, 1985). The bonding forces are applied by atmospheric shear stress, diffusion along, vapour pressure gradients, drifting snow impact, and the impact of animals' walking over the snow (Schmidt, 1980; Pruitt, 1990; Kotlyakov, 1961; Sommerfeld and LaChapelle, 1970). The intake and release of energy throughout the year makes snow a variable habitat.
2. Radiation shield: cold snow reflects most shortwave radiation and absorbs and reemits most long-wave radiation (Male, 1980). Its reflectance of shortwave radiation is a critical characteristic of the global climate system. As snowmelt progresses, the snow cover reflects less shortwave radiation because of a change in its physical properties (O'Neill and Gray, 1973). This reflectance can be additionally reduced about 50 percent by in situ life forms such as populations of snow algae (Kohshima et al., 1994).
3. Insulator: as a porous medium with a large air content, snow has a high insulation capacity and plays an important role protecting microorganisms,

45

plants, and animals from wind and severe winter temperatures (Palm and Tveitereid, 1979). Its insulation can result in strong temperature gradients that fundamentally restructure the snow composition and provide opportunities and constraints for organisms that live in the snow cover (Colbeck, 1983).

4. Reservoir: snow is a reservoir of water that provides habitat and food sources for various life stages of microbes, invertebrates, and small mammals. The physical properties of snow – especially radiation penetration, density, gas content, temperature, wetness, and porosity – control intra-nivean biological activity and in turn may be influenced by the behaviour of supranivean organisms.

5. Transport medium: snow moves as a particulate flux as it is relocated by the wind in open environments or intercepted by vegetation in forests (Budd, Dingle, and Radock, 1966; Schmidt and Gluns, 1991). This flux is influenced to a great degree by macrophyte vegetation. Snow is transformed to a vapour because of sublimation, resulting in transport to colder surfaces or to the atmosphere (Schmidt, 1991; Santeford, 1979). During melt, snow moves as meltwater in preferential pathways within the snowpack to the soil or directly to streams and lakes (Marsh and Woo, 1984a).

2.1.1 Unique Physical Properties of Snow

Snow covers are the milieu or habitat of unique ecosystems partly because of the distinctive physical properties of snow compared with other environments on the Earth's surface. These properties are intrinsic to the snow ecosystem, and all organisms living in the snow ecosystem must contend with or take advantage of them in order to survive and prosper. The unique physical properties of snow are as follows:

1. At $0°C$ water may exist as a solid, liquid, or vapour on the Earth's surface; below this temperature water exists as primarily as ice (snow) and vapour supplemented by thin liquid-like layers on the edge of snow crystals. Above $0°C$, water exists as liquid or vapour. Because of diurnal and annual temperature variations, most of the Earth's snow is seasonal in that it melts (to liquid) or sublimates (to vapour) on an annual basis and interacts with vapour and liquid-like phases of water at temperatures below zero. The liquid water content of a snowpack declines rapidly as the snow temperature drops below $0°C$. However, other factors are also important, such as the

rate of melt within the snow cover, rate of rainfall to the snow-covered surface, liquid retention capacity of snow, and rate of drainage of water from the pack. The combination of these factors can cause rapid variations in the liquid water content of snowpacks.

2. The latent heat of vaporisation is extremely large, approximately 2.83 MJ kg^{-1} of snow. The energy required to sublimate 1 kg of snow therefore is equivalent to that required to raise the temperature of 10 kg of liquid water 67°C. Vaporisation is reversible, as this energy is released to the environment upon recrystallisation of vapour to ice.

3. The latent heat of fusion is large, approximately 333 kJ kg^{-1} of snow. The energy required to melt 1 kg of snow (already at 0°C) therefore is equivalent to that required to raise the temperature of 1 kg of water 79°C. Latent heat is released to the environment during freezing, when liquid water crystallises.

4. The thermal conductivity of a snow cover is low compared with soil surfaces and varies with the density and liquid water content of the snow cover. A typical thermal conductivity for dry snow with a density of 100 kg m^{-3} is 0.045 W m^{-1} K^{-1}, over six times less than that for soil. This means that snow can insulate over six times more effectively than soil for equivalent depths. The total insulation provided by snow strongly depends on its depth.

5. The proportion of shortwave (solar) radiation incident upon a snow cover and then reflected (albedo) is high compared with soil and vegetation and varies over the winter. A fresh, continuous snow cover has an albedo of 0.8–0.9; as a snow cover ages and becomes patchy and wet the snow albedo can drop to 0.5 with areal albedo dropping even further as vegetation and soil become partly exposed. Bare soil and vegetation therefore will absorb as much as eight times the shortwave radiation as a fresh, continuous snow cover. The shortwave radiation that is not reflected by a snow cover is absorbed largely in the top 30 cm of the snowpack. The degree to which this radiation penetrates varies with wavelength – in general, the shorter wavelengths penetrate further than longer wavelengths.

6. Snow cover behaves almost as a blackbody; hence, the long-wave (thermal infrared) radiation incident on a snow cover is absorbed and reradiated as thermal radiation. The wavelength of emission depends on the surface temperature of the snow cover.

47

7. Snow covers are aerodynamically smooth compared with most land surfaces. Snow surfaces have aerodynamic roughness heights (z_0) of 0.01 to 0.7 mm, except during drifting snow when z_0 increases substantially. Land surfaces typically have aerodynamic roughness heights several orders of magnitude greater than this. The result is that, for a constant geostrophic wind speed, the wind speed is usually greater over snow cover than over vegetated surfaces, and turbulent transfer of sensible and latent heat between the atmosphere and the surface is smaller for a snow surface than for adjacent vegetated surfaces.

2.2 Accumulation

Snow accumulation is the first part of any snow ecology study, as the snow cover is the abiotic core of any snow ecosystem. Accumulation is traced from snow formation in the atmosphere and precipitation as snowfall, to wind transport, interception, and deposition and from large scales to small scales with reference to the important role of overwinter sublimation in reducing the final accumulation of snow in many snow environments.

2.2.1 Snowfall

An understanding of snowfall formation helps the snow ecologist to better understand the structure of snow crystals and snow covers and to interpret the spatial distribution of the snow environment. Snowfall formation is also important to snow chemistry, as the incorporation of chemical species in the forming snow crystals leads to wet deposition of chemical species (see Tranter and Jones, Chapter 3). Snow crystals derive from clouds, which form when the atmosphere is supersaturated in that the water vapour pressure exceeds the saturation level. Snow forms in clouds when the temperature is less than $0°C$ and supercooled water and suitable aerosols (cloud condensation nuclei; CCN) are present. Growth of ice crystals from vapour results in snow crystals, and growth from collision of supercooled water droplets produces hail and graupel particles. Ice crystals form around CCN and rapidly grow through aggregation of small ice crystals and riming from water droplets into the familiar snowflake form. The most effective CCN are large aerosol particles with high water solubility. As marine air masses contain higher concentrations of large CCN than do continental air masses, precipitation is more likely from marine air masses with other factors constant (Wallace and Hobbs, 1977). For snowfall to occur there must be sufficient

48

depth of cloud to permit the growth of snow crystals and sufficient moisture and aerosol nuclei to replace those removed from the cloud in falling snowflakes (Schemenauer, Berry, and Maxwell, 1981).

Snowfall in meteorological records is the depth of fresh snow that falls to the ground during a given period (see Groisman and Davies, Chapter 1). In many countries this is measured with a snowfall gauge, which is an open-top cylinder that is exposed to snowfall with resulting accumulations measured periodically and expressed as millimetres of *snow water equivalent* (SWE) or the equivalent depth of water on the ground if all fresh snowfall melted. Snowfall gauges should be shielded from wind exposure to reduce undercatch due to wind, but even better shields such as the Nipher or Tretyakov shields are subject to undercatch of about 25 percent at wind speeds of $7\,\mathrm{m\ s^{-1}}$. Pomeroy and Goodison (1997) report upward revision of published annual snowfall quantities of 31 percent in the Canadian Prairies and from 64 to 161 percent in the Canadian Arctic when snowfall records are corrected for gauge undercatch and other losses. Snowfall in some countries is expressed as a depth (centimetres) of new snow on the ground, assuming a density of $100\,\mathrm{kg\ m^{-3}}$, or as a depth of equivalent water on the ground in millimetres as SWE. If these measurements are reported from areas with open exposure to strong winds (farmland, prairies, tundra) it is almost impossible to relate them to the original snowfall quantity because of redistribution by wind. Even the density of fresh snow (ρ_s, $\mathrm{kg\ m^{-3}}$) that is deposited without strong wind speeds varies considerably from the $100\,\mathrm{kg\ m^{-3}}$ assumption; for instance, Goodison, Ferguson, and McKay (1981) found that fresh snow density in Canada varied from 70 to $165\,\mathrm{kg\ m^{-3}}$. The variation of fresh, dry snow density, ρ_s, with air temperature (T_a, $^\circ$C) can be estimated with an equation developed by Hedstrom and Pomeroy (1998) from measurements reported by the U.S. Army Corps of Engineers (1956) in the Central Sierra Mountains of California and by Schmidt and Gluns (1991) in British Columbia and Colorado, where

$$\rho_s = 67.9 + 51.25 e^{\frac{T_a}{2.59}} \tag{2.1}$$

The relationship suggests fresh snow densities of $143\,\mathrm{kg\ m^{-3}}$ at air temperatures of 1°C declining to $68\,\mathrm{kg\ m^{-3}}$ for temperatures below about -10°C.

2.2.2   Distribution of Snowfall and Snow Cover

The distribution of snowfall may be distinguished into two scales, macro- and mesoscale, with local-scale effects more strongly associated with redistribution after snowfall. Macroscale snowfall distribution includes distances of 100 to 1,000 km and varies with latitude, physiographic province, and proximity to large bodies of water.

At this scale, dynamic meteorological effects such as flow deviations due to the Coriolis effect, standing waves in the atmosphere, flow around mountain ranges, latitudinal temperature changes, and regional moisture sources such as oceans are important. Mesoscale snowfall distribution includes distances of 1 to 100 km, over which changes in orographic cooling or warming, deviation of flow due to sharp topographic change, or convective precipitation in response to addition of heat and water vapour from lakes occur. Groisman and Davies (Chapter 1) discuss snow climatology and the large-scale factors that influence snowfall and distributions of snow cover on a global scale.

Figure 2.1 provides an example of the mesoscale distribution of snowfall in the Great Lakes region of North America (Norton, 1991). Most of the Great Lakes remain open throughout much of the winter, providing a water vapour source to the atmosphere. As topographic influences alone are small in this area, the major influences on snowfall are the dominant paths of winter storms over the Great Lakes and the decrease in temperature with latitude and from water to land. Intensive local convection occurs over relatively warm lake water, producing cloud and snowfall, which move inland with the prevailing wind. As the air mass rises upon reaching the lakeshore, it cools, saturates, and precipitates, producing large amounts of snowfall. These influences cause an increase in annual snowfall from 100 cm deep in southwestern Ontario, where relatively few storms arrive from the Great Lakes, to 320 cm just east of Lake Huron, where a northwesterly flow over the lake dominates after passage of a cold front. Intensive areas of snow accumulation also occur in the fabled snowbelt southeast of Lake Erie and Lake Ontario, where heavy snowfalls occur from Cleveland to Buffalo.

Where vegetation and micro-relief do not vary with elevation, the depth of seasonal snow cover usually increases with elevation because of the increasing number of snowfall events and the decrease in melt as elevation increases. Thus, at a specific location in a mountainous region, a strong linear association is often found between seasonal SWE and elevation within a selected elevation band (U.S. Army Corps of Engineers, 1956). As demonstrated by Meiman (1970), however, even along specific transects the rate of increase in SWE with elevation may vary widely from year to year. In northern latitudes, relatively small differences in evaporation and melt over an elevation transect suppress the observed increase in snow depth with elevation. In western Canadian mountains, the influence of elevation on snow accumulation is most apparent at elevations above 600 m. The increase in accumulation with elevation is more dramatic where wind flows ascend mountains rather than where a descending flow occurs, as shown in Figure 2.2 for the lee side of the Monashee Mountains and the windward side of the Selkirk Mountains of British Columbia (Auld, 1995).

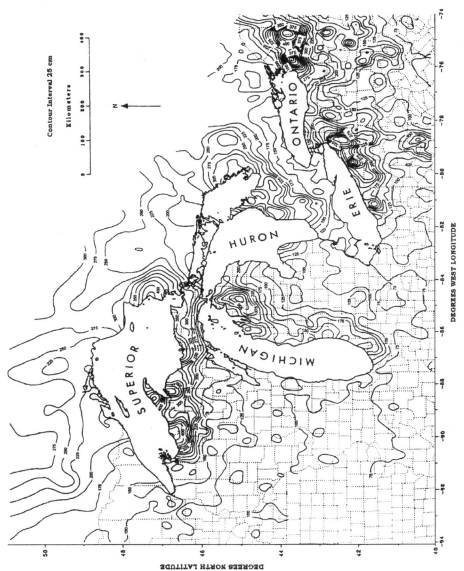

Figure 2.1. Mesoscale distribution of mean seasonal snowfall (1951–1980) in the Great Lakes region of North America, showing the influence of water bodies on snowfall over downwind uplands (after Norton, 1991).

51

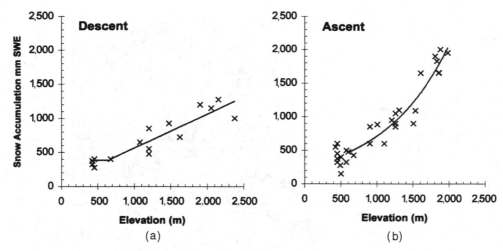

Figure 2.2. Variation in snow accumulation with elevation as mountain ranges are ascended and descended by air masses, which generate snowfall. (a) Descending air mass is shown for the Monashee Mountains. (b) Ascending air mass is shown on the right for the Selkirk Mountains, British Columbia (after Auld, 1995).

Elevation alone is not a causative factor in snow cover distribution; a host of other variables such as slope, aspect, vegetation, wind, temperature, and weather systems must be considered to interpret distribution patterns accurately. Orographic precipitation can be more related to terrain slope and wind flow than to elevation in mountain environments with complex topography. Rhea and Grant (1974) studied the effects of large-scale vertical air mass movement, convective activity, and orographic cooling on Rocky Mountain snowfall. From an analysis of Colorado winter precipitation they found that the long-term average snowfall at a point was strongly correlated with the topographic slope located 20 km upwind. They also found that long-term average precipitation is not well correlated with station elevation except for points along the same ridge.

2.2.3   Wind Redistribution

The effects of wind on the evolution of a snow cover are most evident in open environments. Wind redistribution of snow involves *erosion* of snow cover by the shear force of the wind, *transport* of blowing snow from exposed sites with low aerodynamic roughness, *sublimation* of blowing snow in transit, and *deposition* of snow to sites with higher aerodynamic roughness or less exposure to wind. Dyunin et al. (1991) noted that wind redistribution of snow is the primary process of desertification in steppe environments, and they associate the phenomenon of "northern desertification" with suppression of vegetation in the Russian steppes and forests and the resulting increase in frequency of blowing snow. Dyunin recalls that the Soviet administration attempted to manage snow ecology in the Russian steppes through "Stalin's Nature Transformation Plan," which involved increasing vegetation on the steppes to improve regional water supplies in the 1930s. Walker, Billings, and de Molenaar (Chapter 6) discuss the implications of snow redistribution on vegetation and soils of alpine ecosystems. Estimates of maximum snow cover erosion due to blowing snow are similar in continental steppe or prairie environments; Komarov (1957) estimates that up to 70 percent of annual snowfall is eroded by blowing snow from open terrain in Siberia. In North America the estimate for highland plains in Wyoming is up to 75 percent (Tabler and Schmidt, 1986) and for fallow fields in the southern Canadian prairies it is up to 77 percent (Pomeroy and Gray, 1995). Tabler et al. (1990a) found that about 86 Mg of snow per metre width was transported by wind on the arctic coast of Alaska. Estimates of sublimation increase with downwind fetch but can be up to five times larger than the amount of snow transported off a fetch for long fetches. Tabler (1975) found that 25 percent of eroded snow sublimated in transport over fetches of 1 km, increasing to 75 percent for a 6-km fetch. Snow accumulation in wind-swept environments in the absence of concurrent melt therefore cannot be determined directly from snowfall but from the following mass balance:

$$Q_{\text{surface}} = Q_{\text{snowfall}} - \frac{dQ_{\text{T}}}{dx}(x) - Q_{\text{E}} \qquad (2.2)$$

where Q_{surface} is snow accumulation (kg m^{-2} s^{-1} = mm of SWE s^{-1}), Q_{snowfall} is the snowfall rate (kg m^{-2} s^{-1} = mm of SWE s^{-1}), Q_{T} is the blowing snow horizontal transport flux (kg m^{-1} s^{-1}), x is some distance along a fetch (m), and Q_{E} is the sublimation flux (kg m^{-2} s^{-1}). Erosion of snow cover is caused by the downwind increases in transport rate and the sublimation flux. Deposition of snow cover is caused by snowfall and downwind decreases in transport rate. Pomeroy, Marsh, and Gray (1997a) found that, in the Low Arctic of northwestern Canada, snow accumulation varied from 54 to 419 percent of snowfall, depending on vegetation and exposure. Similar degrees of variation in snow accumulation are found in prairie environments (Gray and others, 1979).

Blowing snow transport involves three modes of movement:

1. *Creep*: the rolling movement of those particles that are too heavy to be lifted by the wind.

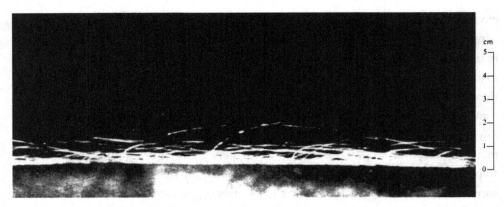

Figure 2.3. Photograph of trajectories of saltating snow particles, with flow direction from left to right (Kobayashi, 1972).

2. *Saltation*: the movement of snow particles by a "skipping" or "jumping" action along the snow surface.
3. *Turbulent diffusion* (suspension): the movement of snow particles in suspended flow at a mean horizontal velocity close to that of the moving air.

In the first few centimetres above the snow surface, particles move primarily in saltation (Figure 2.3). Saltation is the source of all other snow particle movement and is itself derived from eroded snow. Erosion of snow particles occurs when the shear stress exerted on the snow surface by the wind exceeds the sum of snow surface strength and any shear force exerted on vegetation and rocks. The partitioning of atmospheric shear stress by moving snow, surface snow, and vegetation is critical to understanding the control exerted on blowing snow by vegetation and wind exposure. The shear stress itself is approximately proportional to the square of the wind speed; the proportionality is controlled by the aerodynamic surface roughness.

The atmospheric shear stress directly applied to the snow surface does not contribute to supporting the weight of the saltating snow, but it breaks snow crystal bonds and overcomes cohesive forces due to wetness in the snow cover. By breaking bonds and overcoming cohesion, this stress makes wind erosion of snow possible. The shear stress is related to the threshold wind speed, which is the wind speed at the termination or initiation of snow transport. Higher threshold wind speeds are associated with stronger particle bonds and cohesion. Kotlyakov (1961) shows a correlation between threshold wind speed and surface snow hardness in Antarctica. Kind (1981) reports data from several authors who note snow surface hardness of 100 N m^{-2} for a threshold wind speed of 4.3 m s^{-1} rising to 100 kN m^{-2} for a threshold wind speed of 9 m s^{-1} and

$250 \, \text{kN m}^{-2}$ for $18 \, \text{m s}^{-1}$. By describing the resistance of the snow surface to shattering, the threshold wind speed defines the relationship between saltation transport rate and wind speed (Pomeroy and Gray, 1990). Li and Pomeroy (1997) found that for northern prairie environments wet or icy snow has a higher threshold wind speed ($9.9 \, \text{m s}^{-1}$ at 10-m height) than does dry snow ($7.5 \, \text{m s}^{-1}$ for fresh dry snow, $8.0 \, \text{m s}^{-1}$ for aged dry snow). Threshold wind speeds of dry snow are lowest at temperatures near $-25°C$ ($7 \, \text{m s}^{-1}$), increase slightly at colder temperatures, and increase more dramatically as the temperature approaches $0°C$ ($9.4 \, \text{m s}^{-1}$).

Pomeroy and Gray (1990) have modeled saltation transport as a function of the threshold wind speed, exposure of vegetation, and wind speed. The implications of their model are that although saltation transport starts at lower wind speeds over fresh, loose snow, it is relatively inefficient over these snow surfaces. At high wind speeds, greater transport rates may be achieved over the high threshold wind-hardened snow. Exposed vegetation rapidly diminishes the wind energy available to transport snow; as a result, little or no snow erosion can occur from within stands of dense vegetation until the snow depth is within a few centimetres of the vegetation height.

Suspended snow is supported by turbulent diffusion and carried along by moving air at a velocity approximately equal to that of the mean horizontal wind. The source of all suspended snow (in the absence of concurrent snowfall) is saltating snow, so suspended transport can proceed only when saltation transport is occurring. Instead of bouncing along the snow surface, suspended snow is lifted from the top of the saltation layer by turbulent eddies in the atmosphere. Suspended snow exists in a layer from a few centimetres above the snow surface to hundreds of metres above the ground; its maximum height is determined by the duration of the blowing snow storm, intensity of turbulence, and upwind distance of open, snow-covered terrain. As shown in Figure 2.4, the concentration of suspended snow above continuous snow covers reaches a maximum just above the saltation layer and decreases with height at a rate that depends on the wind speed. The peak mass flux of suspended blowing snow may be responsible for abrasion of stems from exposed plants (Bégin and Boivin, Chapter 7) and occurs 10–25 cm above the snow surface, being higher with stronger wind speeds. Pomeroy and Male (1992) show that 77 percent and 40 percent of this total transport would occur below a height of 1 m at 10-m wind speeds of $10 \, \text{m s}^{-1}$ and $30 \, \text{m s}^{-1}$, respectively. Although mass concentration of suspended snow is usually smaller than that of saltating snow, the total mass flux of snow moving in suspension may be large because of the great vertical thickness of the layer and higher wind speeds away from the snow surface.

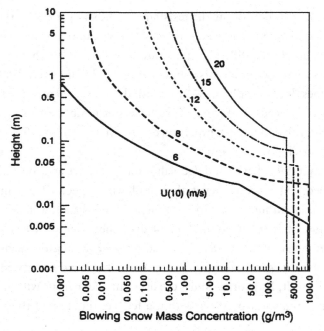

Figure 2.4. Vertical profiles of blowing snow mass concentration as a function of wind speed (after Pomeroy and Male, 1992).

The total snow transport rate q_T can be estimated by summing the saltation and suspension components. Figure 2.5 shows the relative contribution of saltation and suspension transport for a threshold wind speed corresponding to the 10-m wind speed, $u_{10} = 5.5$ m s^{-1} (dry snow of moderate hardness), over an unvegetated, level, and extensive snow field. Suspension dominates over saltation transport, especially at high wind speeds. Alternatively, empirical expressions are available that may be used to estimate q_T directly from wind speed. Pomeroy, Gray, and Landine (1991) recommend

$$q_T = 0.0000022 \, u_{10}^{4.04} \tag{2.3}$$

where q_T is blowing snow horizontal transport (kg s^{-1}) through a 1-m-wide column extending over a range of height from 0 to 5 m and u_{10} (m s^{-1}) is the wind speed measured at a 10-m height.

Sublimation from blowing snow is a significant loss to the surface snow pack during dry winter weather. The theory of sublimation of blowing snow has been described (Schmidt, 1972; Male, 1980; Pomeroy, Gray, and Landine, 1993). The high rates of sublimation of blowing snow particles with respect to the surface of a snow cover are

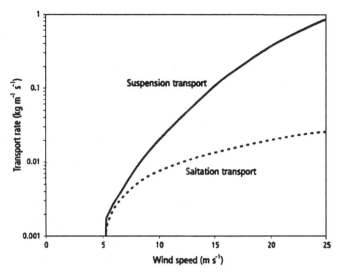

Figure 2.5. Transport rate of blowing snow by saltation and suspension as a function of wind speed (after Pomeroy and Male, 1992).

due to much higher turbulent exchange for suspended snow and a much higher ratio of surface area to mass for a snow particle removed from the pack. For instance, a unit mass of blowing snow has about 3,000 times more surface area exposed to the atmosphere than does an equivalent mass of snow on the ground. The sublimation rate can be calculated with a heat and mass transfer equation developed by Schmidt (1972) and implemented for standard meteorological measurements by Pomeroy (1989). The magnitude of sublimation loss depends on blowing snow mass concentration, wind speed, air temperature, relative humidity, and, less importantly, solar radiation.

Daily variations in temperature or humidity through typical ranges cause at least an order of magnitude change in the sublimation rate, whereas variations in daily radiation input cause only a small change in this rate. For example, the sublimation rate increases 25-fold when air temperature increases from $-35°C$ to $-1°C$. For a blowing snow storm with wind speed at 10-m height equal to $20 \, m \, s^{-1}$, temperature of $-15°C$, and relative humidity of 70 percent, the sublimation loss from blowing snow particles is equivalent to about 1 mm of SWE per hour and will produce an erosion rate of this magnitude where fetches are sufficiently long. Pomeroy and Gray (1995) have calculated the annual losses of snow to wind transport (to field edges, hedgerows, gullies, etc.) and to sublimation during blowing snow on the Canadian Prairies. The average annual blowing snow losses on 1-km fetches of grain stubble (stalk height = 25 cm)

57

and fallow (surface roughness equivalent to grain stalk height = 1 cm) at four locations in the province of Saskatchewan show:

1. On average at least 8 to 11 percent of annual snowfall is removed from a 1-km fetch of 25-cm stubble by wind transport and redeposited. On the southern prairies this percentage doubles (19 percent).
2. Blowing snow fluxes are greater on fallow than stubble. The largest increases with a change in land use from stubble to fallow occur on the southern prairies where the annual transport at least doubles and the annual sublimation losses increase by 7 percent.
3. The percentage of annual snowfall lost to sublimation from a 1-km fetch ranges from 23 to 41 percent of annual snowfall, an amount equal to or greater than (ranging up to 2.6 times) the amount of snow transported and redeposited. The results from 16 locations in the Canadian Prairies show that the average annual amount of snow water lost to sublimation from a 1-km fetch may be up to five times the amount of snow transported to the field edge.

In the prairies of North America, blowing snow erosion generally tends to decrease with increasing latitude from the grassland to boreal forest because of climatic differences and changes in vegetation. Lower wind speeds, air temperatures, and higher humidities prevail at the northerly stations and the vegetation changes from open grassland to mixed grassland, deciduous, and boreal forest. In Arctic regions, blowing snow fluxes are also large, despite low temperatures, because of open exposure and high winds. Pomeroy and Marsh (1997) estimated that 27 percent of annual snowfall sublimated as blowing snow in tundra locations along the subarctic/arctic transition with 31 percent sublimating from tundra surfaces well north of the treeline in northwestern Canada.

Pomeroy and Gray (1994) compared monthly snow losses to transport (to the field edge where redeposition occurs) and sublimation with mean monthly climate variables in order to describe the climatology of blowing snow over the prairie provinces; the results are briefly discussed below.

1. Wind speed: mean monthly blowing snow transport and sublimation tend to increase linearly with increasing monthly mean of wind speed.
2. Temperature: mean monthly blowing snow transport and sublimation increase with decreasing monthly minimum or maximum temperature (below 0°C). The trend for sublimation is opposite the effect of temperature on

the sublimation rate for an ice sphere. However, low air temperatures result in snow covers with low cohesion and low transport thresholds and hence more frequent transport. The increased frequency of blowing snow events more than compensates for lowering of the sublimation rate due to a decrease in temperature. The effect of decreasing air temperature on snow transport is most important at low wind speeds. For a mean monthly wind speed of 5 m s^{-1}, transport quadruples as the monthly mean of daily maximum air temperature drops from $-2°$C to $-25°$C. Mean monthly snow transport on stubble displays a strong temperature dependence because of the effect of midseason melts in reexposing buried vegetation. The frequency of these melts increases as temperature increases; hence, transport increases dramatically as monthly mean daily maximum temperature drops below $-10°$C.

3. Relative humidity: there is a slight increase in snow transport on fallow and stubble land with increasing relative humidity. This trend is presumed to be due to higher humidities that suppress sublimation, leaving more snow available for transport.

4. Snowfall and snow depth: on fallow land, the transport and sublimation fluxes increase with increasing mean monthly snowfall. Sublimation rate exhibits the greatest increase per unit increase in snowfall. On stubble land mean monthly snow transport and sublimation increase exponentially with increasing depth of snow on the ground. The largest increase occurs when the snow cover water equivalent is less than 30 mm because stalks of vegetation are exposed to the wind. At greater depths of snow cover the vegetation is inundated and snow transport will be similar to that on a field of fallow.

In open environments vegetation and terrain features can produce dramatic variations in snow accumulation by trapping wind-transported snow. The effect of a hedgerow in the centre of a field and varying crop vegetation heights is shown in Figure 2.6. Snow accumulation near the hedge is over 10 times greater than that in adjacent fields, and snow accumulation in the fields depends on wheat stubble height. The effect of a sharp topographic change is presented in Figure 2.7, which shows a SWE transect at Trail Valley Creek, 60 km north of Inuvik, Northwest Territories, Canada. The upper tundra plateau drops sharply to the valley bottom, inducing formation of a major side drift. Slightly higher lowland vegetation on the valley bottom holds more snow than the short vegetation on the upland plateau. Mountain ridges are a special case, as data of Föhn and Meister (1983) from near Davos, Switzerland,

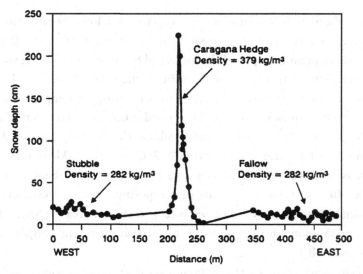

Figure 2.6. Transects of snow depth and density along a fallow field and stubble field divided by a Caragana hedge, measured east of Saskatoon, Canada, in late winter (after Pomeroy et al., 1993).

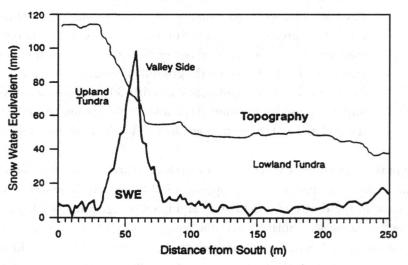

Figure 2.7. Transects of SWE and topography from an upland tundra plateau down a valley side to a lowland tundra valley bottom, north of Inuvik, Canada, measured in late winter (after Pomeroy and Gray, 1995).

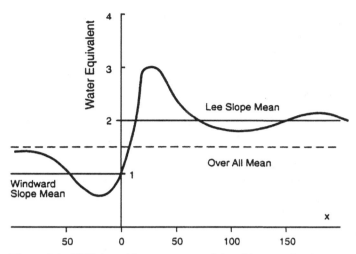

Figure 2.8. SWE deposition patterns on alpine ridges as related to windward and lee slope accumulation, near Davos, Switzerland (after Föhn and Meister, 1983).

show in Figure 2.8. For a snowdrift formed from a prevailing wind direction perpendicular to a sharp crest, the windward and lee slopes have distinctive patterns. The mean snow accumulation on the lee slope is about twice that formed on the windward slope, with a maximum accumulation in the lee slope cornice about six times that of the minimum accumulation on the windward slope just below the crest. The drifts formed near hedges, mountain crests, and valley sides persist much later than adjacent snow covers during melt and provide an important source of late spring meltwater.

The landscape-wide effects of the small-scale snow cover response to vegetation and terrain are particularly evident in prairie and arctic landscapes where a long and windy winter season with a mix of open short vegetation terrain, brushland, pockets of trees, plateaux, and gullies produces distinctive snow accumulations in various landscape types. Snow accumulations in a prairie environment are shown in Table 2.1, which lists the relative amounts of snow water retained by various landscapes at a time near maximum accumulation. It is evident that the combination of terrain and vegetation types provide landscape classes with distinctive influences on snow accumulation.

In Arctic environments similar types of snow accumulation patterns occur with snow relocated to sites of high vegetation and sharp topography. Pomeroy et al. (1997a) divided a forest-tundra transitional landscape north of Inuvik, Northwest Territories, Canada into *sources* (frozen water, tundra, sparse shrub tundra, exposed soil) and *sinks* (open and closed shrub tundra, sparse forest, drifts) of blowing snow based on

61

Table 2.1. *Relative snow water retention on various landscape types in an open grassland environment (Gray et al., 1979).*

Landscape type	Relative accumulation[a]
Level plains	
fallow	1.00
stubble	1.15
pasture (grazed)	0.60
Gradual hill and valley slopes	
fallow, stubble, hayland	1.0–1.10
pasture (ungrazed)	1.25
Steep hill and valley slopes	
pasture (ungrazed)	2.85
brush	4.20
Ridge and hilltops	
fallow, ungrazed pasture	0.40–0.50
stubble	0.75
Small shallow drainageways	
fallow, stubble, pasture (ungrazed)	2.0–2.15
Wide valley bottoms	
pasture (grazed)	1.30
Farm yards (mixed trees)	2.40

[a]Snow water equivalent normalized to snow water equivalent monitored on level plains under fallow.

vegetation and topographic slope criteria. Drifts were indicated for incised stream channels or slopes exceeding 9° adjacent to source areas. They then simulated snow cover development with a blowing snow model for each of the landscape types. The results are shown for one seasonal simulation in Figure 2.9, and indicate drastically different accumulation regimes for drift and tundra sites. The tundra reached its maximum snow accumulation after five months, where as the drift continued to accumulate snow to the eighth month of winter with a maximum measured accumulation about nine times greater than that on the tundra. The shrub-tundra accumulation regime more closely followed snowfall inputs because exposed vegetation restrained blowing snow.

Snow management is practised to increase snow accumulation in windswept areas, to provide water for crops and livestock in many agricultural environments, and to control snow drifting onto highways and buildings. The primary principle used is to increase the roughness of the landscape to trap blowing snow. Commonly used

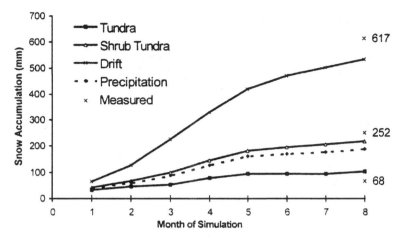

Figure 2.9. Monthly SWE accumulation for different landscape types in a lowland tundra environment, calculated by using a distributed blowing snow model. Modeled period is from October 1992 through May, 1993. Measured SWE is based on long snow surveys of depth and density conducted in early spring (Pomeroy et al., 1997a).

practises are windbreaks of snow fences, woody vegetation (caragana, ash, maple, or other hedgerows), or nonwoody vegetation (strips of sudan grass, tall wheatgrass, uncut strips of grain); crop stubble management (tall and alternate height stubble, trap strips); and snow ridges (Pomeroy and Gray, 1995). Tall stubble will generally induce snow accumulation to near the height of the stubble. An example of snow accumulation in alternative height stubble is shown in Figure 2.10, where snow depth and stubble height are plotted along a transect; this technique is quite effective and still provides straw for farmers. Pomeroy and Gray recommend the following equation to predict the maximum mean depth of snow, d_s (m),

$$d_s = H - v\frac{A_s}{S_d} \tag{2.4}$$

where H is the average plant height (m), A_s is the average ground surface area occupied by each plant (m^2), and S_d is the stalk diameter or width of plant (m). For dense vegetation (wheat stalks, grass) the dimensionless snow trapping coefficient $v = 0.03$, whereas Tabler and Schmidt (1986) recommend $v = 0.01$ for sparse vegetation (sagebrush). Spacing of hedgerows and grass barriers of from $10H$ to $20H$ where H is the windbreak vegetation height have proven useful in a variety of steppe and prairie environments. Properly designed porous (50 percent) snow fences will trap snow up to $15H$ upwind and $35H$ downwind and capture a volume of snow that increases with the

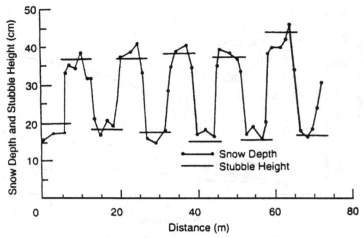

Figure 2.10. Snow depth patterns and stubble height on a wheat field harvested with alternate-height stubble management practice, southern Saskatchewan, Canada (after Pomeroy and Gray, 1995).

square of fence height (Tabler, Pomeroy, and Santana, 1990b). The ecological implications of deep snowdrifts formed by snow fences on tundra vegetation are discussed by Walker et al. (Chapter 6).

2.2.4   Interception by Vegetation

Interception of snowfall by vegetation plays a major role in the snow ecology of coniferous forests, as it results in characteristic snow distributions and winter microclimate. Snow interception is controlled by accumulation of falling snow in the canopy. This snow is subsequently affected by sublimation, melt, and unloading of snow by canopy branches and wind redistribution. Intercepted snow receives snow from snowfall and snow unloaded from upper branches and, of less importance, drip from melting snow on upper branches and vapour deposition during supersaturated atmospheric conditions. Intercepted snow can sublimate to water vapour or become suspended by atmospheric turbulence, followed by further sublimation or deposition to surface snow. Deposition to the surface also may occur by melt and drip to the surface or by directly unloading from branches. Hence, intercepted snow may reach the ground as a solid, liquid, or vapour, but not all intercepted snow eventually reaches the ground (Figure 2.11). The importance of these specific processes in governing snow cover development varies with climatic region, local weather pattern, tree species, and canopy density.

64

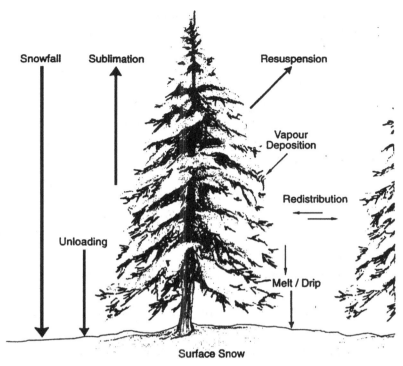

Figure 2.11. Mass fluxes associated with the disposition of winter snowfall in a boreal forest (after Pomeroy and Schmidt, 1993).

The *interception efficiency* of a canopy is the ratio of snowfall intercepted to the total snowfall; this efficiency is an integration of the collection efficiencies for individual branches that compose the canopy. At the branch scale, the *collection efficiency* is the ratio of snow retained by a branch to that incident on the horizontal area of the branch. In early stages of snowfall, snowflakes fall through the spaces between branches and needles, lodging in the smallest spaces until small bridges form at narrow openings. These snow bridges increase the collection area and hence the efficiency with which a branch accumulates snow; further snow is retained on the bridges by cohesion. *Cohesion* of snow results from the formation of microscale ice bonds between snow crystals shortly after contact because of the movement of a thin, liquid-like layer surrounding the crystals or because of the small-scale vaporisation and condensation of water vapour (Langham, 1981). Shortly after initial contact, cohesion increases rapidly; the rate of cohesion accelerates as temperatures approach the melting point. Hence, other factors being equal, the collection efficiency of a branch due to the cohesion of snow increases with increasing temperature.

65

The collection efficiency of a branch is limited by three primary factors (Pomeroy and Gray, 1995):

1. *Elastic rebound* of snow crystals falling onto branch elements and onto snow held by the branch. Rebound is most pronounced below temperatures of $-3°C$ and declines rapidly as temperature rises from $-3°C$ to $0°C$ (Schmidt and Gluns, 1991). Rebound from the branch occurs most effectively near the branch edge; hence large branches lose proportionately less snow to rebound than do small branches.

2. *Branch bending* under a load of snow. Bending decreases the horizontal area of the branch and increases the vertical slope, thereby increasing the probability that falling snow crystals will rebound. The degree to which a branch will bend under a given load increases with branch elasticity. Branch elasticity at subfreezing temperatures is related to the ice crystal content of the branch and increases linearly with increasing temperature (Schmidt and Pomeroy, 1990).

3. *Strength* of the snow structure. As snow accumulates on a branch the degree to which it holds together and to the branch is related to the degree of bonding or strength of the interlocking snow crystals. As temperature increases, the rate at which the snow structure "simplifies" (reduced number of bonds) due to metamorphism increases. Hence, snow strength will decrease with increasing temperature and, when accumulations are large, this may lead to decreased interception efficiency (Gubler and Rychetnik, 1991).

High winds may induce snow redistribution from conifer branches during snowfall, reducing the collection efficiency during a storm. The lowest branch collection efficiencies in the measurements of Schmidt and Gluns (1991) occurred during the highest wind speeds. More effective snow particle rebound during high winds may be the reason for reduced interception efficiency in these cases, though release of accumulating snow, triggered by branch vibrations in the wind may play a role as well.

There are differences in the relationships between canopy-scale interception efficiency and branch-scale collection efficiency with snowfall. Working at small scales, Satterlund and Haupt (1970) and Schmidt and Gluns (1991) found collection efficiencies low for low snowfall, high for medium snowfall, and low for high snowfall amounts. At the canopy scale, Strobel (1978) and Calder (1990) found that interception efficiency declined with snowfall amount. Hedstrom and Pomeroy (1998) developed a physically based model of snow interception from the assumption that the incremental

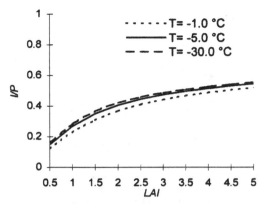

Figure 2.12. Modeled interception efficiency (interception/snowfall) for a snowfall of 10 mm SWE, closed canopy pine forest with initial snow-free conditions, as a function of canopy leaf area index and air temperature (after Hedstrom and Pomeroy, 1998).

interception efficiency is proportional to the snow storage capacity in the canopy (e.g., as the canopy fills with snow the interception efficiency diminishes). The snow storage capacity increases with canopy leaf area and diminishes with existing canopy snow load. Snowfall amount, temperature, canopy closure, wind speed, and time since snowfall also influence interception in this model. The modeled interception efficiency as a function of leaf area index (LAI) and air temperature is shown in Figure 2.12. A doubling of LAI from 2 to 4 causes the interception efficiency to increase from approximately 0.4 to 0.5. Interception is slightly more efficient at low temperatures because of lower fresh snow density in cold conditions and hence better retention in the canopy.

Sublimation of snow can consume considerable amounts of energy from a forest environment in winter. However, the vertical distribution of intercepted snow through the lower 10 to 20 m of the atmosphere results in a large volume of air exposed to snow. This, in addition to the large surface area to mass ratio of intercepted snow and low canopy reflectance of shortwave radiation, results in the potential for notable rates of turbulent transfer of atmospheric heat to the intercepted snow and concomitant removal of water vapour produced at the intercepted snow surface (Harding and Pomeroy, 1996). Figure 2.13 shows snow intercepted in an evergreen canopy at both large and microscale, illustrating the distribution and exposure of snow in the canopy. Pomeroy and Schmidt (1993) provide data suggesting that, for a unit mass of snow,

(a)

(b)

Figure 2.13. Photographs of snow-covered spruce canopy (a) and a snow-covered branch (b), Prince Albert National Park, Saskatchewan, Canada.

68

the surface area of intercepted snow exposed to the atmosphere is 60 to 1,800 times greater than that of snow on the ground. This high surface area and the long period of exposure of snow-covered canopies during northern winters thus provide ample opportunity for sublimation to occur, even when cold temperatures suppress the rate.

Measurements in Prince Albert National Park, Saskatchewan, during December 1992 and January 1993 show that 9 kg of snow accumulated on a single black spruce tree in a cold period ($-25°$C to $-45°$C). In late January ($-20°$C to $-2°$C) about 3 kg of this snow sublimated, whilst the rest fell to the ground; the evolution of mass of tree and snow during this period is shown in Figure 2.14. Landscape-based snow surveys over the winter showed that 31 percent of the annual snowfall sublimated from intercepted snow in black spruce and jack pine canopies. A review of the loss of intercepted snow by sublimation (Schmidt and Troendle, 1992) suggests that approximately one-third of annual snowfall is lost to sublimation from intercepted snow in dense coniferous canopies in the region. Extrapolating this figure to the boreal forest of western Canada (Northwestern Ontario to Great Slave Lake) and correcting for spatial distributions of coniferous canopy density leads to an average loss of 46 mm SWE in a region that receives from 100- to 200-mm water equivalent of annual snowfall. Hence

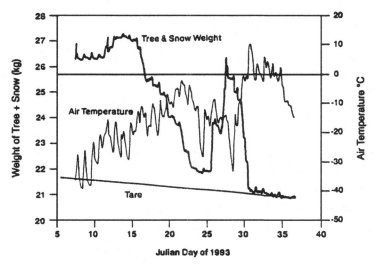

Figure 2.14. Combined weight of a 12-m black spruce tree and its intercepted snow load along with air temperature during a sequence of sublimation, snowfall, and sublimation, Prince Albert National Park, Saskatchewan (after Pomeroy and Gray, 1995).

sublimation of intercepted snow has important effects on the amount of snow that accumulates on the ground and subsequently water supply to the terrestrial and aquatic ecosystems. At local scales sublimation losses vary with canopy cover and provide a mechanism for forest structure to influence water supply through the medium of snow.

The result of interception processes is that, under a forest canopy, both snow depth and water equivalent vary:

1. In relation to the distance to trees, decreasing with decreasing distance to a coniferous tree trunk and slightly increasing with decreasing distance to a deciduous tree trunk; and
2. Between stands of different tree species, with higher accumulations under deciduous trees and smallest accumulations under coniferous trees.

Sturm (1992) developed the following model for snow depth d_s (cm) as a function of distance x (m) from a coniferous tree trunk in a boreal forest, using data from Alaska, USA, and northern Ontario, Canada:

$$d_s(x) = d_s(3) + [d_s(0) - d_s(3)]e^{-(\frac{x}{k})^2} \tag{2.5}$$

where $d_s(x)$, $d_s(0)$, and $d_s(3)$ are the snow depths at distances of x, 0, and 3 m from the central trunk, respectively, and k (m) is a tree size parameter that varies from 0.2 for stunted black spruce in northern boreal forests to 5 for larger white spruce or fir trees in montane or southern boreal forests. Woo and Steer (1986) present measurements showing $d_s(0)$ is approximately one-half of $d_s(3)$. Figure 2.15 shows the change in SWE and snow depth with distance from a single 10-m tall white spruce in a predominantly aspen forest near Waskesiu Lake, Saskatchewan, Canada. Note that SWE undergoes a greater change with distance from the trunk than does depth, because snow density is lower under the spruce branches than beyond the branches. Therefore, Sturm's model for depth cannot be directly modified to calculate SWE with an assumed constant density.

At the forest stand scale, the influence of winter leaf area index (horizontal area of needles and stems over a unit area of ground) on snow accumulation is shown in Figure 2.16, as measured in midwinter in the southern boreal forest of Saskatchewan, Canada. Wind relocation and midwinter melts are not important factors in these relatively calm, cold forests. Leaf + stem area index was measured for the canopy above each snow measurement point. There is a roughly linear decrease in SWE with increasing leaf area, although individual tree species have distinctive effects and there

70

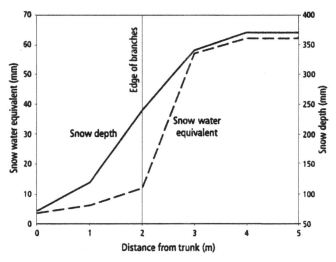

Figure 2.15. Change in midwinter snow depth and SWE with distance from a white spruce in a stand of trembling aspen, Prince Albert National Park, Saskatchewan, Canada (after Pomeroy and Goodison, 1997).

is a variation of SWE and leaf area within a species type for conifers that departs from the larger trend somewhat.

The winter evolution of average SWE by stand species type over time in the same boreal forest is examined in Figure 2.17. It is evident that throughout the winter the aspen stand and small clearing accumulated more snow than did the jack pine or black

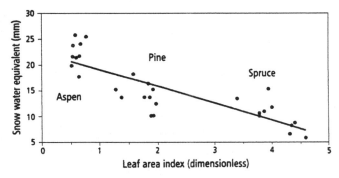

Figure 2.16. Relationship between LAI and SWE in midwinter, southern boreal forest, Prince Albert National Park, Saskatchewan, Canada (after Pomeroy and Goodison, 1997).

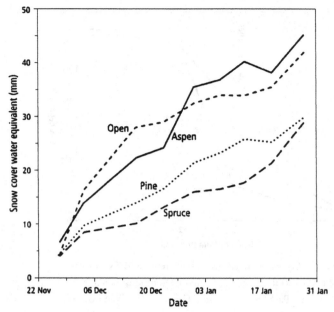

Figure 2.17. Winter evolution of SWE under various forest stands and in a small clearing, southern boreal forest, Prince Albert National Park, Saskatchewan (after Pomeroy and Gray, 1995).

spruce stands. A warming trend at the end of January removed all intercepted snow from the canopy so the differences in SWE at the end of that month represent differential losses due to sublimation of intercepted snow and amount to about 35 percent of the snowfall to that date. Sublimation directly from the snow cover on the ground is negligible in this environment (Harding and Pomeroy, 1996).

Logging operations are changing the vegetation of many forests throughout the world, and some research on snow cover distribution in forested environments has concentrated on the differences between the amounts of snow collected in forests and in openings. The results demonstrate larger amounts of snow in clearings, although the degree of difference is related to the size of the clear-cut. In southeastern British Columbia, Canada, Toews and Gluns (1986) report that snow accumulation in clear-cuts ranged from 4 to 118 percent more than that in adjacent coniferous forests, with a mean difference of 37 percent. In the foothills of southern Alberta, Canada, Golding and Swanson (1986) report snow accumulation increased from 20 to 45 percent from forest to clearing.

Kuz'min (1960) suggests that the SWE in a forest (SWE_f) and in a clearing (SWE_c) can be empirically related to forest canopy cover density p (expressed as a decimal

72

fraction) as:

$$SWE_f = SWE_c(1 - b_f p) \tag{2.6}$$

where for a Russian fir forest $b_f = 0.37$. Harestead and Bunnell (1981) develop a similar form of equation and find b_f varies from 0.24 to 1.4 for coniferous forests in California and Montana in the United States. This expression compares forest to open area accumulation without reference to the size of the clearing. The average depth of snow in a clearing, however, can vary with the size of the opening; increasing the size decreases the average depth in windy, cold snow environments. Troendle and Leaf (1981) and others note that an opening of diameter $5H$ (H = height of the surrounding forest canopy) accumulates the maximum amount of snow in the high mountains of Colorado. Beyond diameters of $12H$, the clearing retains less snow than the adjacent coniferous forest. The reduction in snow cover in large clearings is caused by the increasing likelihood that wind transport erodes snow from the clearing as clearing size increases. This feature of snow accumulation in forest clearings has important implications for the effects of clear-cut blocks on snow accumulation in formerly forested regions. In more humid and temperate regions such as southern British Columbia, blowing snow is suppressed by warm temperatures and SWE does not vary substantially with clear-cut size (Toews and Gluns, 1986).

2.3　　Energetics of Snowpacks

2.3.1　　Atmospheric Boundary

The energy and mass exchanges between the snow cover and the atmosphere involve various interacting processes, primarily sensible and latent turbulent heat transfer, radiative energy transfer, and phase changes. The energy flow associated with these processes is often expressed as a flux of energy (W m^{-2}) or daily energy flux (kJ m^{-2}) with respect to a unit area of the snow surface. Hence sensible heat flux involves a convective (turbulent) flow of heat, latent heat flux is an equivalent heat exchange due to phase change (evaporation, sublimation, refreezing), and radiative flux is due to absorption of shortwave or long-wave radiation and emission of long-wave radiation. By convention, most snow scientists classify the energy used in melting snow as a separate flux from latent (evaporative) heat. Shook and Gray (1997) found for continuous snowpacks in open environments that more than 90 percent of total phase change energy went to melt rather than evaporation. Complete reviews of energy exchanges during winter and snowmelt seasons are provided by Male (1980), Male and

73

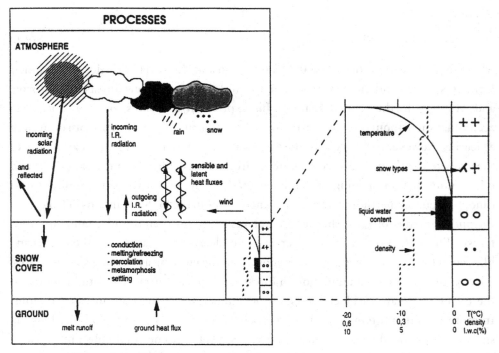

Figure 2.18. Conceptual schematic of mass and energy fluxes controlling the energetics of a snow cover and their relation to snowpack structure, properties and processes, the atmosphere, and the ground. I.R., infrared; l.w.c, liquid water content.

Gray (1981), and Pomeroy and Goodison (1997). These exchanges and other important snowpack energetic processes are shown in Figure 2.18. Their relative importance depends on:

1. Weather conditions: air temperature, humidity, wind speed, incoming shortwave and long-wave radiation and precipitation; and
2. State of the snow cover: albedo, snow surface temperature, surface roughness.

Specific weather and snow cover conditions may favour one or more of these processes in supplying energy to the snow cover for a period of time, while at another time the same processes may remove energy. In most places, incoming shortwave (solar) radiation is the principal source of energy, but other processes can involve energy exchanges of similar orders of magnitude (i.e., between 100 and 500 W m^{-2}). Processes such as photosynthesis, snowmelt, and evaporation driven by solar energy follow a marked diurnal cycle, and the daily average of the resulting fluxes or transformations may be much smaller than peak values.

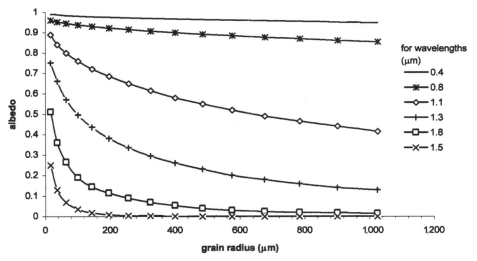

Figure 2.19. Diffuse albedo versus snow crystal radius for six wavelengths (after Warren, 1982).

As solar radiation is a major component of the energy flux to snow, the time of the year and latitude have strong effects on the snow energy balance. Much of the shortwave radiation incident on a snow surface is reflected, with albedos as high as 0.9 for compact, dry, clean snow, dropping to 0.5–0.6 for wet snow and to 0.3–0.4 for porous, dirty snow. A comprehensive review of measurements and mathematical models of snow reflectance and transmission of radiation is given by Warren (1982). Figure 2.19 shows the model results of Warren and Wiscombe (1980) for diffuse light albedo of pure, dry, deep, level snow as a function of wavelength and snow grain radius. Although the albedo is a bulk shortwave reflectance, the spectral reflectance of snow differs substantially over the range of wavelengths from 200 to 2,800 nm. From 200 to 800 nm, reflectance is very high and relatively independent of grain size. Above 1,900 nm, reflectance is very low. Between 800 and 1,900 nm, reflectance declines sharply and is very sensitive to snow grain size. The larger and more rounded the grains, the lower the reflectance in this range. Variation of albedo from 0.9 to 0.5 is an important factor for the surface energy balance for plants and animals.

Male (1980) notes that in field conditions the albedo is primarily a function of snow grain size, whether the surface is illuminated by diffuse radiation or direct sunlight, contamination, and the roughness of the surface. In general the spectral reflectance of fresh snow is high, and once snow has thawed reflectance decreases. Refreezing of snow cover causes no notable change in albedo although contaminants such as soil dust can lower albedo (Conway, Gades, and Raymond, 1996). Warren (1982) calculated that

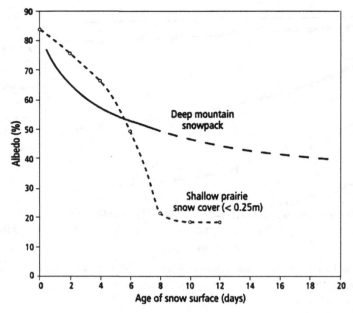

Figure 2.20. Variation in albedo, expressed as percent, over time for melting snowpacks in the mountains (deep) and prairies (shallow) (O'Neill and Gray, 1973).

0.1 percent of volcanic ash in snow will decrease peak albedo in the visible wavelengths from 0.95 to 0.4 for 1-mm-radius snow grains. Albedo in polar climates can increase with time as blowing snow shatters snow crystals and reduces grain size on the surface (Liljequest, 1956). However, albedo in subpolar climates decays with time since the last snowfall, with very different decay rates for shallow and deep snowpacks as shown in Figure 2.20. As snow covers less than 30 cm deep melt, the underlying ground begins to influence the areal albedo, as shown in the data of O'Neill and Gray (1973) for Canadian Prairie snowpacks; the decrease in the actual snow albedo, however, is much less than that of the areal value as wet, bare ground has an albedo of about 0.1. For shallow snow covers in continental environments, a seasonal decrease in albedo of 0.0061 per day has been observed during the premelt period (1 February to melt) and 0.071 per day during melt, dropping to an average value of 0.17 when the surface is free of snow (Granger and Gray, 1990). Snow intercepted in coniferous canopies has no effect on the areal albedo of forests despite the high reflectance of individual intercepted snow clumps (Pomeroy and Dion, 1996) because these canopies act as light traps, retaining most incoming solar radiation and therefore warming above ambient air temperatures when the incoming solar radiation flux is notable. Under coniferous

76

forest canopies albedo is strongly controlled by accumulation of leaf litter on the snow surface and on reduction in snow-covered area during melt.

Over snow surfaces exposed to the atmosphere, outgoing long-wave radiation is generally larger than incoming radiation, leading to a net loss of long-wave radiative energy from the snow cover (Male and Granger, 1979). However, under forest canopies, or when warm air masses enter a previously cold area, a downward net long-wave flux can develop. Incoming long-wave radiation depends on the temperature of the emitting surface (upper atmosphere, cloud layer, or plant canopy) and generally is greater on cloudy days and in forested environments where conifers absorb shortwave radiation and radiate long-wave radiation (Male, 1980). The downward long-wave flux from canopies increases with increasing solar angle because shortwave radiation penetrates further into the canopy and therefore warms the lower branches, which then emit long-wave radiation toward the underlying snowpack (Pomeroy and Dion, 1996). Emitted long-wave radiation is proportional to the fourth power of surface temperature and may be calculated from the Stefan–Boltzmann law

$$L\uparrow = \varepsilon_s \sigma T_s^4 \tag{2.7}$$

where ε_s is the surface emissivity, σ is the Stefan–Boltzmann constant (5.67×10^{-8} W m^{-2} K^{-4}), and T_s is the absolute temperature of the surface (K). The emissivity of snow is generally taken as 0.98, although Warren (1982) shows that it varies from 0.975 to 0.995 with radiation wavelength over the range 3 to 15 μm and varies from 0.96 to 0.99 with snow grain size at wavelengths above 15 μm.

The radiation balance over snow shows wide variations because of cloud cover, forest canopy, and topography. The simple situation of melting snow over an open area of low relief is shown in Figure 2.21 (Granger and Male, 1978). The first two days are clear, followed by continuous cloud on the third day. Fog occurs for a period on the first day. The high snow albedo causes the reflected shortwave flux to be the same magnitude as incoming shortwave radiation. Incoming and outgoing long-wave radiation are also of similar magnitude. Outgoing long-wave radiation remains nearly constant because of the limited range of surface temperatures during melt.

The example shown in Figure 2.21 contrasts with nonmelting snow where long-wave cooling at night and the low thermal conductivity of the snowpack cause the snow surface temperature and hence outgoing long-wave radiation to drop dramatically. Under coniferous forests, the net shortwave flux to snow decreases sharply with increasing canopy density and can be an order of magnitude less than incoming radiation to the canopy top under mature coniferous canopies. The daily mean net radiation flux decreases with increasing canopy density up to canopy densities of about

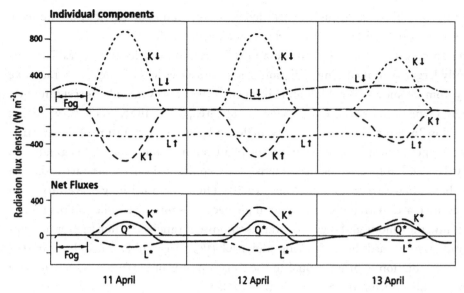

Figure 2.21. The component fluxes of incoming ($K\downarrow$, $L\downarrow$) and outgoing ($K\uparrow$, $L\uparrow$) shortwave and long-wave radiation, respectively, and the radiation balance (net) of all-wave (Q^*), shortwave (K^*), and long-wave (L^*) radiation during 3 days over a melting prairie snow cover. Day 1, fog then clear; day 2, clear; day 3, cloud cover (from Granger and Male, 1978).

60 percent, whence the flux rises. Net radiation at the snow surface can be significantly greater in a deciduous forest than in an open area because of emission of long-wave radiation from the canopy.

The turbulent fluxes of sensible (Q_H) and latent (Q_E) heat are also important to the energy balance of a snow cover. Sensible heat is derived from atmospheric heat energy, and its flux is the rate at which heat is carried to or from the snow surface by atmospheric turbulence. Latent heat is that released by sublimation or evaporation at the snow surface, and its flux is the rate at which water vapour is carried to or from the snow surface by atmospheric turbulence. The fluxes of sensible and latent heat depend on the respective vertical gradients of temperature and humidity in the atmosphere and turbulent transfer of heat and water vapour at the snow surface. Wind, temperature, and humidity gradients above snow covers are normally not available, requiring the use of bulk transfer calculations to estimate Q_H and Q_E. Male (1980) reports that bulk transfer formulae are generally poor estimators of the latent heat flux over snow, as the atmospheric water vapour pressure at one level above the snow surface provides insufficient information to accurately define a gradient and therefore the water vapour flux. Male and Granger (1979) do report correlations between daily totals of sensible heat estimated by profile and bulk transfer approaches; however, significant differences can exist on any single day.

Bulk transfer equations for estimating turbulent fluxes take the form

$$Q_H = D_H u_z(T_a - T_s), \quad Q_E = D_E u_z(e_a - e_s) \tag{2.8}$$

where T is air temperature (°C); u_z is wind speed at height z; e is water vapour pressure (mb), subscripts a and s denote air and snow surface, respectively; D_H is the convective heat bulk transfer coefficient (kJ m^{-3} °C^{-1}); and D_E is the latent heat bulk transfer coefficient (kJ m^{-3} Pa^{-1}). In eastern Canada, Gold and Williams (1961) found D_H to be 0.015 and D_E to be 0.025 for u_z at 2 m, T_a at 1.2 m, and e_a measured independently. For melting snow in western Canada, D_H is approximately 0.007 and D_W is approximately 0.0052 for 1-m measurement heights (z and a); notable variation from these values may be expected depending on stability, fetch distance, and uniformity of terrain.

For continuous snow covers in open environments, the direction of latent heat flux is controlled by the direction of net radiation with cycles of condensation at night and evaporation during the day. Male and Granger (1979) report daily net evaporation rates (less condensation) in central Saskatchewan ranging from 0.02 to 0.3 mm of SWE/day with a mean of 0.1 mm of SWE/day. These fluxes removed for 14 to 22 percent of the incoming energy to the snow cover.

For discontinuous snow covers in open environments, local advection of sensible energy from bare ground to snowpatches becomes an important component of snowpack energy balance. Gray and O'Neill (1974) report that sensible heat transfer supplied 44 percent of the incoming energy to an isolated snowpatch; when surrounded by continuous snow cover this value drops to 7 percent. The model developed by Liston (1995) shows a 30 percent increase in energy inputs to melting snow as the area of exposed ground increases. The energy inputs also increased with increasing patchiness of snow. Shook and Gray (1996) show that the patchiness of snow during melt can be described mathematically by the mean and coefficient of variation of snow depth before melt begins. Areas of deciduous vegetation (grain stubble, tall grasses, shrubs) collect deep, even snow covers with low coefficient of variation of SWE. Snow in such taller vegetation will remain less patchy during melt than snow over short vegetation, prolonging the melt period compared with poorly vegetated sites with similar snow depth. Local advection of energy driven by turbulent transfer from very tall (2–3 m), brushy vegetation stalks to underlying snowpacks caused snowpacks in clear-cuts to melt five times faster than that in adjacent mature forest canopies in the southern boreal forest of western Canada and resulted in a 2-week shorter snow season in cleared lands compared with natural forest (Pomeroy and Granger, 1997).

2.3.2   Soil Boundary

The ground heat flux to snow on a daily basis is typically considered to be a small component of the energy balance. However, because it is persistent, it can have an important cumulative effect early in the melt season in retarding or accelerating the time of melt and in affecting the intranivean environment. Incorrect calculation of ground heat flux can therefore significantly change the estimated timing and magnitude of snowmelt. In locations with incompletely frozen soils, the ground heat flux is positive; Gold (1957) measured up to 860 kJ m^{-2} per day moving from warm soils to melting snow near Ottawa. However, in the Arctic the flux is negative in late winter; recent measurements showed a peak of -900 kJ m^{-2} per day during snowmelt over frozen permafrost soils near Inuvik, Northwest Territories. Marsh and Woo (1987) report similarly elevated levels from permafrost near Resolute, Northwest Territories. The large negative ground heat flux plays a major role in delaying the snowmelt season in northern Canada despite high sensible and radiant heat fluxes.

The ground heat flux is not completely due to the temperature gradient and thermal conductivity of soils; infiltration of meltwater into soils can involve a significant energy flow, especially if the meltwater refreezes in frozen soils, releasing latent heat. Infiltration may result in a warming of upper soil layers, an increase in apparent thermal conductivity, and further melting of frozen soil water or refreezing of meltwater in the frozen soil (Zhao, Gray, and Male, 1997). The refreezing of meltwater at the soil–snow interface, forming a basal ice layer, releases a significant quantity of latent heat and is a distinctive characteristic of melting Arctic snow covers (Woo, Heron, and Marsh, 1982). These phenomena change the ground heat flux by transferring sensible heat to the soil and absorbing or releasing latent heat. Hence, in an unsaturated frozen soil, infiltration and refreezing of meltwater in soils may reduce the ground heat flux from large negative to very small values (smaller than -10 W m^{-2}). Rouse (1990) divides the ground heat flux for frozen soil into that due to the change in storage of sensible heat and the change in storage of the latent heat of fusion in ground ice; however, relatively few values for these individual components are available. Male and Granger (1979) measured net daily ground heat flux from 454 to -378 kJ m^{-2} per day over 3 years in central Saskatchewan during melt over frozen soils, with strong variation from year to year. Granger and Male (1978) note that the soil heat flux on the prairies shows a diurnal cycle that lags the radiation cycle and is of opposite sign. As shown in Figure 2.22, when snowmelt begins, this

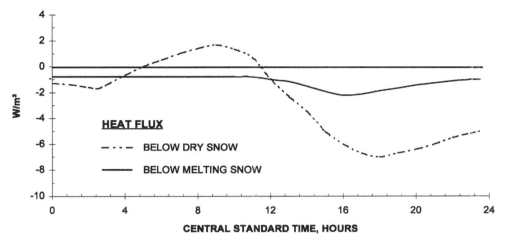

Figure 2.22. Soil heat flux below a shallow prairie snowpack for dry snow and for melting (wet) conditions. Note that, after the meltwater reaches the soil and is refrozen, the soil heat flux is dramatically dampened (after Granger and Male, 1978).

diurnal cycle is almost completely eliminated because of meltwater's refreezing in frozen soils.

2.3.3   Internal Energetics

Most of the energy exchanges between snow and its environment occur at the atmosphere or ground interfaces; however, because snow is porous, some radiation and convective fluxes occur within the top few centimetres of the snowpack. The important fluxes that can directly penetrate the snowpack are radiation, conduction, convection, and meltwater or rainwater percolation. Phase change within the snowpack complicates the discussion of internal energetics but can be extremely important for life forms requiring liquid water (see Hoham and Duval, Chapter 4) or as media for chemical reactions (see Tranter and Jones, Chapter 3).

Whilst most shortwave radiation is reflected by snow, part of the solar radiation flux penetrates the snowpack and is absorbed a few centimetres beneath the surface. The penetration and extinction of incoming shortwave radiation is of prime importance to the development of life within the snowpack. Sergent et al. (1987) measured the spectral extinction of radiation as a function of snow grain size and density; the results permit the theoretical description of radiation penetration through snow. Figure 2.23 shows irradiance penetration of natural light as a function of wavelength for various depths in a snowpack with 100-μm-radius spherical grains and a density of 200 kg m^{-3}.

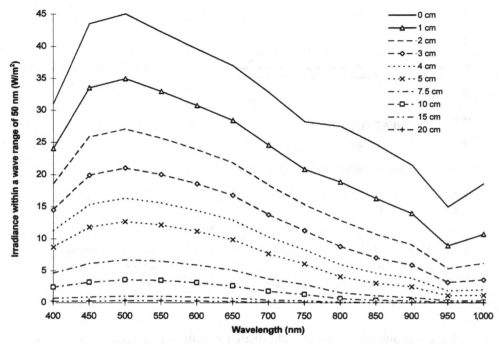

Figure 2.23. Solar irradiance within the snowpack as a function of wavelength for various depths in a clean snowpack composed of uniform spherical snow grains of 200 μm diameter and a density of 200 kg m^{-3}.

Within the wavelength range of 400–1000 nm, significant radiation penetrates to almost 20 cm below the surface. The maximum irradiance at depth occurs for a wavelength of about 500 nm because of the combined effect of the distribution of the solar spectrum and the spectral absorption characteristics of ice. Figure 2.24 shows the differing effects of heavy, wet and loose, dry snow on irradiance for two different wavelength bands. The larger grain size associated with wet snow (500-μm radius) permits deeper penetration in the visible light wavelengths; however, the additional energy is relatively small at less than 3 W m^{-2} for most depths and an insignificant difference near the surface and near 20 cm below the surface. The coarser grains found in preferential meltwater flow paths promote greater irradiance at depth in these paths; this irradiance is the basis of a technique to map macro flow paths in snowpacks (McGurk and Marsh, 1995).

Dry snow undergoes thermal conduction, thermal convection (where temperature gradients are extreme), and windpumping (in windswept areas), which govern the heat flux between the bottom and top of the snow cover (Sturm, 1991; Colbeck, 1989).

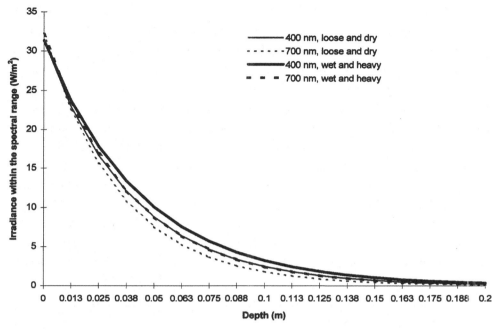

Figure 2.24. Solar irradiance within the snowpack as a function of snow structure. Two structure types are shown: heavy and wet (density = 400 kg m^{-3}, grain size = 500-μm radius), and loose and dry (density = 200 kg m^{-3}, grain size = 1-mm radius) for two wavelengths of 50-nm bandwidth.

Temperature gradients in dry snow covers induce water vapour gradients and consequently diffusion of water vapour from warmer parts to colder parts. When temperature gradients exceed 10°C m^{-1}, complete "destructive" ice crystal sublimation and recrystallisation occur as water vapour moves along the gradient (Colbeck, 1987). The corresponding sublimation or condensation necessary to maintain a thermodynamic equilibrium involves fluxes of latent heat, which account for between 10 and 40 percent of heat transport in snow (Yosida et al., 1955). These fluxes are generally taken into account as an "effective" thermal conductivity that combines all heat transfer mechanisms.

The low thermal conductivity of snow, compared with other types of materials, is of primary importance in the protection of intranivean and subnivean life from severe winter conditions (Aitchison, Chapter 5). The thermal conductivities of air and ice at −15°C are about 0.023 and 2.2 W m^{-1} K^{-1}, respectively; that of snow must lie between these value. Thermal insulation by the snow cover depends strongly on its depth but also on the crystal structure and density of the surface layers. The dependence on density is due to the effective thermal conductivity of snow, which increases approximately with the square of density except for low–density snow reformed by kinetic temperature gradient metamorphism. The nonlinear relationship between thermal conductivity and density causes the thermal insulation of an inhomogeneous, layered snowpack to be higher than that of a uniform snowpack of the same depth and density. Sturm et al. (1997) have reviewed thermal conductivity studies over the last century and supplemented these with a new large set of observations for a variety of snow types. They found that effective thermal conductivity (K_{eff}) increased from $0.05\ \mathrm{W\ m^{-1}\ K^{-1}}$ for low–density fresh snow (density $\approx 100\ \mathrm{kg\ m^{-3}}$) to $0.6\ \mathrm{W\ m^{-1}\ K^{-1}}$ for dense drifted snow (density $\approx 500\ \mathrm{kg\ m^{-3}}$). The data and confidence intervals indicated by Sturm et al. are shown in Figure 2.25 and substantiate that there is little relation between thermal conductivity and density for low–density snow (density $< 200\ \mathrm{kg\ m^{-3}}$), which is dominated structurally by depth hoar. Sturm et al. also fitted an empirical regression equation to their data points. The equation is consistent with the earlier work of

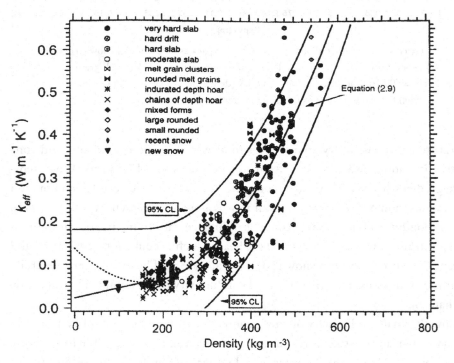

Figure 2.25. Effective thermal conductivity of snow as a function of snow crystal type and snow density, along with Equation 2.9 plotted with its confidence intervals (CL). For a discussion of snow crystal classification, see Section 2.4 in this chapter (after Sturm et al., 1997).

Yosida et al. (1955) and is

$$k_{\text{eff}} = 0.138 - \frac{\rho}{990} + \frac{\rho^2}{309} \qquad \text{for } 156 \leq \rho \leq 60$$

$$k_{\text{eff}} = 0.023 + \frac{\rho}{4273} \qquad \text{for } \rho < 156 \text{ kg}$$

(2.9)

Temperature regimes in dry snowpacks are exceedingly complex and are controlled by a balance of the energy regimes at the top and bottom of the snowpack, radiation penetration, effective thermal conductivity of snow layers, water vapour transfer, and latent heat exchange during metamorphism. Finite difference, energy, and mass balance models (e.g., Anderson, 1976) have been used to simulate many of the important energy transfers. SNTHERM (Jordan, 1991) contains a relatively complete and accurate physical description and is increasingly popular. Beyond the scope of regimes predicted by point models, horizontal and vertical inhomogeneities in natural snow cover cause substantial spatial variation in the thermal regime of snowpacks. For instance, lower accumulation because of snow interception occurs under isolated conifers, resulting in shallow, cold snowpacks near these trees (Sturm, 1992). Temperature stratification within dry snowpacks is usually unstable (warm temperatures lying below cold temperatures) from formation until late winter and spring, as energy inputs from the soil boundary exceed those from the atmosphere and upper layers. As a result temperatures become warmer with depth, with gradients as high as about $50°C \text{ m}^{-1}$ in shallow subarctic and arctic snowpacks during early midwinter. In cold climates with frozen soils, an inversion can develop in late winter where the upper snowpack warms to higher temperatures than the lower layers; this reflects higher energy inputs from the atmosphere (often due to long sunlit periods in the northern spring) than from the frozen soil.

For a given climate, the thermal regime in the snowpack strongly depends on the amount of snowfall early in the winter season. This is illustrated in Figure 2.26 with the evolution of the temperature throughout a snowpack over a winter period in an alpine environment. The temperature is simulated by the snow model CROCUS from meteorological conditions observed in the French Alps during 1995–1996 at an altitude of 2,400 m above sea level. Figure 2.26b shows the simulation with recorded snowfall, and Figure 2.26a and 2.26c correspond to simulations with 50 percent less and 50 percent more snowfall, respectively. In the period from mid-November to mid-December, the 50-cm-deep snowpack efficaciously keeps the bottom snow layer near the melting point (Figure 2.26c), whilst a 30-cm snowpack (Figure 2.26b) does not prevent the temperature from dropping as low as $-5°C$ in the bottom layer. For the 20-cm-deep snowpack the bottom layer reaches $-8°C$ (Figure 2.26a). Other factors

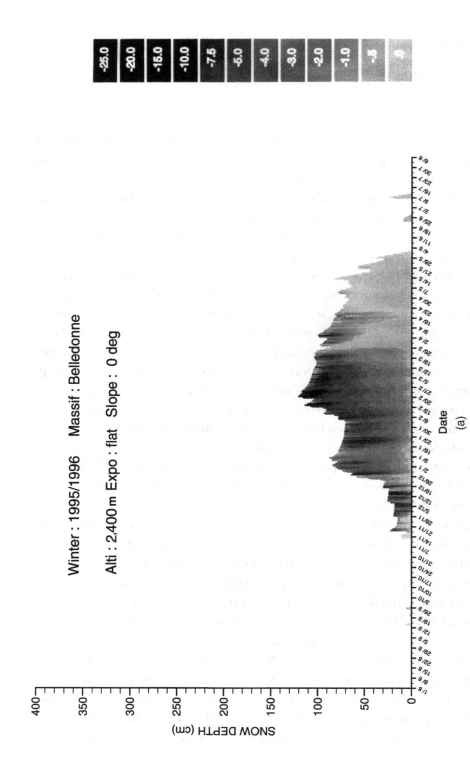

Figure 2.26. Seasonal snow temperature regime in an alpine snowpack modeled by CROCUS using meteorological data from 2,400 m above sea level on Belledonne Massif, France, in winter 1995–96. (a) 50 percent less snowfall than recorded; (b) recorded snowfall at 2400 m above sea level French Alps; (c) 50 percent more snowfall than recorded.

86

Winter : 1995/1996   Massif : Belledonne

Alti : 2,400 m Expo : flat   Slope : 0 deg

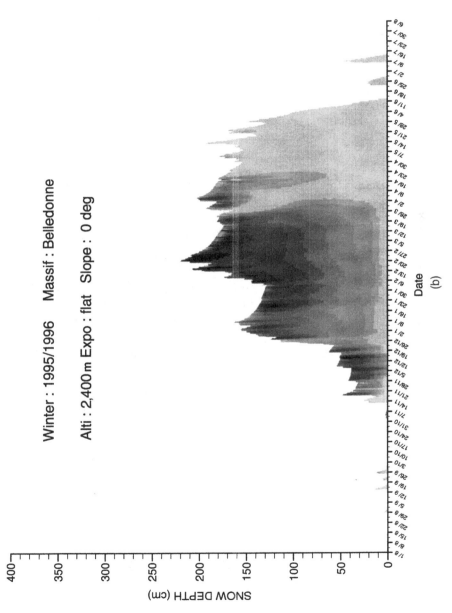

(b)

Figure 2.26. (*Continued*)

87

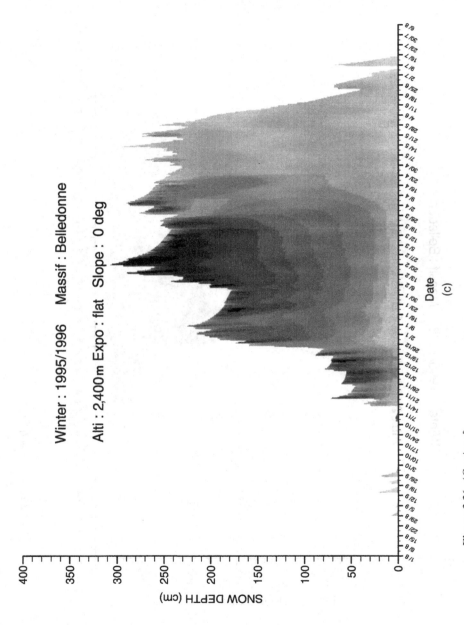

Winter : 1995/1996   Massif : Belledonne

Alti : 2,400m Expo : flat   Slope : 0 deg

Figure 2.26. (*Continued*)

88

can strongly influence the thermal regime in this environment. For a deep snowpack a midwinter rainfall would increase density and decrease depth. A doubling of density in such circumstances would reduce the thermal resistivity of the snowpack by a factor of eight!

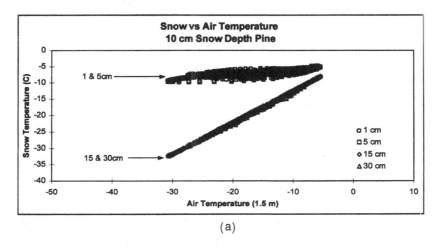

(a)

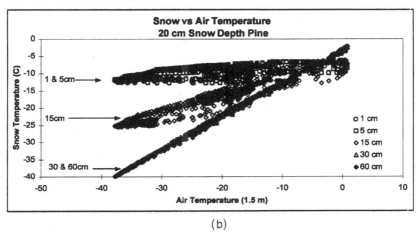

(b)

Figure 2.27. Thermal regime of a boreal forest snowpack. Temperatures measured half-hourly at various heights above the ground in Prince Albert National Park, Saskatchewan, Canada. Note that temperatures measured at heights greater than the snow depth listed for that period are air temperatures slightly above the snow surface rather than snow temperatures. (a) Early winter, 10 cm depth, 100 kg m^{-3} density; (b) early midwinter, 20 cm depth, 150 kg m^{-3} density; (c) midwinter, 30 cm depth, 180 kg m^{-3} density; (d) late winter, 30 cm depth, 200 kg m^{-3} density. Depths and densities varied over the period and values stated are approximate.

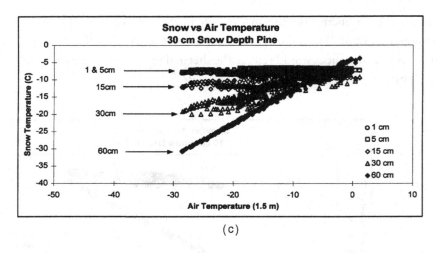

(c)

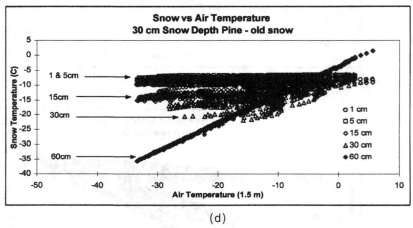

(d)

Figure 2.27. (*Continued*)

The progression of the thermal regime in a continental boreal forest snowpack is shown in Figure 2.27. Air temperatures, snow temperatures, and snow depth were measured half-hourly over a winter under a pine canopy in central Saskatchewan, Canada (Pomeroy et al., 1997b). The strong diurnal variation in air temperature permits an assessment of the thermal regime of natural snowpacks in severe environments. In early winter an extended period occurred where the snowpack depth and density were approximately 10 cm and 100 kg m^{-3}, respectively; snow temperatures and air temperatures for this period are shown in Figure 2.27a. The temperatures at 1 and 5 cm above the ground varied from $-5°C$ to $-10°C$, whilst air temperatures just above the snow surface varied from $-8°C$ to $-33°C$, a substantial moderation and an

90

indication that thermal moderation can develop under very shallow snowpacks. Later in winter, a 20-cm-deep 150 kg m^{-3} snow cover developed. Snow temperatures in the bottom 5 cm varied in a similar manner to the earlier snow ($-7°$C to $-13°$C) and in the layer 15 cm above the ground from $-6°$C to $-26°$C, whilst air temperatures varied from $-1°$C to $-40°$C (Figure 2.27b). Clearly the thermal moderation increases with depth from the snow surface and decreases with distance from the ground. When a 30-cm-deep snowpack develops, the degree of thermal moderation at the lowest layers increases, with snow temperatures at 1 and 5 cm of $-7°$C to $-9°$C for an air temperature range from $-4°$C to $-30°$C (Figure 2.27c). However, at air temperatures greater than $-15°$C, the temperature regimes at various depths become difficult to distinguish and snow temperatures can be *exceeded* by air temperatures. This reduced thermal stratification is indicative of colder soils and a smaller soil heat flux as winter progresses. The thermal stability that develops due to cold soils in later winter can promote snow temperatures consistently below that of air temperatures as shown in Figure 2.27d for a 30-cm-deep snowpack with a density of 200 kg m^{-3}. Points to the right of a 1:1 line between air and snow temperature indicate conditions when the snowpack is colder than the air. Until melt occurs, the thermal insulation of the snowpack will cause the intranivean and subnivean environments to be sometimes colder than the supranivean environment.

The alpine snow temperature simulation shown in Figure 2.26 showed intranivean temperatures near the melting point from March until the end of melt in June. In this situation the energy inputs to the snowpack have raised its temperature to the melting point and further energy inputs are used to satisfy the latent heat associated with phase change. However, unsteady weather conditions, diurnal variations in the energy status of a snowpack, and stratification of a snowpack cause the internal energetics during melt to be quite complex. Subarctic and arctic snowpacks can undergo melt in upper layers whilst sustaining snow temperatures significantly below freezing in the lower layers. Internal heat fluxes in wet snow, or in partially wet snow, are principally controlled by conduction and by latent heat release due to refreezing of liquid water. In most snow covers, flow initiates along preferential flow paths, in which "flow fingers" may penetrate dry snow several days before the more general "matrix flow" reaches a layer (Marsh and Woo, 1984a). When percolating meltwater reaches cold internal snow layers, a part of this water refreezes and the latent heat produced warms the layer very quickly toward the melting point. When the surface of a wet snow cover cools, it induces an internal heat flux by conduction, which is balanced by refreezing of the liquid water present in the underlying wet layers. At night, this refreezing is usually confined to a depth of less than 20 cm below the surface. Matrix flow in completely wet snow can transmit liquid water produced at the surface by rain or melting through the internal layers without any related energy exchange.

2.3.4 Snowmelt

The rate of snowmelt is primarily controlled by the energy balance near the upper surface, where melt normally occurs. Shallow snowpacks may be considered as a "box" to which energy is transferred by radiation, convection, and conduction and across whose boundaries mass fluxes of solid, liquid, and vaporous water occur. Male (1980) indicates that, for deep or cold snows, the energetics of layers within the pack must be considered discretely.

In temperate climates, snowpacks tend to be uniformly close to the melting temperature (isothermal) when melt commences at the surface. However, in cold climates, the change in internal energy of the snowpack dU/dt can be a significant term in the energy balance during melt of shallow snowpacks. Meltwater is released from the pack in a diurnal cycle in response to cycling of energy inputs; the nighttime energy deficit must be compensated for the next day before the pack can return to $0°C$ and release water.

The melting point for snow may differ slightly from $0°C$ because of impurities in ice, but typically once the snow layer reaches $0°C$ any further energy inputs are directed toward snowmelt and $dU/dt = 0$. The energy balance for melting snow is expressed in terms of Q_m, the energy available for melting a unit volume of snow,

$$Q_m = K^* + L^* + Q_H + Q_E + Q_G + Q_R - \frac{dU}{dt} \qquad (2.10)$$

where K^* is the net shortwave radiation flux, L^* is the net long-wave radiation flux, Q_H is the sensible heat flux, Q_E is the latent heat flux, Q_G is the ground heat flux, and Q_R is the energy flux derived from precipitation.

The amount of meltwater can be calculated from Equation 2.10 by the expression

$$SWE_m = \frac{Q_m}{(\rho_w L_f B)} \qquad (2.11)$$

where SWE_m is the snow meltwater (in mm), ρ_w is the density of water, L_f is the latent heat of fusion, and B is the fraction of ice in a unit mass of wet snow. B is usually 0.95 to 0.97.

Net radiation and sensible heat largely govern the melt of shallow snowpacks in open environments (Male and Gray, 1981). At the beginning of melt, radiation is the dominant flux with sensible heat growing in contribution through the melt. The relative importance of some of the energy balance terms during melt of a High Arctic

92

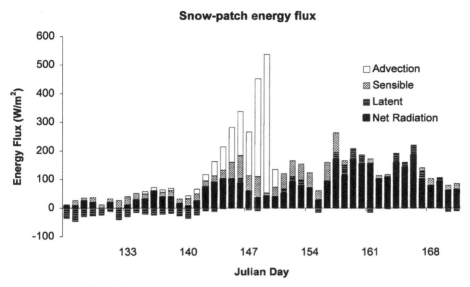

Figure 2.28. Energy fluxes for an isolated melting snowpatch in the Arctic (after Marsh and Pomeroy, 1996).

snowpack is shown in Figure 2.28, which presents daily fluxes to an individual snow-patch measured by Marsh and Pomeroy (1996) near Inuvik, Northwest Territories. In an alpine, midlatitude region, these terms have very different relative importance (Brun et al., 1989) as shown in Figure 2.29. Various mathematical models are available to calculate the energy fluxes for melting snow from standard meteorological variables. Over a patchy snow cover, the classic formulae to calculate Q_H and Q_E underestimate the turbulent heat fluxes because advection of energy from bare areas to snowpatches plays an important role that cannot be neglected. Over nonisothermal snow covers, the calculation of the different fluxes is quite difficult and requires the use of sophisticated snow models (Anderson, 1976; Brun et al., 1989, 1992; Jordan, 1991; Bader, 1992; Yamazaki and Kondo, 1992).

If a complete set of meteorological measurements is not available, then temperature index models may be used to predict snowmelt (Rango and Martinec, 1987; Braun, 1991). Despite their poor performance for open, continuous snow covers, considerable success has been achieved in using index models to relate melt to air temperature in mountainous areas, forested areas, and for patchy snow cover where advection becomes a dominant term. Shook, Gray, and Pomeroy (1993) have shown that the potential for advection of sensible heat increases dramatically as basin snow cover decreases from 70 to 60 percent and remains relatively high as basin snow cover further

93

decreases. Where sensible heat is a dominant term, application of temperature index models may be expected to be relatively successful. Temperature index calculations employing melt factor M_f have the form

$$SWE_m = M_f(T_A - T_B) \qquad (2.12)$$

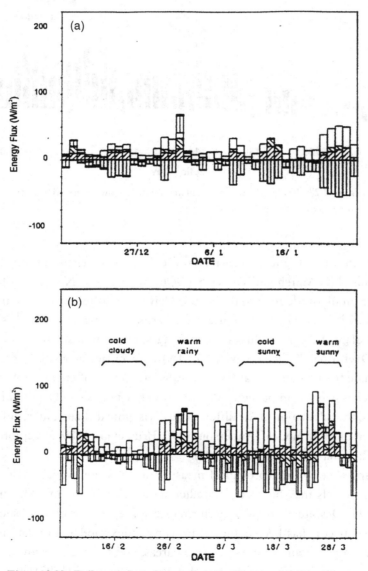

Figure 2.29. Daily variation of energy balance components of an alpine snow cover in the French Alps during three periods: (a) midwinter, (b) late winter, and (c) spring.

94

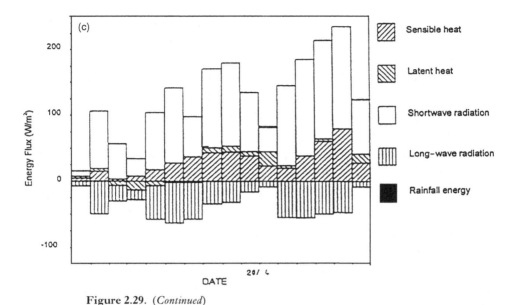

Figure 2.29. (*Continued*)

where T_A is the mean air temperature over a given time period and T_B is a base temperature below which melt does not occur (usually taken as $0°C$). Granger and Male (1978) found the melt factor to vary from 6 to 28 mm $°C^{-1}$ day^{-1} for snowmelt on the prairies. Exceptionally high values of M_f (50 mm $°C^{-1}$ day^{-1}) were occasionally measured in April and associated with high radiation fluxes. The use of melt factors in calculating snowmelt should be done with caution because of the considerable variation in their value from year to year and with location.

2.4   Snow Cover Structure

An important characteristic of snow covers is the stratification of their physical structure. In analogy to geological strata, snow stratification results from successive snowfalls over the winter and processes that transform the snow cover between snowfalls. A new snowfall creates a new surface layer that will undergo various transformations before being buried by the next snowfall. Snow crystals undergo the most important transformation by changing their original shapes (stars, needles, dendrites, etc.) under the effect of thermodynamics; this phenomenon is called metamorphism. Crystal transformations due to metamorphism and other transformations of the snow cover result in snow properties that are distinctive enough to permit easy identification of stratigraphic layers over most of the snow season.

95

Transformations of snow cover are controlled by two interacting processes: snow settling and snow metamorphism. They cannot be completely separated; settling of a snow layer is principally due to the weight of upper layers and often is represented as a Newtonian viscosity. Snow metamorphism also contributes to settling, especially for fresh snow layers. Furthermore, the type of crystal strongly influences snow viscosity and so affects the settling rate. Transformations by settling and metamorphism are fundamental phenomena because most of the physical properties of snow strongly depend on snow density and on the type and size of grains comprising a snow layer. Snow metamorphism depends on several variables, the principal ones being temperature, temperature gradient, and liquid water content. Liquid water content is quite influential and a distinction between dry and wet snow is always necessary to describe the behaviour of snow layers. As a result of metamorphism, snow grains present various shapes and sizes, which have been distinguished and classified in an international classification created by a working group of the International Commission on Snow and Ice (Colbeck et al., 1990). This classification describes the different types of snow crystals, their size, and their formation. The classification is process oriented in that it remarks on the physical processes involved in creating snow crystal forms and therefore is useful in snow ecological studies where the snow processes also affect ecological processes. The dry and wet snow processes that lead to these crystal forms are discussed in the following sections. A brief summary of the International Classification is shown in Table 2.2, the main classes being precipitation (1), dry snow (2, 3, 4, 5, 7), wet snow (6), ice (7), and surface-generated features (9). Some examples of snow crystal forms, as investigated by microphotography, are shown in Figure 2.30.

2.4.1   Dry Snow

Dry snow is characterised by the relative absence of liquid water. This absence is not complete, because snow crystals may be surrounded by a liquid-like layer that is thin enough to neglect when considering metamorphic processes, but is important to snow chemistry and possibly to microbiology. Snow is necessarily dry when its temperature is below the melting point, but dry snow also may exist in melting snow covers as, for instance, in a dry layer persisting between wet snow layers.

2.4.1.1   *Dry Snow Metamorphism*

Dry snow metamorphism is principally controlled by the vertical temperature gradient across the snow layer. This gradient develops because of radiative cooling from the snow surface, warm and/or wet soils, and the low thermal conductivity of

96

Table 2.2. *International classification of snow crystals (adapted from Colbeck et al., 1990).*

Basic classification	Shape	Process classification
1. Precipitation particles	a) Columns b) Needles c) Plates d) Stellar dendrites e) Irregular crystals f) Graupel g) Hail h) Ice pellets	a–g) Cloud-derived falling snow h) frozen rain
2. Decomposing and fragmented precipitation particles	a) Partly rounded particles b) Packed shard fragments	a) Freshly deposited snow b) Wind-packed snow
3. Rounded grains	a) Small rounded particles b) Large rounded particles c) Mixed forms	a–c) Dry equilibrium forms
4. Faceted crystals	a) Solid hexagonal b) Small faceted c) Mixed forms	a) Solid kinetic growth form b) Early kinetic growth form c) Transitional form
5. Cup-shaped crystals and depth hoar	a) Cup crystal b) Columns of depth hoar c) Columnar crystals	a) Hollow kinetic crystal b) Columns of a) c) Final growth stage
6. Wet grains	a) Clustered rounded grains b) Rounded polycrystals c) Slush	a) No melt–freeze cycles b) Melt–freeze cycles c) Poorly bonded single crystals
7. Feathery crystals	a) Surface hoar crystals b) Cavity hoar	a) Kinetic growth in air b) Kinetic growth in cavities
8. Ice masses	a) Ice layer b) Ice column c) Basal ice	a) Refrozen water above less-permeable layer b) Frozen flow finger c) Frozen ponded water
9. Surface deposits and crusts	a) Rime b) Rain crust c) Sun crust, firn-spiegel d) Wind crust e) Melt–freeze crust	a) Surface accretion b) Freezing rain on snow c) Refrozen sun-melted snow d) Wind-packed snow e) Crust of melt–freeze grains

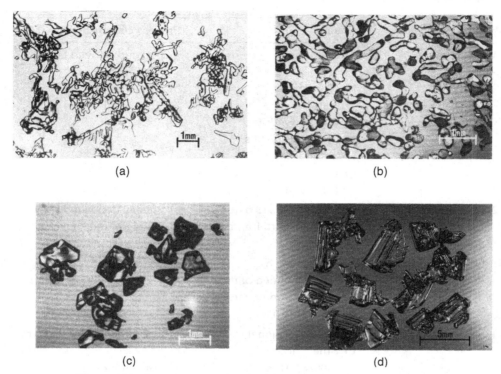

(a)

(b)

(c)

(d)

Figure 2.30. Examples of snow crystals as classified in Table 2.2 (from Colbeck et al., 1990): (a) partially decomposed fresh snow, 2a (by E. Akitaya); (b) large rounded dry grains, 3b (by E. Akitaya); (c) solid hexagonal faceted crystals, 4a (by E. Akitaya); (d) cup-shaped depth hoar crystal, 5a (by K. Izumi).

the snow cover. Low thermal conductivity permits the soil to stay relatively warm, even when the snow surface temperature is very cold. The temperature gradient in the snow cover induces high local temperature gradients between adjacent crystals. Because the saturation water vapour pressure over a flat ice surface is an exponential function of temperature, the temperature gradients induce water vapour pressure gradients and consequently vapour diffusion from the warmest crystals toward the coldest ones. Warm crystals are destroyed and re-formed on the cold side, resulting in a change of crystal shape and size. The rate of diffusion is balanced by sublimation and solid condensation processes.

Water vapour pressure is also a function of the radius of curvature of the ice face, so that the sharpest parts of snow crystals tend to sublimate most rapidly, consequently becoming rounded and diminished. Concave parts of crystals and the largest crystals tend to grow because of this effect. There is a competition between the temperature gradient and curvature effects because condensation on the coldest crystals tends to produce the sharp edges typical of ice crystallisation. Consequently, there are some threshold conditions that separate the dry metamorphism producing faceted and angular crystals and that producing rounded crystals. These thresholds are not absolute but can be described as follows:

1. Low temperature gradient metamorphism. When the temperature gradient is lower than $5°C\,m^{-1}$, it produces small, rounded grains (class 3 of the International Classification) characterised by an efficient settling rate and good cohesion due to the growth of ice bonds between the grains by sintering. They are typically encountered in temperate regions with heavy snowfalls or strong redistribution due to the wind.

2. Medium temperature gradient metamorphism. When the temperature gradient is between $5°C\,m^{-1}$ and $15°C\,m^{-1}$, it drives the formation of faceted crystals (class 4). The increase in grain size is slow because gradient effects are partially balanced by curvature effects.

3. High temperature gradient metamorphism. When the temperature gradient is higher than $15°C\,m^{-1}$, it drives the formation of depth hoar crystals (class 5). These large, plate-like crystals grow quickly. Such snow layers generally are characterised by a very low settling rate and weak cohesion. They are typically encountered in cold and dry climate regions.

Quantitative empirical laws to describe snow metamorphism have been determined by Marbouty (1980) and Brun et al. (1992). In natural snow, the temperature gradient is not constant and often passes through several thresholds amongst metamorphism types over a season. A dry snow layer therefore may undergo successive types of metamorphism that are partially reversible. The results are often intermediate or mixed snow types.

2.4.1.2 *Grains and Pores*

The size, type, and bonding of snow crystals are responsible for the pore size and permeability of the snowpack. Depth hoar layers are characterised by very large grains (2–10 mm). They also include large pores with sizes similar to that of the crystals. These pores partially survive the spring wetting due to rain or snowmelt. Because of their kinetic growth, depth hoar crystals are not principally linked by ice bonds but exist as piles linked by a few strands of brittle chains. Sometimes large voids of several centimetres diameter form because of a previous collapse of grains, especially under ice crusts or close to rocks or tree branches.

Faceted crystals are formed by medium temperature gradient metamorphism or as a first stage of high temperature gradient metamorphism. Their grain size is between 0.3 and 1 mm. In most cases, the structure is a combination of ice bonds and piled crystals, so that most of the properties of faceted snow layers are intermediate between those of depth hoar and of rounded crystals.

Rounded grains are very small (0.1–0.4 mm) and are principally formed from fresh snow drifted by strong winds. Their small radius of curvature leads to very efficient sintering between such crystals. The result is strong bonding, with the ice bonds relatively large compared with grain size. Pore size is accordingly very small with no large voids because the structure is not subject to collapse.

An example of pores in snow is shown in Figure 2.31; the photograph is taken from a thin section of a specially treated snowpack sample from a former National Hydrology Research Institute laboratory in the Canadian Rockies (Dozier, Davis, and Perla, 1987). Dozier et al. note from analysis of such photographs that the ratio of ice–pore interface surface area to mass is initially high for fresh snow (100 m^2 kg^{-1}),

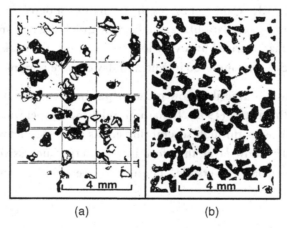

(a) (b)

Figure 2.31. Snow structure collected at middepth from a subalpine snowpit near Banff, Alberta, by Ron Perla and photographed at a former National Hydrology Research Institute snow laboratory. Sample density is 375 kg m^{-3}. (a) Photograph of disaggregated grains from the snow sample, displaying both faceted and rounded structure; (b) photograph of a thin-section plane prepared from the sample and showing the cross-sectional snow structure. Dark areas represent snow/ice and white areas represent pores in the snowpack (Perla, 1978)

100

declining by a factor of 10 for old seasonal snow. These surface areas presumably form a habitat for microorganisms that require ice, water, and air to survive (see Chapter 4).

2.4.1.3 *Density and Hardness*

In low wind speed environments, fresh snowfall has very low hardness (ram hardness < 10 N; surface hardness < 1 kN m^{-2}) and a density range from 50 to 120 kg m^{-3}. High, medium, and low temperature gradient metamorphism modify this snow, resulting in various snow sintering and densification rates. Goodison et al. (1981) measured rates of snow density increase of 7 kg m^{-3} hour^{-1} for the first 6 hours after snowfall. The ranges of density and bonding induce a large range of hardness. The cohesion and hardness of depth hoar or faceted grain layers are normally very low. If these layers are formed from light, fresh snow, they reach a maximum density varying from 150 to 325 kg m^{-3}. The corresponding ram hardness is often lower than 10 N m^{-2} (Brun and Rey, 1987). Thick depth hoar layers can be an impediment to the movement of large animals (and humans) on snow cover that do not have special walking features to decrease the pressure applied by their feet or hooves. The low shear strength (less than 5 kN m^{-2}) permits relatively easy movement of large animals inside or through depth hoar layers. These mechanical properties may be partially conserved during a rapid melt in spring.

In contrast, small rounded grain layers formed from wind redistribution or low temperature gradient metamorphism present a high density (250 to 500 kg m^{-3}), high hardness (ram hardness larger than 50 N), and strong cohesion (larger than 10 kN m^{-2}). Movement over these snow covers is facilitated but movement inside is impeded. The density of dry rounded grain layers is sufficiently high for these grains to retain their high hardness and strong cohesion during the melting period.

Exposure to wind hardens and densifies snow; for example, the surface hardness of freshly fallen snow varied from 2.5 to 1,000 kN m^{-2} in one winter (1986–1987) near Saskatoon, on the Canadian Prairies (Pomeroy, 1988). The higher hardness values are associated with conditions of high wind speeds, greater age, and concurrent blowing snow during deposition. Hardness increases with the wind speed required to sustain blowing snow. Li and Pomeroy (1997) have shown for prairie environments that these wind speeds increase with the logarithm of time since fresh snowfall. This suggests an initial rapid increase in hardness after snowfall with an asymptotic increase as elapsed time becomes large. Snow deposited within exposed vegetation cover has much lower hardness and density. When snow is transported by wind, the crystals undergo changes to their shape, size, and other physical properties, and upon redeposition form drifts

101

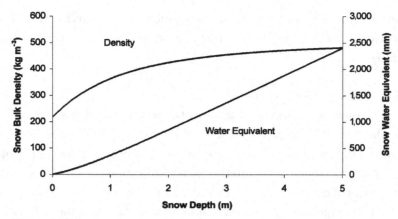

Figure 2.32. Increase in mean density of snowdrifts with increasing depth; trend from extensive measurements in Wyoming, USA, by Tabler et al. (1990b).

and banks of higher density than the parent material. Gray, Norum, and Dyck (1971) measured a sixfold increase in 24 hours, from 45 kg m^{-3} to 250 kg m^{-3}, in the density of freshly fallen snow during blowing snow on the Canadian Prairies. When drifted snow becomes very deep, compression further controls the snow density. As shown in Figure 2.32, Tabler et al. (1990b) report that the mean density of deep snow drifts ρ_s (kg m^{-3}) increases with depth d_s (cm) as

$$\rho_s = 522 - \frac{20470}{d_s}\left(1 - \exp\left[-\frac{d_s}{67.3}\right]\right) \tag{2.13}$$

This relationship should be applied only to snowdrifts deeper than 60 cm, as there is no strong relationship between depth and density for shallow windblown snow covers. Figure 2.33 (Pomeroy and Gray, 1995) shows the seasonal increase in snow density determined for various snow environments from data sources around the world.

2.4.2 Wet Snow

2.4.2.1 Irreducible Water Content and Water Flow

Wet snow is characterised by a significant amount of liquid water in snow. Significant, in this instance, means an amount larger than that in the liquid–like layer that still exists at subfreezing temperatures on the edges of ice crystals (Hobbs, 1974). Consequently, wet snow is usually encountered only at mean snow temperatures equal to the melting point. An exception is where zones of preferential meltwater flow into the normally dry and cold internal pack layers (Marsh and Woo, 1984a). Wet snow can also exist in small areas of the snowpack at below 0°C due to freezing point depression

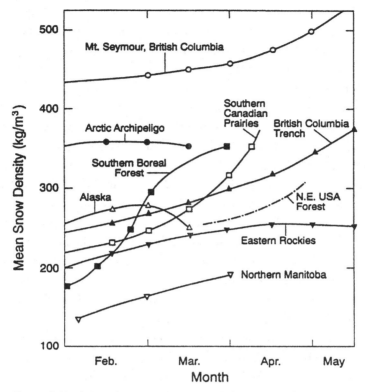

Figure 2.33. Seasonal variation in average snow cover density, drawn from many sources (Pomeroy and Gray, 1995)

by chemical impurities and pressure exerted by overlying snow layers. Freezing point depression is not normally more than a few tenths of a degree Celsius. Dry snow also may exist at the melting point. A deep, isothermal snow cover typically retains dry snow in internal layers when the surface begins to melt. Consequently, it is necessary to measure the liquid water content of a snow layer to determine whether it is wet or dry snow.

Liquid water content may be measured with reference to the snow mass, snow volume, or the volume of pores. The typical range of liquid water content in well-drained snow layers is from 0.1 to 8 percent of the snow mass. Liquid water percolates downward when its content exceeds the irreducible water content, which is that held within a snow layer by capillary forces (Colbeck, 1973; Denoth et al., 1979). When the water flux is high, or when a snow layer lies upon an impermeable layer of snow, ice, or soil, the snow may become saturated. Saturated layers may also form when a snow layer lies upon a permeable ground or snow layer with larger grain size than the overlying layer. In this last case, snow is saturated over a depth ranging from 3 to 10 cm; the depth depends on grain size. Smaller grains result in a deeper saturated layer. Field measurements of snow liquid water content are accurately and easily done with dielectric sensors and a density kit (Denoth et al., 1984).

Water flow through snow is not stable and is strongly affected by flow instabilities such as relatively impermeable layers, zones of preferential flow called flow fingers, and large meltwater drains. Flow fingers and wet snow above less-permeable layers in cold snow are shown in Figure 2.34. Meltwater drains are large and often extend directly to the bottom of the pack (Wakahama, 1968), whilst flow fingers usually initiate and terminate at snow layer boundaries (U.S. Army Corps of Engineers, 1956). A conceptual cross-sectional representation of water movement around impermeable layers and in flow fingers is shown in Figure 2.35 (adapted from Marsh and Woo, 1984a). Water dammed at these boundaries moves rapidly down to the next one via the flow fingers until it refreezes or reaches the soil. In this manner, flow instabilities concentrate water within a melting pack, resulting in a greater flux of water at the leading edge of the wetting front than if that front were from uniform flow. Liquid water therefore can quickly reach snow layers far below the surface whilst the dry zones in the upper pack remain so for some time. Marsh and Woo (1984a) report a mean flow finger diameter of 3.6 cm and spacing of 13.1 cm, with fingers occupying 22 percent of the horizontal area of the pack in the Canadian High Arctic. In the subarctic/boreal forest transition Marsh (1988) noted finger diameters of 5 cm and similar spacing resulting in a horizontal area of 27 percent. For a warm, isothermal snowpack with a 1.2-m depth, the meltwater model of Marsh and Woo (1984b) suggests that the flow finger wetting front can reach the ground 4 days before the background wetting front, whilst for a cold snow pack (-12°C at bottom) the wetting fronts reach the bottom at approximately the same time, because the flow finger water is refrozen as it penetrates the cold layers of the pack. On slopes, meltwater may be diverted downslope by ice layers so that it reaches the snowpack base well downhill of its origin (English et al., 1987). In general, peak discharges from a snowpack with flow instabilities are much higher and faster than if the pack were homogeneous. The inhomogeneities of flow paths, water content, and water flux during melt have important implications for the chemistry and microbiology of snow covers as discussed in Chapters 3 and 4.

2.4.2.2 *Grain Growth*

Liquid water in snow drastically affects the evolution and physical properties of snow crystals and layers, primarily by wet snow metamorphism. The melting point at an ice–water interface depends on radius of curvature of the interface. Hence,

104

Figure 2.34. Photograph showing dye injection into developing flow fingers and wet layers in cold arctic snow (by Phil Marsh, NHRI).

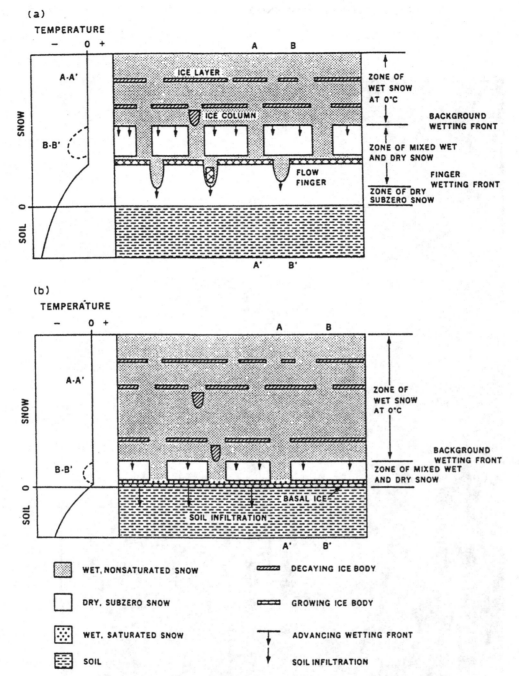

Figure 2.35. Conceptual representation of water flow, flow finger development, temperature change, and ice layer growth during meltwater percolation into an initially cold snowpack (after Marsh and Woo, 1984b). (a) Nonisothermal, before meltwater reaches the base; (b) Isothermal, after meltwater reaches the base but before the pack is fully wetted.

106

the sharpest parts of the snow crystals can be melting when the concave or flattest parts are refreezing. This induces a microflow of water from sharp points to flat or concave points on a snow crystal, rounding the crystals into a classic wet grain shape. Furthermore, larger grains are growing at the expense of the rapidly melting smaller ones to form a snow layer of large, rounded grains. Wet snow metamorphism is very efficient when the liquid water content is larger than 5 percent of mass, because the presence of liquid veins among the grains facilitates the heat and mass transfer necessary for the phase changes (Brun, 1989). In wet snow under constant conditions, the growth rate of the grain diameter varies with the cubic root of time (Raymond and Tusima, 1979; Brun, 1989). However, for natural snow under liquid water contents that vary with diurnal and daily energy input changes, the grain growth rate is more linear with time (Marsh, 1987). Wet snow grains rapidly round and grow up to a diameter of about 0.5 mm in a few hours. However, in natural conditions it takes several days to reach a diameter of 1 mm and several weeks to reach 2 mm. An example of wet snow grain structure is shown in Figure 2.36.

2.4.2.3   *Density and Hardness*

Wet snow metamorphism at first induces a decrease in cohesion and hardness as liquid bonds form between the crystals at the expense of solid ones. This phenomenon is most likely due to higher concentrations of impurities in ice bonds than in surrounding crystals with subsequent depression of the melting point. The decrease of cohesion in wet snow is well known through its inducement of avalanche

Figure 2.36. Microphotograph of wet, clustered, rounded snow grains, showing the snow grain, wet snow bonds, and surrounding liquid water layer, 6a (from Colbeck et al., 1990).

activity and plays an important role in snow ecology by changing the ability of animals to move upon the snow surface or inside the pack.

The presence of liquid water increases the compaction rate because of the decrease in viscosity of wet snow and because the metamorphism directly transforms faceted or plate-like crystals into spherical ones. Denser, more cohesive, and harder snow covers form in a few days even when wetting affects fresh, low-density snow. If wet conditions prevail for from one to several weeks, the density can reach 350–550 kg m^{-3}, with strong diurnal variations as the liquid water content of the snow changes with meltwater production. It is possible during active snowmelt to measure a density in the morning of 350 kg m^{-3} and a density in late afternoon (when the snowpack is primed with meltwater) of 550 kg m^{-3}.

2.4.2.4   *Refreezing, Ice Layers, Basal Ice Growth*

When a wetting front is stationary, the heat flux from wet to dry snow may be sufficient to refreeze water and form ice bodies within the snowpack. As this happens, flow fingers form ice columns, snow layer boundaries form ice layers, and the snow–soil boundary forms basal ice (Marsh, 1991). An example of a large ice column capped by an ice layer is shown in Figure 2.37. When refreezing, ice columns develop very large, rounded, clustered crystals (Kattelmann, 1990). Horizontal ice layers often develop where a flow finger spreads the flowing water over a cold internal layer made from small rounded grains and overlying a layer made up of larger crystals (for instance, at the wind-packed to depth hoar layer transition). Ice layers are typically 0.1 to 20 cm thick and extend horizontally from 0.2 to more than 3 m (Langham, 1974). Basal ice forms where the meltwater flux exceeds the infiltration rate of frozen soils and there is a strong negative heat flux from the snow to the soil. Basal ice layers may become quite thick (up to 70 cm) and persist after the snow cover has melted, but they are most prevalent in permafrost regions (Woo et al., 1982).

Initially refrozen ice bodies within a snow cover are formed of large polycrystalline melt-freeze particles up to 5 mm in diameter, which are bonded in a honeycomb structure. As freezing continues this structure becomes a hard, rigid, relatively imper-meable ice body with a density greater than 800 kg m^{-3} (Marsh and Woo, 1984a). As the temperature rises above -0.1°C, the permeability of ice layers increases rapidly, hard-ness decreases, and both parameters may change diurnally during melt. The hardness of refrozen meltwater in surface or subsurface layers is remarkably high and usually sufficient to support a human or large animal. However, it is subject to strong diurnal fluctuations during melt, along with temperature and net radiation. Basal ice prevents

108

Figure 2.37. Photograph of a large ice column, capped by an ice layer that had been removed from a cold snow cover in the Arctic (Phil Marsh, NHRI). The column was formed by meltwater refreezing where ponded along a horizontal structural divide in the snowpack and in a vertical preferential flow path.

access to ground-hugging vegetation by grazing herbivores during snowmelt in the High Arctic.

Rain-on-snow events play a major role in wetting the snow surface and in forming ice crusts, especially during midwinter when the energy balance is generally unable to generate sufficient snow melt to wet deep snow layers. In regions where such events are very rare under the present climate, a climate change could increase their frequency and consequently change drastically the structure of the snow cover and the ecology.

2.4.3   Snow Cover Classification

Snow classification is an extremely complex subject because classification schemes tend to be most useful for the purposes of their derivation and no scheme has received widespread use in the study of snow. Snow covers evolve over time and vary at both small and meso spatial scales in association with vegetation and topography and at very large scales with climate. One might argue that, with a good understanding of the physical characteristics of a snow cover, static classification of such a dynamic phenomenon is somewhat misleading and probably unnecessary to gain an understanding of snow ecology.

Pruitt (1984) and McKay and Gray (1981) have presented schemes that classify snow by vegetation zone, presuming a strong association between natural vegetation cover and regional climate. Although these schemes correspond with broad ecozone categories and hence are easy to apply, they have not been rigorously defined in terms of snow properties, are not sensitive to land use or climate change, and offer incomplete physical descriptions of snow covers.

A scheme that has attempted to classify global snow cover using physically based measures is that by Sturm, Holmgren, and Liston (1995); elements of this scheme and global snow maps have been presented in Chapter 1. Sturm et al.'s climatic scheme is based on three major variables (temperature, wind speed, and snowfall) and discriminates snow cover into seven classes (*tundra, taiga, alpine, prairie, maritime, mountain,* and *ephemeral*). Climatic rather than land cover indicators are used despite the naming of classes by associated ecoregion and biome; hence, a tundra snow can develop in a prairie environment if the climate falls within the tundra snow class and alpine snow can develop in a transitional forest/prairie (parkland) region. The strict use of climatic indicators can result in confusing classifications when applied outside the region of development, as the microclimate that influences snow cover formation is strongly associated with both climate and vegetation cover (biome) as described earlier in this chapter. Despite the difficulties that can arise in using a climatic classification for ecological purposes, the scheme of Sturm et al. provides rigorous quantitative descriptions of snow characteristics in various climates. If used with discretion it provides the most objective physical classification currently available.

In Figure 2.38 snow stratigraphy using the crystal classification in Table 2.2 is demonstrated for class averages in mid to late winter. Tundra and prairie snow is shallow and dominated by wind slab and either depth hoar (tundra) or fresh to

110

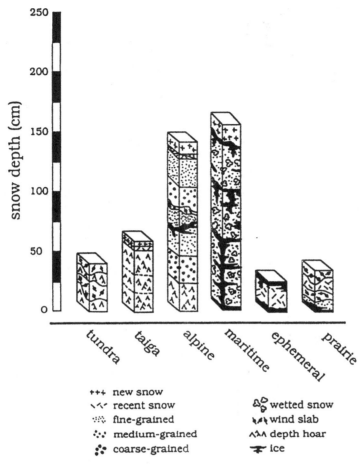

Figure 2.38. Basic stratigraphy and texture for each snow class in Sturm et al.'s climatic classification, as found in mid-to-late winter (after Sturm et al., 1995).

fine-grained snow (prairie); taiga snow is deeper and dominated by depth hoar; alpine and maritime snow are the deepest and structurally the most complex, with layers of varying grain size, wetness, and ice bodies. Boundaries between tundra, taiga, alpine, and maritime classes along with measured data from Alaska are shown in Figure 2.39 for average winter air temperature, soil heat flux, snow depth, snow–soil interface temperature, and vertical temperature gradient. These values provide representative ranges for these snow properties. The vertical temperature gradient shows that only tundra and taiga snow develop gradients sufficient for kinetic metamorphism; the gradients are controlled by cold air temperatures and shallow snow depth and result

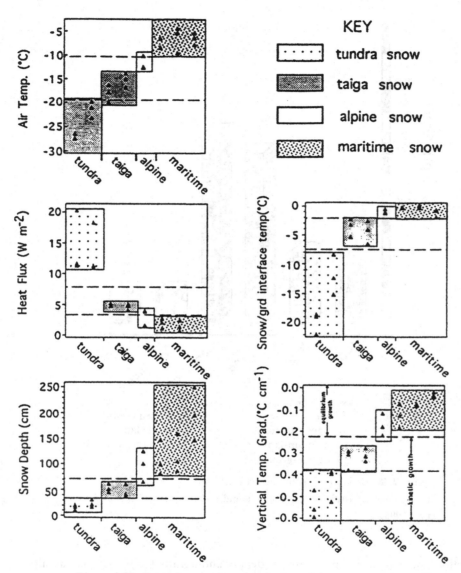

Figure 2.39. Snow classes and measured values of air temperature, ground heat flux, snow–soil interface temperature, snow depth, and vertical temperature gradient for tundra, taiga, maritime, and alpine snow (after Sturm et al., 1995).

in large soil heat fluxes and cold snow–soil interface temperatures. The snow properties shown in Figures 2.38 and 2.39 can be used to anticipate conditions in the appropriate environments where measurements are lacking. A map of average snow cover classes for much of the globe is shown in Chapter 1.

2.5   Impact of Snow on Soil and Aquatic Ecosystems

Snow cover provides an insulating blanket over soil and lake ice for the winter period and provides an important episodic flux of latent heat, water, and chemicals into soils and water bodies during the spring melt. The important role of snow covers on lakes in controlling lake ice thickness and strength is reviewed by Adams (1981). The insulation of soils by snow cover is discussed in the preceding sections with typical snow–soil interface temperatures shown in Figure 2.39. In northern prairie environments, snow accumulation is managed to provide a deep insulating snow cover for winter wheat crops (Steppuhn, 1981). In many high latitude and altitude environments, snowfall accounts for over 40 percent of annual precipitation, and the release as meltwater over the course of a few days can be the single most important hydrological event in these environments (Woo, 1990).

2.5.1   Infiltration of Meltwater into Frozen and Unfrozen Soils

Meltwater infiltration into unfrozen soils usually occurs in environments with deep snowpacks or maritime climates. Soils frozen to depths less than 150 mm behave as unfrozen soils with respect to infiltration. The proportion of snowmelt water that infiltrates to unfrozen soils depends on the application rate of meltwater from the snowpack, the hydraulic conductivity of soil layers, and the water-retention characteristics of the soil. Snowmelt discharge rates less than the saturated hydraulic conductivity of the soil will completely infiltrate. Discharge rates in excess of the saturated hydraulic conductivity will infiltrate until the soil becomes saturated and ponding of water at the base of the snowpack occurs. When water is ponded on the soil surface, procedures such as the Philip infiltration equation (Philip, 1957) are used to calculate the infiltration rate. Infiltration in this case depends on the hydraulic conductivity of the soil, suction at the wetting front in the soil, soil porosity, and initial soil water content. A complete discussion of techniques for calculating infiltration into unfrozen soils is provided by Rawls, Brakensiek, and Miller (1993).

Frozen soils develop where snow cover is thin or extremely cold winter temperatures prevail. Frozen soil indices can be used to estimate the occurrence of soils with frozen infiltration characteristics; these indices are exponential functions of the mean daily air temperature and depth of snow on the ground (Molnau and Bissell, 1983). One index developed for the northwestern United States is

$$FSI_i = A(FSI_{i-1}) - T_m e^{(-0.4Kd_s)} \tag{2.14}$$

113

where FSI is the frozen soil index, i is the day of the snow year, T_m is mean daily air temperature, d_s is the depth of snow on the ground (cm), A is a daily decay coefficient (0.97), and K is a snow depth reduction coefficient (0.57 cm^{-1}). FSI values between $56°C$ and $83°C$ days correspond to the transition between frozen and unfrozen infiltration characteristics in soils of the northwestern United States.

Frozen soils normally have lower infiltration capacities than do unfrozen soils of similar saturation levels because the presence of ice from initial water or refreezing reduces the effective porosity of the soils. Saturated frozen soils lacking macropores have a negligible infiltration capacity and are considered impermeable. Unsaturated frozen soils have snowmelt infiltration amounts that vary with the SWE of the snow cover and the frozen water content of the upper 300 mm of soil. Granger, Gray, and Dyck (1984) provide the following equation to calculate snowmelt infiltration into frozen prairie soils:

$$INF = 5(1 - \theta_p)SWE^{0.584} \tag{2.15}$$

where INF is infiltration (mm of water) and θ_p is the degree of soil pore saturation ($\text{mm}^3 \text{ mm}^{-3}$) that can be measured in the autumn. Pomeroy et al. (1997b) provide a somewhat different equation for southern boreal forest snowmelt infiltration, where

$$INF = (1 - 1.36\,\theta_p)\frac{SWE^{2.41}}{364} \tag{2.16}$$

Equation 2.16 normally should be applied only to snowpacks of less than 100 mm SWE. The higher exponent for the forest compared with the prairie reflects the role of deep snowpacks in insulating soil, so that soils are not always completely frozen at the time of melt. Where snow was deep (normally clear-cuts or burns), Pomeroy et al. found the soils to be incompletely frozen; hence infiltration was high. Frozen soils that contain large cracks or macropores normally will be able to accommodate infiltration of all snowmelt water.

2.5.2 Runoff and Streamflow Generation

Streamflow is generated by snowmelt water that directly runs off rather than infiltrating or from water that infiltrates and then moves downslope through a shallow subsurface soil of high permeability. During snowmelt, frozen or saturated soils restrict infiltration and evaporation is relatively low; this promotes a water excess over a basin and permits relatively large runoff generation for the amount of water applied to the ground. As a result, peak annual streamflows often occur directly after

114

snowmelt. The constituent waters of this freshet comprise both snowmelt water and water expelled from soils by infiltrating snowmelt water, with important implications for stream chemistry (see Tranter and Jones, Chapter 3). For point scales, the influence of SWE on infiltration and runoff generation is shown, presuming negligible evaporation, for a variety of soil types in prairie, arctic, and boreal forest environments in Figure 2.40. The effect of deep forest environment snowpacks in promoting warm soils causes forest runoff to drop with increasing SWE for deep snow and dry soils.

In northern forests (boreal, northern hardwoods), from 40 to 60 percent of annual streamflow is derived from snowmelt, with increases in snowmelt runoff of from 24 to 75 percent when the forest is removed by harvesting or fire (Hetherington, 1987). In cold, semiarid environments (arctic, northern prairies, steppes) greater than 80 percent of annual streamflow is derived from snowmelt, even though snowfall accounts for less than 50 percent of annual precipitation (Gray, 1970). Snowmelt in the western cordillera of North America and mountain systems of central Asia is the major source of water when carried as streamflow to semiarid regions downstream. Snowmelt water sustains arctic, alpine, prairie, and boreal forest lakes and wetlands, which are primary aquatic habitats in their respective ecosystems. Prairie pothole ponds are part of an interesting symbiotic relationship between snow, pond/wetland, and trees. The poplar trees surrounding these ponds induce large snowdrifts, which upon melt replenish the pond water supply. The high water table surrounding the pond in turn sustains the poplar trees.

Attempts to modify streamflow generation through snow management in the United States have involved streamflow enhancement in the upper Colorado River basin by selective clear-cutting of coniferous forests at high elevations and in the upper Missouri River basin by placement of arrays of tall snow fences in high-elevation steppe basins. In the Canadian Prairies, snow management is practised by leaving tall wheat stubble strips (one per swath) perpendicular to the primary wind direction to trap small snowdrifts. Snowpack can more than double that of adjacent fallow fields but, when infiltration capacity is low, results in increased runoff from the field. Snow management can increase water supplies deriving from snowmelt on the prairies. Estimates by de Jong and Kachanoski (1987) suggest that if snow management were universally adopted across the Canadian Prairies, water supply in the region would increase by 2–4×10^9 m^3 of water per year, more than the amount of water currently used for irrigation. Inadvertent modification of the snow environment by clear-cutting forests or agricultural practises can cause unanticipated changes in downstream runoff, increased flooding, and damage to the aquatic ecosystem, but it

115

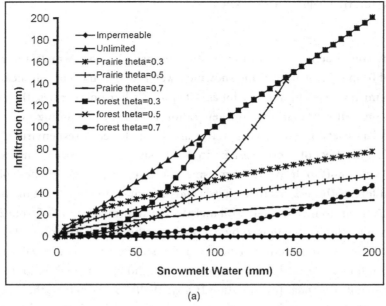

(a)

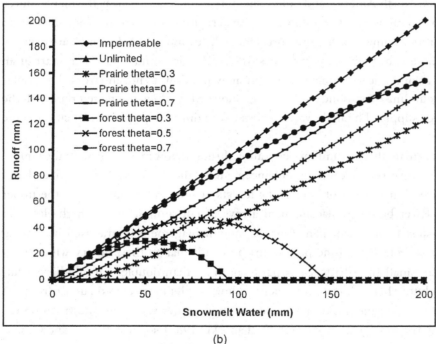

(b)

Figure 2.40. Disposition of snowmelt in various environments with frozen soils (prairie, arctic, southern boreal forest), presuming negligible evaporation during melt. Soils coated with a basal ice layer will be impermeable (typical of High Arctic); those with extensive soil macropores will be unlimited (typical of organic terrain and some heavy clays). Theta refers to θ_p, the degree of pore saturation. (a) Infiltration to frozen soils; (b) local runoff generation.

116

may also provide a means to ameliorate the impacts of drought associated with climate warming.

2.6 Conclusions

Snow is one of the most complex physical materials on Earth and therefore provides a challenging habitat for life. Its presence is often ephemeral, governed by weather, climate, topography, and vegetation cover. As a substance, it is crystalline at small scales and porous at larger scales. It is highly reflective and uniquely undergoes phase change to both liquid and vapour forms of water at temperatures that are normally encountered in the winter and under conditions that may be manipulated by life forms. Snow is also one of the lightest natural materials, such that it is relocated by wind and vegetation and can be burrowed in or stepped through by mammals. The porous feature of snow means that it is an excellent insulator and is capable of remaining wet and percolating meltwater over periods of several days. The crystalline nature of snow sometimes results in poor strength, so that it cannot support heavy animals or remain in the branches of trees when bond strength weakens upon warming. Snowmelt water – whether retained within the snowpack or delivered to soils and streams – is a critical supporter of life and medium for nutrients. The following chapters will detail the chemical properties of snow and how various plant and animal communities live in, on, and beneath this fascinating material.

2.6.1 Future Research

Research into snow physical processes will continue to provide information of benefit to snow ecologists. Often physical research has been conducted to solve applied problems (e.g., avalanche control, remote sensing) but has then been found useful in biological fields (e.g., snow microbiology). Scientists conducting research on snow properties in relation to ecology should note that because of the complex multi-phase physical nature of snow, relationships developed must be consistent with conservation of mass and energy. The inter-relationships between snow properties and vegetation have been addressed in this chapter, by Walker, Billings and de Mollenaar (Chapter 6) and by Bégin and Boivin (Chapter 7) and involve a fascinating series of feedbacks between vegetation structure and snow structure. In this sense, the snowpack not only affects ecology but also can be controlled by the resulting vegetation. No other feedback between physical and biological systems is so strong, evident, and persistent around the world. The relationships between snow and

117

vegetation are complex and vary with biome and climate; they will require a specialized research effort focusing on process-based interactions in both open and forested environments.

2.7   References

Adams, P. 1981. Snow and ice on lakes. In *Handbook of Snow, Principles, Processes, Management and Use*, D.M. Gray and D.H. Male (Eds.). Pergamon Press, Toronto, pp. 437–474.

Anderson, E. 1976. A point energy and mass balance model of snow cover. *NOAA Technical Report NWS 19*, U.S. National Weather Service, Dept. of Commerce, Washington, DC.

Auld, H. 1995. Dependence of ground snow loads on elevation in western Canada. *Proc. Western Snow Conf.*, 63, 143–146.

Bader, H.-P. 1992. Modelling temperature distribution, energy and mass flow in a phase-changing snowpack. I. Model and case studies. *Cold Regions Sci. Technol.*, 20, 157–181.

Braun, L. 1991. Modelling the snow water equivalent in the mountain environment. In *Snow, Hydrology and Forests in High Alpine Areas*, H. Bergmann, H. Lang, W. Frey, D. Issler, and B. Salm (Eds.). IAHS Publication No. 205, IAHS Press, Wallingford, UK, pp. 3–17.

Brun, E. 1989. Investigation on wet snow metamorphism in respect of liquid water content. *Ann. Glaciol.*, 13, 22–26.

Brun, E., and Rey, L. 1987. Field studies on snow mechanical properties. In *Avalanche, Formation, Movement and Effects*, B. Salm and H. Gubler (Eds.). IAHS Publication No. 162, IAHS Press, Wallingford, UK, pp. 183–193.

Brun, E., Martin, E., Simon, V., Gendre, C., and Col, ou C. 1989. An energy and mass model of snow cover suitable for operational avalanche forecasting. *J. Glaciol.*, 35(121), 333–342.

Brun, E., David, P., Sudal, M., and Brunot, G. 1992. A numerical model to simulate snow-cover stratigraphy for operational avalanche forecasting. *J. Glaciol.*, 38(128), 13–22.

Budd, W.F., Dingle, W.R.J., and Radok, U. 1966. The Byrd snow drift project: outline and basic results. *Studies Antarctic Meteorol. Am. Geophys. Union Antarctic Res. Ser.*, 9, 71–134.

Calder, I.R. 1990. *Evaporation in the Uplands*. Wiley, Chichester, p. 144.

Colbeck, S.C. 1973. *Theory of Metamorphism of Wet Snow*. CRREL Research Report 313, U.S. Army Cold Regions Research and Engineering Laboratory, Hanover, NH.

Colbeck, S.C. 1983. Theory of metamorphism of dry snow. *J. Geophys. Res.*, 88(C9), 5475–5482.

Colbeck, S.C. 1987. Snow metamorphism and classification. In *NATO ASI Series C: Mathemat. Physic. Sci.*, Vol. 211, *Seasonal Snowcovers: Physics, Chemistry, Hydrology*, H.G. Jones and W.J. Orville-Thomas (Eds.). Reidel, Dordrecht, The Netherlands, pp. 1–35.

Colbeck, S.C. 1989. Air movement in snow due to windpumping. *J. Glaciol.*, **35**(120), 209–213.

Colbeck, S.C., Akitaya, E., Armstrong, R., Gubler, H., Lafeuille, J., Lied, K., McClung, D., and Morris, E. 1990. *The International Classification for Seasonal Snow on the Ground.* International Association of Hydrological Sciences, International Commission on Snow and Ice, Wallingford, Oxfordshire.

Conway, H., Gades, A., and Raymond, C.F. 1996. Albedo of dirty snow during conditions of melt. *Water Resources Res.*, **32**(6), 1713–1718.

de Jong, E., and Kachanoski, R.G. 1987. The role of grasslands in hydrology. *Can. Bull. Fish. and Aquatic Sci.*, **215**, 213–241.

Denoth, A., Seidenbusch, W., Blumthaler, M., Kirchlechner, P., Ambach, W., and Colbeck, S.C. 1979. *Study of Water Drainage from Columns of Snow.* CRREL Report 79–1, U.S. Army Cold Regions Research and Engineering Laboratory, Hanover, NH.

Denoth, A., Foglar, A., Weiland, P., Matzler, C., Aebischer, H., Tiuri, M., and Sihvola, A. 1984. A comparative study of instruments for measuring the liquid water content of snow. *J. Appl. Phys.*, **56**, 2154–2159.

Dozier, J., Davis, R.E., and Perla, R.I. 1987. Snow microstructure measurements using stereology. In *Snow Property Measurement Workshop.* Technical Memorandum No. 130, National Research Council of Canada, Ottawa, pp. 60–70.

Dyunin, A.K., Kvon, Ya.D., Zhilin, A.M., and Komarov, A.A. 1991. Effect of snow drifting on large-scale aridization. In *Glaciers-Ocean-Atmosphere Interactions*, V.M. Kotlyakov, A. Ushakov, and A. Glazovsky (Eds.). IAHS Publication No. 208, IAHS Press, Wallingford, UK, pp. 489–494.

English, M.C., Semkin, R.G., Jeffries, D.S., Hazlett, P.W., and Foster, N.W. 1987. Methodology for investigation of snowmelt hydrology and chemistry within an undisturbed Canadian shield watershed. In *NATO ASI Series C: Mathemat. Physic. Sci., Vol. 211, Seasonal Snowcovers: Physics, Chemistry, Hydrology*, H.G. Jones and W.J. Orville-Thomas (Eds.). Reidel, Dordrecht, The Netherlands, pp. 467–500.

Föhn, P.M.B., and Meister, R. 1983. Distribution of snow drifts on ridge slopes: measurements and theoretical approximations. *Ann. Glaciol.*, **4**, 52–57.

Gold, L.W. 1957. The Influence of Snowcover on Heat Flow from the Ground. In *Snow and Ice*, IAHS Publication No. 46, IAHS Press, Wallingford, UK, pp. 13–21.

Gold, L.W., and Williams, G.P. 1961. Energy balance during the snowmelt period at an Ottawa site. In *Snow and Ice*, IAHS Publication No. 54, IAHS Press, Wallingford, UK, pp. 288–294.

Golding, D.L., and Swanson, R.H. 1986. Snow distribution patterns in clearings and adjacent forest. *Water Resources Res.*, **22**(13), 1931–1940.

Goodison, B.E., Ferguson, H.L., and McKay, G.A. 1981. Measurement and data analysis. In *Handbook of Snow, Principles, Processes, Management and Use*, D.M. Gray and D.H. Male (Eds.). Pergamon Press, Toronto, pp. 191–274.

Granger, R.J., and Gray, D.M. 1990. A net radiation model for calculating daily snowmelt in open environments. *Nordic Hydrol.*, **21**, 217–234.

Granger, R.J., and Male, D.H. 1978. Melting of a prairie snowpack. *J. Appl. Meteorol.*, **17**(12), 1833–1842.

Granger, R.J., Gray, D.M., and Dyck, G.E. 1984. Snowmelt infiltration into frozen Prairie soils. *Can. J. Earth Sci.*, **21**, 669–687.

Gray, D.M. (Ed). 1970. *Handbook on the Principles of Hydrology*. Canadian National Committee for the International Hydrological Decade, Ottawa.

Gray, D.M., and O'Neill, A.D.J. 1974. Application of the energy budget for predicting snowmelt runoff. In *Advanced Concepts in the Study of Snow and Ice Resources*, H.S. Santeford and J.L. Smith (Eds.). National Academy Press, Washington, DC, pp. 108–118.

Gray, D.M., Norum, D.I., and Dyck, G.E. 1971. Densities of prairie snowpacks. *Proc. Western Snow Conf.*, **39**, 24–30.

Gray, D.M., and others. 1979. Snow accumulation and redistribution. In *Proceedings, Modeling of Snow Cover Runoff*, S.C. Colbeck and M. Ray (Eds.). U.S. Army Cold Regions Research and Engineering Laboratory, Hanover, NH, pp. 3–33.

Gubler, H. 1985. Model of dry snow metamorphism by interparticle vapor flux. *J. Geophys. Res.*, **90**(C8), 8081–8092.

Gubler, H., and Rychetnik, J. 1991. Effects of forests near the timberline on avalanche formation. In *Snow, Hydrology and Forests in High Alpine Areas*, H. Bergman, H. Lang, W. Frey, D. Issler, and B. Salm (Eds.). IAHS Publication No. 205, IAHS Press, Wallingford, UK, pp. 19–38.

Harding, R.J., and Pomeroy, J.W. 1996. The energy balance of the winter boreal landscape. *J. Climate*, **9**, 2778–2787.

Harestead, A.S., and Bunnell, F.L. 1981. Prediction of snow-water equivalents in coniferous forests. *Can. J. Forest Res.*, **11**, 854–857.

Hedstrom, N.R., and Pomeroy, J.W. 1998. Accumulation of intercepted snow in the boreal forest: measurements and modelling. *Hydrol. Processes*, **12**, 1611–1623.

Hetherington, E.D. 1987. The importance of forests in the hydrological regime. *Can. Bull. Fish. and Aquatic Sci.*, **215**, 179–211.

Hobbs, P.V. 1974. *Ice Physics*. Clarendon Press, Oxford.

Jordan, R. 1991. *A One-Dimensional Temperature Model for a Snow Cover*. Cold Regions Research and Engineering Laboratory Special Report 91-16, U.S. Army Research and Engineering Laboratory, Hanover, NH.

Kattelmann, R. 1990. Liquid water at the snowpack surface [Abstract]. *Eos*, **71**(43), 1328.

Kind, R.J. 1981. Snow drifting. In *Handbook of Snow, Principles, Processes, Management and Use*, D.M. Gray and D.H. Male (Eds.). Pergamon Press, Toronto, pp. 338–359.

Kobayashi, D. 1972. Studies of snow transport in low-level drifting snow. *Contrib. Inst. Low Temp. Sci. Ser. A*, **24**, 1–58.

Kohshima, S., Yoshimura, Y., Seko, K., and Ohata, T. 1994. Albedo reduction by biotic impurities on a perennial snowpatch in the Japan Alps. In *Snow and Ice Covers: Interactions with the Atmosphere and Ecosystems*, H.G. Jones, T.D. Davies, A. Ohmura, and E. Morris (Eds.). IAHS Publication No. 223, IAHS Press, Wallingford, UK, pp. 323–330.

120

Komarov, A.A. 1957. Some rules on the migration and deposition of snow in western Siberia and their application to control measures. *Trudy Transportno-Energetichesko-go Instituta*, **4**, 89–97 (National Research Council of Canada Technical Translation 1094, 1963).

Kotlyakov, V.M. 1961. Results of study of the ice sheet in Eastern Antarctica. In *Antarctic Glaciology*. IAHS Publication No. 55, IAHS Press, Wallingford, UK, pp. 88–99.

Kuz'min, P.P. 1960. Snowcover and snow reserves. *Gidrometeorologicheskoe, Izdatelsko*, Leningrad. (Translated by US National Science Foundation, Washington, DC, 1963).

Langham, E.J. 1974. Phase equilibria of veins in polycrystalline ice. *Can. J. Earth Sci.*, **11**, 1280–1287.

Langham, E.J. 1981. Physics and properties of snowcover. In *Handbook of Snow, Principles, Processes, Management and Use*, D.M. Gray and D.H. Male (Eds.). Pergamon Press, Toronto, pp. 275–337.

Li, L., and Pomeroy, J.W. 1997. Estimates of threshold wind speeds for snow transport using meteorological data. *J. Appl. Meteorol.*, **36**, 205–213.

Liljequest, G.H. 1956. Energy exchange of an Antarctic snow field: short wave radiation (Maudheim 71°03'S 10°56'W). In *Norwegian-British-Swedish Antarctic Expedition, 1949–52, Scientific Results*. Vol. 2, Part 1A. Norsk Polarinstitutt, Oslo.

Liston, G.E. 1995. Local advection of momentum, heat and moisture during the melt of patchy snow covers. *J. Appl. Meteorol.*, **34**(7), 1705–1715.

Male, D.H. 1980. The seasonal snowcover. In *Dynamics of Snow and Ice Masses*, S. Colbeck (Ed.). Academic Press, Toronto, pp. 305–395.

Male, D.H., and Granger, R.J. 1979. Energy and mass fluxes at the snow surface in a prairie environment. In *Proceedings of the Modelling of Snow Cover Runoff*, S. Colbeck and M. Ray (Eds.). U.S. Army Cold Regions Research and Engineering Laboratory, Hanover, NH, pp. 101–124.

Male, D.H., and Gray, D.M. 1981. Snowcover ablation and runoff. In *Handbook of Snow, Principles, Processes, Management and Use*, D.M. Gray and D.H. Male (Eds.). Pergamon Press, Toronto, pp. 360–436.

Marbouty, D. 1980. An experimental study of temperature-gradient metamorphism. *J. Glaciol.*, **26**(94), 303–312.

Marsh, P. 1987. Grain growth in a wet Arctic snow cover. *Cold Regions Sci. Technol.*, **14**, 23–31.

Marsh, P. 1988. Flow fingers and ice columns in a cold snowcover. *Proc. Western Snow Conf.*, **56**, 105–112.

Marsh, P. 1991. Water flux in melting snow covers. In *Advances in Porous Media, Vol. 1*, M.Y. Corapcioglu (Ed.). Elsevier Science, Amsterdam, pp. 61–124.

Marsh, P., and Pomeroy, J.W. 1996. Meltwater fluxes at an arctic forest-tundra site. *Hydrol. Processes*, **10**, 1383–1400.

Marsh, P., and Woo, M-K. 1984a. Wetting front advance and freezing of meltwater within a snowcover 1. Observations in the Canadian Arctic. *Water Resources Res.*, **20**(12), 1853–1864.

Marsh, P., and Woo, M-K. 1984b. Wetting front advance and freezing of meltwater within a snowcover 2. A simulation model. *Water Resources Res.*, **20**(12), 1865–1874.

Marsh, P., and Woo, M-K. 1987. Soil heat flux, wetting front advance and ice layer growth in cold, dry snow covers. In *Snow Property Measurement Workshop*. Technical Memorandum No. 140, Snow and Ice Subcommittee, Associate Committee on Geotechnical Research, National Research Council of Canada, Ottawa, pp. 497–524.

McGurk, B.J., and Marsh, P. 1995. Flow-finger continuity in serial thick-sections in a melting Sierra snowpack. In *Biogeochemistry of Seasonally Snow-Covered Catchments*, K.A. Tonnessen , M.W. Williams, and M. Tranter (Eds.). IAHS Publication No. 228, IAHS Press, Wallingford, UK, pp. 81–88.

McKay, G.A., and Gray, D.M. 1981. The distribution of snowcover. In *Handbook of Snow, Principles, Processes, Management and Use*, D.M. Gray and D.H. Male (Eds.). Pergamon Press, Toronto, pp. 153–190.

Meiman, J.R. 1970. Snow accumulation related to elevation, aspect and forest canopy. *Proc. Workshop Seminar on Snow Hydrology*. Queen's Printer for Canada, Ottawa, pp. 35–47.

Molnau, M., and Bissell, V.C. 1983. A continuous frozen ground index for flood forecasting. *Proc. Western Snow Conf.*, **51**, pp. 109–119.

Norton, D.C. 1991. Digital investigation of Great Lakes regional snowfall, 1951–1980. *Proc. Eastern Snow Conf.*, **48**, 67–80.

O'Neill, A.D.J., and Gray, D.M. 1973. Spatial and temporal variations of the albedo of a prairie snowpack. In *The Role of Snow and Ice in Hydrology: Proceedings of the Banff Symposium, Vol. 1*, Sept. 1972, UNESCO-WMO-IAHS, Geneva-Budapest-Paris, pp. 176–186.

Palm, E., and Tveitereid, M. 1979. On heat and mass flow through dry snow. *J. Geophys. Res.*, **84**(C2), 745–749.

Perla, R.I. 1978. *Snow Crystals*. NHRI Paper No. 1, National Hydrology Research Institute, Environment Canada, Ottawa.

Philip, J.R. 1957. The theory of infiltration: 1. The infiltration equation and solution. *Soil Sci.*, **83**(5), 345–357.

Pomeroy, J.W. 1988. *Wind Transport of Snow*. Thesis in partial fulfilment of the requirements for the Degree of Doctor of Philosophy in the College of Graduate Studies and Research, University of Saskatchewan, Saskatoon.

Pomeroy, J.W. 1989. A process-based model of snow drifting. *Ann. Glaciol.*, **13**, 237–240.

Pomeroy, J.W., and Dion, K. 1996. Winter radiation extinction and reflection in a boreal pine canopy: measurements and modelling. *Hydrol. Processes*, **10**, 1591–1608.

Pomeroy, J.W., and Goodison, B.E. 1997. Winter and snow. In *The Surface Climates of Canada*, W.G. Bailey, T.R. Oke, and W.R. Rouse (Eds.). McGill-Queen's University Press, Montreal, pp. 68–100.

Pomeroy, J.W., and Granger, R.J. 1997. Sustainability of the western Canadian boreal forest under changing hydrological conditions – I – snow accumulation and ablation. In

122

Sustainability of Water Resources under Increasing Uncertainty, D. Rosjberg, N. Boutayeb, A. Gustard, Z. Kundzewicz, and P. Rasmussen (Eds.). IAHS Publication No. 240, IAHS Press, Wallingford, UK, pp. 237–242.

Pomeroy, J.W., and Gray, D.M. 1990. Saltation of snow. *Water Resources Res.*, **26**(7), 1583–1594.

Pomeroy, J.W., and Gray, D.M. 1994. Sensitivity of snow relocation and sublimation to climate and surface vegetation. In *Snow and Ice Covers: Interactions with Atmosphere and Ecosystems*, H.G. Jones, T.D. Davies, A. Ohmura, and E.M. Morris (Eds.). IAHS Publication No. 223, IAHS Press, Wallingford, UK, pp. 213–226.

Pomeroy, J.W., and Gray, D.M. 1995. *Snow Accumulation, Relocation and Management.* NHRI Science Report No. 7. National Hydrology Research Institute, Saskatoon.

Pomeroy, J.W., and Male, D.H. 1992. Steady-state suspension of snow: a model. *J. Hydrol.*, **136**, 275–301.

Pomeroy, J.W., and Marsh, P. 1997. The application of remote sensing and a blowing snow model to determine snow water equivalent over northern basins. In *Applications of Remote Sensing in Hydrology*, G.W. Kite, A. Pietroniro, and T.J. Pultz (Eds.). NHRI Symposium No. 17, National Hydrology Research Institute, Saskatoon, pp. 253–270.

Pomeroy, J.W., and Schmidt, R.A. 1993. The use of fractal geometry in modelling intercepted snow accumulation and sublimation. *Proc. Eastern Snow Conf.*, **50**, 1–10.

Pomeroy, J.W., Gray, D.M., and Landine, P.G. 1991. Modelling the transport and sublimation of blowing snow on the prairies. *Proc. Eastern Snow Conf.*, **48** 175–188.

Pomeroy, J.W., Gray, D.M., and Landine, P.G. 1993. The Prairie Blowing Snow Model: characteristics, validation, operation. *J. Hydrol.*, **144**, 165–192.

Pomeroy, J.W., Marsh, P., and Gray, D.M. 1997a. Application of a distributed blowing snow model to the Arctic. *Hydrol. Processes*, **11**, 1451–1464.

Pomeroy, J.W., Granger, R.J., Pietroniro, A., Elliott, J., Toth, B., and Hedstrom, N. 1997b. *Hydrological Pathways in the Prince Albert Model Forest: Final Report*, NHRI Contribution Series No. CS–97007, National Hydrology Research Institute, Saskatoon, Environment Canada.

Pruitt, W.O. 1984. Snow and living things. In *Northern Ecology and Resource Management*, University of Alberta Press, Edmonton, pp. 51–77.

Pruitt, W.O. 1990. Clarification of some Api characteristics in relation to Caribou *(Rangifer tarandus)*. *Rangifer*, Special Issue No. 3, 133–137.

Rango, A., and Martinec, J. 1987. Large-scale effects of seasonal snow cover and temperature increase on runoff. In *Large Scale Effects of Seasonal Snow Cover*, B.E. Goodison, R.G. Barry, and J. Dozier (Eds.). IAHS Publication No. 166, IAHS Press, Wallingford, UK, pp 121–127.

Rawls, W.J., Brakensiek, D.L., and Miller, N. 1983. Green-Ampt infiltration parameters from soils data. *J. Hydraulics Div. Ame. Soc. Civil Eng.*, **109**, 62–70.

Raymond, C.F., and Tusima, K. 1979. Grain coarsening of water-saturated snow. *J. Glaciol.*, **22**(86), 83–105.

Rhea, J.O., and Grant, L.O. 1974. Topographic influences on snowfall patterns in mountainous terrain. In *Advanced Concepts and Techniques in the Study of Snow and Ice Resources*, National Academy Press, Washington, DC, pp. 182–192.

Rouse, W.R. 1990. The regional energy balance. In *Northern Hydrology, Canadian Perspectives*, T. Prowse and S. Ommanney (Eds.). Supply and Services Canada, Saskatoon, pp. 187–206.

Santeford, H.S. 1979. Snow-soil interactions in interior Alaska. In *Proceedings, Modelling of Snow Cover Runoff*, S. Colbeck and M. Ray (Eds.). U.S. Army Cold Regions Research and Engineering Laboratory, Hanover, NH, pp. 311–318.

Satterlund, D.R., and Haupt, H.F. 1970. The disposition of snow caught by conifer crowns. *Water Resources Res.*, **6**, 649–652.

Schemenauer, R.S., Berry, M.O., and Maxwell, J.B. 1981. Snowfall formation. In *Handbook of Snow, Principles, Processes, Management and Use*, D.M. Gray and D.H. Male (Eds.). Pergamon Press, Toronto, pp. 129–152.

Schmidt, R.A. 1972. *Sublimation of Wind-Transported Snow: A Model*. USDA Forest Service Research Paper RM-90, Rocky Mountain Forest and Range Experiment Station, Fort Collins, CO.

Schmidt, R.A. 1980. Threshold windspeeds and elastic impact in snow transport, *J. Glaciol.*, **26**(94), 453–467.

Schmidt, R.A. 1991. Sublimation of snow intercepted by an artificial conifer. *Agric. Forest Meteorol.*, **54**, 1–27.

Schmidt, R.A., and Gluns, D.R. 1991. Snowfall interception on branches of three conifer species. *Can. J. Forest Res.*, **21**, 1262–1269.

Schmidt, R.A., and Pomeroy, J.W. 1990. Bending of a conifer branch at subfreezing temperatures: implications for snow interception. *Can. J. Forest Res.*, **20**, 1250–1253.

Schmidt, R.A., and Troendle, C.A. 1992. Sublimation of intercepted snow as a global source of water vapour. *Proc. Western Snow Conf.*, **60**, 1–9.

Sergent, C., Chevrand, P., Lafeuille, J., and Marbouty, D. 1987. Caractérisation optique de différents types de neige, Extinction de la lumière dans la neige. *J. Phys.*, **48**(C1, suppl. no. 3), 361–367.

Shook, K., and Gray, D.M. 1996. Small-scale spatial structure of shallow snowcovers. *Hydrol. Processes*, **10**, 1283–1292.

Shook, K., and Gray, D.M. 1997. Snowmelt resulting from advection. *Hydrol. Processes*, **11**, 1725–1736.

Shook, K., Gray, D.M., and Pomeroy, J.W. 1993. Temporal variations in snowcover area during melt in Prairie and Alpine environments. *Nordic Hydrol.*, **24**, 183–198.

Sommerfeld, R.A., and LaChapelle, E. 1970. The classification of snow metamorphism. *J. Glaciol.*, **9**(55), 3–17.

Steppuhn, H. 1981. Snow and agriculture. In *Handbook of Snow, Principles, Processes, Management and Use*, D.M. Gray and D.H. Male (Eds.). Pergamon Press, Toronto, pp. 60–125.

Strobel, T. 1978. Schneeinterzeption in Fichten-Bestaenden in den Voralapen des Kantons Schwyz. In *Proc. IUFRO Seminar on Mountain, Forests and Avalanches*. Davos, Switzerland, pp. 63–79.

Sturm, M. 1991. *The Role of Thermal Convection in Heat and Mass Transport in the Subarctic Snow Cover*, CRREL Report 91–19, U.S. Army Cold Regions Research and Engineering Laboratory, Hanover, NH.

Sturm, M. 1992. Snow distribution and heat flow in the taiga. *Arctic Alpine Res.*, **24**(2), 145–152.

Sturm, M., Holmgren, J., and Liston, G.E. 1995. A seasonal snow cover classification system for local to global applications. *J. Climate*, **8**, 1261–1283.

Sturm, M., Holmgren, J., Konig, M., and Morris, K. 1997. The thermal conductivity of seasonal snow. *J. Glaciol.*, **43**(143), 26–41.

Tabler, R.D. 1975. Estimating the transport and evaporation of blowing snow. In *Snow Management on the Great Plains*. Great Plains Agricultural Council Publication No. 73. University of Nebraska, Lincoln, pp. 85–105.

Tabler, R.D., and Schmidt, R.A. 1986. Snow erosion, transport and deposition. In *Proc. Symposium on Snow Management for Agriculture*, H. Steppuhn and W. Nicholaichuk (Eds.). Great Plains Agricultural Council Publication No. 120. University of Nebraska, Lincoln, pp. 12–58.

Tabler, R.D., Benson, C.S., Santana, B.W., and Ganguly, P. 1990a. Estimating snow transport from wind speed records: estimates versus measurements at Prudhoe Bay, Alaska. *Proc. Western Snow Conf.*, **58**, 61–78.

Tabler, R.D., Pomeroy, J.W., and Santana, B.W. 1990b. Drifting snow. In *Cold Regions Hydrology and Hydraulics*, W.L. Ryan and R.D. Crissman (Eds.). American Society of Civil Engineers, New York, pp. 95–146.

Toews, D.A., and Gluns, D.R. 1986. Snow accumulation and ablation on adjacent forested and clearcut sites in southeastern British Columbia. *Proc. Western Snow Conf.*, **54**, 101–111.

Troendle, C.A., and Leaf, C.F. 1981. Effects of timber harvest in the snow zone on volume and timing of water yield. In *Interior West Watershed Management Symposium*, D.M. Baumgartner (Ed.). Washington State University, Pullman, pp. 231–243.

U.S. Army Corps of Engineers. 1956. *Snow Hydrology*. North Pacific Division, Portland, OR.

Wallace, J.M., and Hobbs, P.V. 1977. *Atmospheric Science: An Introductory Survey*, Academic Press, San Diego.

Wakahama, G. 1968. The Metamorphism of Wet Snow. In *Snow and Ice: Reports and Discussions*. IAHS Publication No. 79, IAHS Press, Wallingford, pp. 370–379.

Warren, S.G. 1982. Optical properties of snow. *Rev. Geophys. Space Phys.*, **20**, 67–89.

Warren, S.G., and Wiscombe, W.J. 1980. A model for the spectral albedo of snow, II, snow containing atmospheric aerosols. *J. Atmos. Sci.*, **37**, 2734–2745.

Woo, M-K. 1990. Permafrost hydrology. In *Northern Hydrology: Canadian Perspectives*, T.D. Prowse and C.S.L. Ommanney (Eds.). NHRI Science Report No. 1,

National Hydrology Research Institute, Environment Canada, Saskatoon, pp. 63–76.

Woo, M-K., and Steer, P. 1986. Monte Carlo simulation of snow depth in a forest. *Water Resources Res.*, **22**(6), 864–868.

Woo, M-K., Heron, R., and Marsh, P. 1982. Basal ice in High Arctic snowpacks. *Arctic Alpine Res.*, **14**, 251–260.

Yamazaki, T., and Kondo, J. 1992. The snowmelt and heat balance in snow-covered, forested areas. *J. Appl. Meteorol.*, **31**, 1322–1327.

Yosida, Z., and others. 1955. Physical studies on deposited snow: thermal properties. *Contrib. Inst. Low Temperature Sci., Ser. A*, **27**, 19–74.

Zhao, L., Gray, D.M., and Male, D.H. 1997. Numerical analysis of simultaneous heat and mass transfer during infiltration into frozen ground. *J. Hydrol.*, **200**, 345–363.

126

3 The Chemistry of Snow: Processes and Nutrient Cycling

MARTYN TRANTER AND H. GERALD JONES

3.1 Introduction: Snow Chemistry and Ecology

The previous chapter demonstrated that the physical characteristics of snow are determining factors in the transfer of energy in snow-covered systems (Pomeroy and Brun, Chapter 2). In this chapter, we demonstrate how the chemical composition of snow plays an important role in biogeochemical cycles, particularly in the supply of nutrients to terrestrial and aquatic ecosystems, because where nutrient reserves are limited, biological productivity may depend heavily on snow meltwater inputs to the soil. In other cases, seasonal snowmelt is a major input to the hydrology of catchments. Water quality and aquatic productivity during meltwater runoff can also depend to a large extent on the chemical composition of the meltwaters.

The chemistry of snow has been studied for two main reasons. The first is because of the potential impact of snow meltwaters on the quality of surface waters (Bales, Davis, and Williams, 1993; Peters and Driscoll, 1987; Schöndorf and Hermann, 1987; Tranter et al., 1988a) and the second is to interpret the climatic and pollution records recorded in cold, dry snow cover (Bales and Wolff, 1995). Acidity in snows derived from the scavenging of pollutants (SO_2 and NO_x) from air masses originating in industrial urban areas (Heubert et al., 1983; Barrie and Vet, 1984; Dasch and Cadle, 1985; Delmas, Briat, and Legrand, 1982; DeWalle et al., 1983) is flushed out of the snowpack early in the melt season, giving rise to fluxes of low pH meltwaters to soils and streams (Hendershot et al., 1992; Goodison, Louie, and Metcalfe, 1986; Galloway et al., 1987). Decreases in aquatic productivity and fish kill during the spring freshet occur (Hagen and Langeland, 1973; Muniz, 1991). Earlier work concentrated on the flushing or elution of solute from the melting snow cover. More recent work has concentrated on other chemical processes that occur in snow cover. It has become clear that certain chemical species can be gained or lost during physical changes in the snow cover (Gregor, 1991), can be transformed under certain meteorological conditions (Pomeroy, Davies, and Tranter, 1991), and can be affected by microbiological activity (Jones, 1991). During

127

the 1990s, the synergy between snow chemistry and snow ecology was established (Jones, 1999).

Two main types of processes influence the chemical composition of snow and melt-waters: those involving the heat and mass fluxes that occur during sublimation and melting (Davis, 1991) and those involved with chemical transformations (Tranter, 1991). Chemical migration may take place during the former processes, but, in general, the chemical species are conservative in the sense that no chemical transformation takes place. The physical properties of individual species (e.g., solubility, vapour pressure) play a role in determining any change in the chemical loading of the snow cover (Brimblecombe and Shooter, 1991). Chemical transformations may occur by chemical reactions (e.g., oxidation; Bales et al., 1987) or may arise from microbiological activity (e.g., algal assimilation of nitrogen [N] species; Hoham et al., 1989).

Thus, it has become apparent during the past decade that snow is not a passive reservoir of chemical species. It has also become clear that one cannot study the chemical dynamics of snow without taking into account the physical and biogeochemical characteristics of the snow cover environment. This chapter deals with these aspects of snow and the environment through a discussion of the chemical evolution of snow from the time of formation in the atmosphere to the final runoff of meltwaters to stream channels during the ablation season.

3.2   The Chemistry of Snowfall

3.2.1   Formation of Snow Crystals and the Scavenging of Atmospheric Species

The atmosphere and the surface of the biosphere form a complex recycling system that maintains the composition of the earth's gaseous envelope (Mooney, Vitousek, and Matson, 1987). The atmosphere is a dynamic reservoir where terrestrial and marine emissions from the earth's surface are subjected to oxidation processes before being recycled back to the oceans and continental land masses. The main transfer mechanisms are by direct (dry) deposition of chemical species to surfaces or by precipitation (snow and rain) scavenging (Barrie, 1991; Cadle, 1991).

Snowfall contains crustal elements, such as calcium and magnesium, from terrigenous dust (Delmas et al., 1996; Frazén et al., 1994; Hinkley, 1994), anthropogenic pollutants (Landsberger et al., 1989), weak organic acids (Maupetit and Delmas, 1994), neutral organics from natural sources (Likens, Edgerton, and Galloway, 1983), and trace metals (Thornton and Eisenreich, 1982). Snow formation and snowfall incorporate

128

chemical species from the atmosphere by three main processes: imprisonment during initial formation of ice crystals; capture of gases, aerosol, and larger particulates within clouds; and scavenging of these materials below the cloud layers during snowfall (Barrie, 1991).

Cloud water droplets contain solute as a result of aerosol scavenging and diffusion of atmospheric gases into solution. The soluble species are mainly NH_4^+, SO_4^{2-}, NO_3^-, Ca^{2+}, K^+, and Mg^{2+}, derived from natural and anthropogenic emissions, in addition to Na^+ and Cl^- from sea–salt aerosol. Ice crystals initially form in the atmosphere by the freezing of supercooled cloud water droplets at temperatures $> -40°C$ (see Figure 3.1), which is catalysed by ice nuclei that usually consist of sea–salt aerosols, particulate organic debris, and/or fine particulate clays (Kamai, 1976; Pomeroy and Brun, Chapter 2). Ice crystals may form by direct deposition of water vapour onto ice nuclei and by water droplets' either touching ice nuclei (contact nucleation) or scavenging ice nuclei, so that ice crystals form within the water droplet (immersion–freezing nucleation). Solute does not fit into the ice crystal lattice particularly well (Colbeck, 1981). Hence, most solute is rejected into the outer edges of the ice crystal during freezing.

After nucleation, the ice crystals may grow by direct deposition of water vapour onto the ice crystal surface at the expense of supercooled water droplets in the immediate vicinity. Dendritic, needle and plate–like crystals are the main result of this process (Magono and Lee, 1966). These growing ice crystals contain very low solute concentrations (Scott, 1981; Hewitt and Cragin, 1994), largely limited to the original, imprisoned nuclei. Limited amounts of compounds with relatively high vapour pressure – e.g., HNO_3, HCl – may be incorporated into the growing crystal by absorption or adsorption. In-cloud conditions favouring collisions between growing crystals and supercooled droplets result in the formation of rounded crystals known as rimed snow or soft snow pellets (graupel). These types of crystals contain much larger amounts of solute than crystals formed mainly by vapour transport (Cerling and Alexander, 1987), because much of the solute contained in the supercooled droplets is captured by rapid freezing. For example, at Dye 3 in Greenland, rimed snow crystals compose $\sim$5 percent of the snowpack and contain $\sim$30 percent of the solute (Borys et al., 1993).

Further scavenging of aerosols and other particulates occurs below clouds by adsorption or impaction onto the surface of all types of crystals during snowfall (Scott, 1981). Snow is a better scavenger of particulate material than rain (Raynor and Haynes, 1983; Leuenberger et al., 1988; Nicholson, Branson, and Giess, 1991). The efficiency of scavenging is related to the surface area : mass ratios of the crystals as they sweep out

129

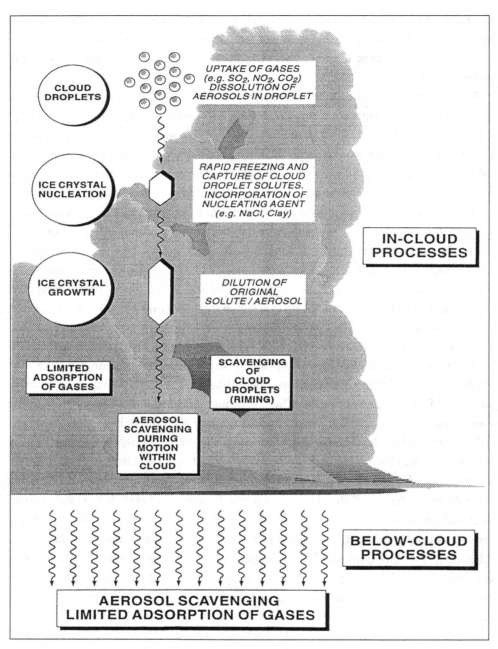

Figure 3.1. The main processes that influence the chemical composition of snow crystals during growth and fall.

the air during fall. The overall scavenging efficiency of precipitation can be expressed either as a washout ratio, W, or as a scavenging coefficient, ω (Misra et al., 1985). The values of these parameters for removal of chemical species from the atmosphere by precipitation are not constant and depend on meteorological conditions and the type of precipitation. The washout ratio is dimensionless and is expressed as the ratio of the amount of a chemical species in precipitation to that in the air,

$$W = \rho_a(C_p/C_a) \qquad\qquad (3.1)$$

where ρ_a is the density of air (g m^{-3}), C_p is the concentration of the species of interest in precipitation (μg g^{-1}), and C_a is the concentration of that species in air (μg m^{-3}).

The scavenging coefficient ω (s^{-1}) can be thought of as a rate constant for the transfer of species from air to precipitation on an event basis,

$$\omega = \ln[C_t/C_0]/t \qquad\qquad (3.2)$$

where C_t is the concentration of the species of interest (μg m^{-3}) in air at time t (s), and C_0 (μg m^{-3}) is the concentration of the species in air at time zero (Raynor and Haynes, 1983).

Table 3.1 shows typical values for W and ω for some of the more common chemical species found in snow.

General relationships between the formation of snow crystals (based on snow crystal type) and the amount of scavenged solute have been found, despite the complexity of physical and chemical processes outlined above, the large range in the concentration of chemical species, and the variable values found for W and ω. Lamb, Mitchell, and Blumernstein (1986) found that the highest concentrations of solute were associated with the smallest unrimed crystals, demonstrating the importance of surface area: mass ratios as a scavenging parameter. Borys et al. (1983) showed that riming was also a factor in the ability of snowfall to scavenge the atmosphere as, in general, the solute content of snow crystals increased with the degree of riming. Hewitt and Cragin (1994) isolated individual snowfall crystals by crystal habit and found that dendritic and stellar plates had similar chemical content for most species, except for Cl$^-$, which showed much higher concentrations in stellar plates than dendritic crystals. The reason for this phenomenon is unknown. The distribution of certain species within, or on, different individual crystals in snowfall has been reported (Kamai, 1976). Rango, Wergin, and Erbe (1996) further suggested that the preferential incorporation of rime on certain snowfall crystal features, such as edges, should be reflected by pollutant distribution at the scale of individual crystals.

131

Table 3.1. *Removal of chemical species from the atmosphere by snow: washout ratios (W; dimensionless) and scavenging coefficients (ω, s⁻¹).*

Species	Washout ratio (W)			Scavenging coefficient (ω, s^{-1})	Reference
	Mean*	Range†	Reference		
NH_4^+	410	40–2,200	1		
SO_4^{2-}	570	20–3,000	1, 2, 3, 4	5×10^{-5}	3
NO_3^-	1,000		1, 3	5×10^{-5}	3
SO_2	200	20–500	1		
HNO_3	3,600–5,000‡		1, 3	25×10^{-5}	3
Ca	2,600	170–8,190	1		
K	450	80–4,600	1		
Cr	530	200–1,280	1		
Cu	1,220	60–3,800	1		
Fe	1,160	10–7,110	1		
Ni	1,190	280–3,350	1		
Pb	240	20–1,350	1		
Zn	2,470	280–1,1500	1		

*Mean values from reference 1.
†Range of all references cited.
‡Means from references 1 and 3.
References: 1, Cadle, Dasch, and Mulawa (1985); 2, Scott (1981); 3, Heubert et al. (1983); 4, Davidson et al. (1987).

3.2.2 Spatial and Temporal Variability of Snowfall Composition

The chemical composition of snowfall depends on factors such as the origin of the air masses that are scavenged, the altitude at which snow is deposited, and the meteorological conditions during snowfall (Colin et al., 1989; Davies et al., 1992). Maritime air masses give rise to snow containing mostly Na^+ and Cl^- (Tranter et al., 1986), whereas polluted air masses from industrial areas deposit snow that is highly acidic because of the presence of strong acid anions (NO_3^-, SO_4^{2-}) from fossil fuel combustion (Davies et al., 1984; Landsberger et al., 1989). Table 3.2 shows the chemical concentrations of snowfalls in different regions of the world.

The chemical composition of snow may show spatial variability on length scales of metres through tens of kilometres (Tranter et al., 1987; Pomeroy, Marsh, and Lesack, 1993; Suzuki, 1987; Laird, Taylor, and Kennedy, 1986), reflecting factors such as the proximity to pollution sources, the impact of wind redistribution (Pomeroy and

Table 3.2. *Chemical composition of snowfall at selected sites throughout the world.*

Location	pH	Ca^{2+}	Mg^{2+}	Na^+	K^+	NH_4^+	NO_3^-	SO_4^{2-}	Cl^-	Source
European Alps	4.4–5.3	18–49	3–15	3–27	1–6	17–60	12–46	28–68	8–32	Puxbaum, Kovar, and Kalina (1991)
Central Asian Mountains	*	19–70	*	1–44	*	*	2.9–60	2.2–51	1–32	Lyons, Wake, and Mayewski (1991)
Turkey Lakes Watershed, SE Canada	4.57	34	0.9	10	0.2	7.5	19	17	3.7	Semkin and Jeffries (1988)
Mid-Wales	3.9–4.5	4–14	4–11	13–30	1–5	*	11–64	16–78	21–69	Reynolds (1983)
Sapporo, Japan	4.4–6.4	13–63	18–67	59–190	2.3–6.4	*	*	70–99	63–310	Suzuki (1987)
Cairngorms, Scotland	4.4	2.5	11	52	2.1	9.8	20	26	91	Davies et al. (1992)
Svalbard	5.4–6.7	0–46	0–200	4–2000	0–96	*	0–7	0–240	0–2400	Hodgkins, Tranter, and Dowdeswell (1997)
South Pole	5.4	*	0.16	0.63	0.03	0.16	1.4	1.5	1.3	Legrand and Delmas (1984)

*Missing values. Single values refer to volume-weighted mean concentrations. All units (except pH) are $\mu Eq/L$.

Brun, Chapter 2), and differential scavenging by vegetation. Snowfall at high altitudes usually contains lower concentrations of chemical species than at lower altitudes because the depth of air column available for aerosol scavenging is smaller. In addition, air masses subjected to orographic rise release relatively concentrated deposition at lower altitudes and relatively clean snow at higher altitudes (Laird et al., 1986; Lyons, Wake, and Mayewski, 1991). However, in the case of some chemical species (e.g., volatile organochlorines), the concentrations in snow can increase with altitude. This is due to volatilisation of the compounds from relatively "warm" snow at low altitudes and subsequent "cold condensation" in snow precipitated at cooler higher altitudes (Blais et al., 1998). Meteorological conditions during snowfall also influence the chemical composition of snow crystals. Pomeroy et al. (1991) demonstrated that the solute content may change during windblown snow events, related to the extent of sublimation and aerosol scavenging.

The temporal variability of the chemistry of individual snowfalls (Colin et al., 1987) is due to the progressive scavenging of the chemical species during the snowfall event (Raynor and Haynes, 1983). Chemical concentrations often decrease exponentially with time. This gives rise to the use of the scavenging coefficient, ω, to characterise the efficiency of precipitation in removing pollutants from the atmosphere. Recently, transfer functions have been used to characterise the relationship between concentrations of chemical species in the atmosphere and those found in snow (Bales and Wolff, 1995). These functions are complex. They account for not only the removal of species from the atmosphere but also for postdepositional changes in cold snow covers. The latter depend on both the reversible physical exchange of species with the atmosphere and chemical reactions at the snow surface and within the snow cover.

3.3   Chemistry of Cold, Dry Snow Cover

Snow on the ground becomes consolidated into snow cover during cold ($< 0°C$) weather. The temporal and spatial variation in the chemical composition of snowfall usually produces a snow cover that is chemically heterogeneous. Cold snow covers are subjected to a number of physical (Pomeroy and Brun, Chapter 2) and chemical processes (Jones and Stein, 1990; Bales, 1991), which further modify the chemical concentrations of individual snow strata and the total chemical load of the snow cover. The main processes (see Figure 3.2) are surface exchange at the snow–atmosphere interface (dry deposition and volatilisation), surface and subsurface chemical reactions, snow-grain metamorphism within the pack and, in the case of

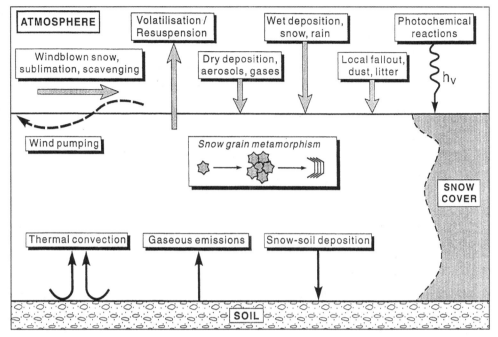

Figure 3.2. The main physical and chemical processes that influence the chemical composition of cold, dry snow cover during the accumulation season.

seasonal ground cover, basal-exchange processes at the snow–soil interface (e.g., gaseous emissions from soil).

3.3.1 Processes at the Atmosphere–Snow Surface Interface

3.3.1.1 *Dry Deposition*

Dry deposition is the direct deposition of chemical species from the atmosphere to the snow surface (Cadle, 1991). Aerosols and particulates may be directly deposited, whereas gaseous species may also be adsorbed (Conklin, 1991). In general, dry deposition to snow cover is much lower than to most other snow- and ice-free terrestrial surfaces, because the surface roughness of snow is much lower (0.01 cm compared with 0.03–100 cm; Cadle, 1991). Hence, the relative magnitude of dry to wet deposition is smaller for snow- and ice-covered terrain than for other terrestrial surfaces.

The net dry deposition flux (F_x, μeq m^{-2}) of a species (X) is expressed as the product of the atmospheric concentration (C_x, μeq m^{-3}) of the species and the deposition velocity ($V_{d(x)}$, m s^{-1}), so that

$$F_x = C_x V_{d(x)} \qquad (3.3)$$

135

Table 3.3. *Dry deposition velocities* (V_d; *cm s*$^{-1}$) *for various gases and aerosols to snow surfaces (after Cadle, 1991).*

Species	Deposition velocity V_d (cm s^{-1})
SO_2	0.02–0.25
NO	<0.03
NO_2	0.005–0.12
HNO_3	<0.02–2.0
NH_4^+	0.08–0.1
O_3	0.01–0.08
Organic carbon aerosols	0.28
Elemental carbon	<0.2–0.55
Crustal aerosols	0.2–3.4
Particulates (0.15–0.3 μm)	0.03
Particulates (0.5–1 μm)	0.02

F_x is negative when there is net deposition to the snow surface. The deposition velocity, $V_{d(x)}$, is usually specified with reference to a height above the snow surface (e.g., 1–2 m), because the flux, rather than the atmospheric concentration, C_x, is independent of height. Deposition velocities vary with meteorological conditions, physical characteristics of the snow surface, and the physical properties and chemical reactivity of the species. Bales et al. (1987) found that dry deposition rates to new snow were higher than those to old snow. This was believed to be the result of a reduction in the area of crystal surfaces during the metamorphism of snow. The importance of crystal form in dry deposition has also been reported by Ibrahim, Barrie, and Fanaki (1983), who suggested that the interception of aerosols by ice needles in relatively fresh snow contributed significantly to the measured rates of dry deposition.

Cadle, Dasch, and Mulawa (1985) calculated that V_d for HNO_3 was approximately one order of magnitude greater than for SO_2. The difference in deposition rates was attributed to the relative solubility of the two gases in the liquid layers around the crystals and also to other factors such as relative diffusion rates into the ice lattice and the rate of oxidation of SO_2 to SO_4^{2-} at the air–crystal interface (Bales et al., 1987; Bales, 1991). The rate of deposition of HNO_3 is much higher to wet snow than to cold, dry snow (Cadle, 1991). Values of V_d for various chemical species (gases and aerosols) to snow are shown in Table 3.3. The values are within the range 0.02–3×10^{-2} m s^{-1}. At these levels of deposition velocity, dry deposition can be a significant factor in the

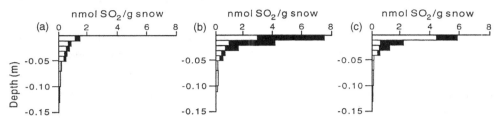

Figure 3.3. Impact of dry deposition on the SO_2 concentration of surface snow (after Valdez et al., 1987). Black bars show increases in SIV (SO_2) concentrations over a 6-hour period, white bars show increases in SVI (HSO_3^-; SO_4^{2-}) concentrations. Conditions are: (a) Temperature, $-2.2°C$; SO_2 mixing ratio, 81 ppbv; age snow, 112 days; density, ρ_s, 400 kg m^{-3}. Measured dry deposition velocity, 2.0×10^{-4} m s^{-1}. (b) Temperature, $0°C$; SO_2 mixing ratio, 70 ppbv; age snow, 9 days; density, ρ_s, 380 kg m^{-3}. Measured dry deposition velocity, 9.3×10^{-4} m s^{-1}. (c) Temperature, $0°C$; SO_2 mixing ratio, 79 ppbv; age snow, 10 days; density, ρ_s, 400 kg m^{-3}. Measured dry deposition velocity, 5.9×10^{-4} m s^{-1}.

chemical composition of snow. For example, a fresh snowfall of 10 mm snow water equivalent (SWE) containing $600 \, \mu g \, L^{-1}$ of HNO_3 (i.e. a load of 6 mg m^{-2}) will double the load of HNO_3 (to 12 mg m^{-2}) within 2 days of being in contact with an air mass containing an average HNO_3 concentration of $2.5 \, \mu g$ m^{-3}. Field studies in North America suggest that the overall contribution of dry deposition to the chemical composition of snow in the case of SO_4^{2-} and NO_3^- is approximately 20–25 percent (Cadle and Dasch, 1987; Barrie and Vet, 1984; Cadle, 1991). Figure 3.3 shows an examples of the impact of dry deposition on SO_2 concentrations in surface snow, which depend on snow temperature ($°C$), age (days) and density of snow (ρ_s), and the mixing ratio of SO_2 in air (ppbv). (Valdez et al., 1987).

Snow covers under forest canopies may receive greater contributions of dry deposition. Dry deposition velocities of gases and aerosols to vegetative surfaces are higher than those to snow (Höfken, Meixner, and Ehhalt, 1983; Dasch, 1987), so the removal of adsorbed species from the canopies by throughfall, wet snow, and/or rain can contribute significantly to the chemical load of the snow cover. Aeolian dust and other particulate matter are continuously being deposited on snow covers. The result of dust deposition usually is to reduce the acidity of snows (Sequeira, 1991), particularly during melt periods (Clow and Ingersoll, 1994; Delmas et al., 1996).

Throughout the winter, cold snow also accumulates nutrients by the deposition of biological debris either as fallout from above the snow surface or by incorporation within the snow cover itself. In forest snow covers much of the deposition arises as litterfall from the canopy (Jones and Debois, 1987) and mammalian excrement (White, Garrolf, and Heisey, 1997). Invertebrate fallout from wind-borne arthropods and winged invertebrates may also contribute nutrients to snow, particularly in more open areas (Edwards and Banko, 1976), whereas vertebrate and invertebrate activity under and within the snow cover (Aitchison, Chapter 5) transforms and redistributes nutrients from the soil upward into the snow matrix. Inorganic N and phosphorus [P] species are readily bioavailable for snow algae, which develop in the latter stages of the spring melt (Hoham, 1987). Organic material, particularly excrement, serves as a substrate for heterotrophic organisms such as fungi (Stein and Amundsen, 1967). Biological deposition thus may be an important component of nival food webs (Hoham and Duval, Chapter 4; Aitchison, Chapter 5).

However, the amount of nutrients deposited by animals and plants on snow that is available to organisms is difficult to evaluate. This is due to the extremely high spatial variability of the sources. Jones (1991) attempted to calculate nutrient inputs to snow from different vertebrates by considering population densities and the amount and chemical composition of animal excrement. In the case of animals that herd (e.g., deer), local (5- to 10-ha scale) deposition rates may be relatively high (0.2 kg N ha^{-1} day^{-1}). On the same scale, solitary mammals, such as moose and hares, deposit only 1.5×10^{-4} to 2.5×10^{-3} kg N ha^{-1} day^{-1}. Although not strictly applicable to the estimations of nutrient contents of cold snow cover, data published by Edwards and Banko (1976) indicate that the pool of particulate nutrient (debris, dead and living invertebrates) mostly blown onto snowpatches from the surrounding cover contains up to 0.045 kg N ha^{-1}.

3.3.1.2 *Volatilisation*

In general, the dry deposition of aerosol species to snow is considered to be irreversible (Bales and Wolff, 1995). However, recent work has suggested that there may be a concomitant loss of some ionic species (in particular, SO_4^{2-}) during sublimation of snow crystals, but there is no known mechanism to explain such a loss (Cragin and McGilvary, 1995). Laird, Buttry, and Sommerfeld (1999) found no loss of HNO$_3$ adsorbed on artificial snow in columns sealed from ambient air. However, they obtained evidence that a small amount (≤ 10 percent) of HNO$_3$ can exist as vapor within the interstitial space of the snow matrix under certain conditions because of volatilisation of HNO$_3$ from the ice surfaces. In open exposed natural sites, wind transport has the potential to change the chemical composition of snow due to three main physical processes: namely sublimation of water vapour, scavenging of aerosols and gases from the atmosphere, and volatilisation (Pomeroy and Jones, 1996). Pomeroy et al. (1991) found evidence that both gains (scavenging) and losses (volatilisation) of NO_3^- and NH_4^+ could take place at different stages during windblown snow events.

Some chemical species, particularly organic species, may volatilise directly from snow to the atmosphere because of their relatively high vapour pressures. For example, Hogan and Leggett (1995) showed that synthetic organic compounds (paranitrotoluene, dinitrotoluene, and nitroglycerine) stored in cold snow cover are lost as the snowpack ripens. Organochlorines (e.g., hexachlorobenzene) may be lost from arctic snow cover by volatilisation during the summer (Gregor, 1991).

3.3.1.3   *Photochemical Reactions in Surface Snow Cover*

Snow is in intimate contact with the atmosphere, and oxidation of certain species by atmospheric oxidants may take place on cold snow grain surfaces, particularly if a liquid film is present (Conklin and Bales, 1993). Bales (1991) has modeled the chemical oxidation of S(IV), SO_2, to S(VI), SO_3^{2-}/SO_4^{2-}, based on the known oxidation rates by H_2O_2, O_3, and O_2. In addition to the presence of liquid films, however, light may also play a key role in the chemistry of snow cover by initiating photochemical reactions.

It is known that ice crystals play an active role in the photochemistry of the upper atmosphere (Molina et al., 1987) and of clouds (Mitra, Barth, and Pruppacher, 1990). Ice crystals on the ground thus have the potential for photochemical reactions. Photoreactive species found commonly in snow include peroxides (Neftel, 1991), trace metals (Rahn and McCaffrey, 1979), and NO_3^-. Although unequivocal evidence of photochemical processes in snow has yet to be obtained, some workers have reported changes in the chemical composition of snow that are consistent with photochemical reactions.

Palmer, Smith, and Neirink (1975) reported the presence of O_3 in the surface snows of a high mountain valley, which suggests that photooxidation reactions may have taken place involving species such as the OH radical, H_2O_2, O_3, and organic peroxides (Gunz and Hoffmann, 1990a, 1990b). Sigg, Neftel, and Zircher (1987) measured the decrease in concentrations of H_2O_2 in surface alpine snows and proposed that photolysis was the primary mechanism. Photochemical reactions in snow are complex and are difficult to distinguish from other processes (e.g., dry deposition) related to exchange at the snow–atmosphere interface. Neubauer and Heumann (1988) suggested that the apparent loss of NO_3^- from Antarctic snow was due either to the photodegradation of HNO_3 to NO_2 by solar radiation and/or to the evaporation of HNO_3 from snow during metamorphism, but they were unable to distinguish between the two mechanisms. Evidence for the photochemical production of NO and NO_2 (NO_x) in Antarctic snow cover has been recorded by Jones et al. (2000). However, they were not able to demonstrate that NO_3^- was the precursor of NO_x. The duration of exposure and the intensity and wavelength of light are principal factors in photochemical reactions. For example, Duchesneau (1993) found that the apparent dry deposition rate of NO_3^- to snow cover on a high-elevation ice cap differed in an area exposed to radiation from that in an area deliberately shaded by a flysheet. It was suggested that the concomitant photodegradation of HNO_3 during the deposition process was responsible for the lower dry deposition rate in the irradiated area. However, a similar experiment over a shorter time period in a subarctic site gave no such result (Jones et al., 1993), and it is believed that both the exposure time and the higher degree of light intensity and ultraviolet radiation at the high-elevation site were responsible for the difference between the two snow covers.

In addition to organic peroxides, other photoreactive organic species are found in snow. Jaffrezo, Clain, and Maschet (1994) reported that the concentrations of benzo[a]pyrene in Greenland snow cover dropped by over 90 percent during post-deposition modification. It was suggested that photochemical degradation of the benzo[a]pyrene took place because of the presence of OH radicals from H_2O_2 decomposition.

3.3.2 In-Pack Processes

3.3.2.1 *Metamorphism*

Metamorphism of ice crystals commences almost immediately after deposition (Colbeck, 1987; Davis, 1991). Metamorphism of dry snow crystals occurs because of redistribution of water vapour between and within crystals as water vapour is transferred from surfaces of high curvature to sites of low curvature. The crystals lose their characteristic dendritic, stellar, or needle structure and are reconstituted as spherical snow grains (Pomeroy and Brun, Chapter 2). Solute may also become redistributed and concentrated on the snow grain surfaces or, in the case of grain aggregations, along the snow grain boundaries. Further metamorphic processes follow that lead to growth of larger snow grains and faceted crystals (Colbeck, 1987). Large rounded snow grains grow at the expense of smaller grains when temperature gradients within the snow cover are small. Grains that lose water vapour should become more concentrated in ionic species if these solutes have negligible vapour pressures.

140

Growth of the larger grains by crystallisation of water vapour on the grain surfaces should lead to more dilute concentrations of solutes. It is not known to what extent this concentration-dilution process occurs in dry snow (Granberg, 1985), as grain clusters, composed of small and large grains physically bound together, are often formed (Colbeck, 1987). In addition, the loss of all, or almost all, the water from a grain would cause this simple concentration-dilution model to break down, as highly concentrated small droplets may be subjected to phoretic transport to grain surfaces. The growth of grains probably is also accompanied by solute exclusion from the growing ice lattice. Thus, the net effect of this type of metamorphism is believed to be a concentration of solute onto or near the surfaces of ice crystals (Bales, 1991; Colbeck, 1981, 1987; Davis, 1991). The solute may be located in a quasi-liquid surface layer that is believed to exist at temperatures $>-35°C$ (Ushakova and Troshkina, 1974), as discrete aerosol or as concentrated, "doped ice" pockets (Davis, 1991).

The growth of large faceted crystals occurs (at the expense of rounded grains) if temperature gradients are large (Colbeck, 1987). Such crystals are known as the kinetic-growth form and are common in depth hoar (Pomeroy and Brun, Chapter 2). Some studies have shown that formation of depth hoar is accompanied by a loss of ions. As ionic species are assumed to be conserved without chemical change in cold dry snow, loss must be due to some form of transport either to the base of the pack or to adjacent snow strata and the atmosphere. Laberge and Jones (1991) found that SO_4^{2-} was lost during depth hoar formation. They suggested that a process that could physically transfer small concentrated SO_4^{2-} particulates away from the depth hoar was responsible for the losses. By contrast, Pomeroy et al. (1993) found that SO_4^{2-} and Cl^- concentrations in depth hoar increased in proportion to the overall loss of water vapour from the depth hoar to adjacent snow layers. However, NO_3^- concentrations remained approximately constant, indicating a concomitant loss of the species. The loss was believed to occur via HNO_3 vapour. In summary, the formation of kinetic-growth crystals seems to lead to the loss of some species, but the precise mechanisms are unknown.

3.3.2.2   *Windpumping and Thermal Convection*

Cyclic air movement between the atmosphere and snow cover occurs when wind conditions are favourable (Colbeck, 1989). The phenomenon is known as windpumping and theoretically could increase or decrease the concentrations of chemical species in the snow cover, because aerosols and gases could be filtered from the air or removed from interstitial air pockets. Gjessing (1977) simulated natural windpumping

141

by forcing air through snow and reported that concentrations of K^+, NO_3^-, and SO_4^{2-} increased in the snow cover.

If temperature gradients increase to a critical threshold level, air movement may be initiated in the snow cover by thermal convection (Pomeroy and Brun, Chapter 2), and convective air currents (Sturm, 1991) could also redistribute chemical species throughout the pack. However, no evidence for such redistribution has been documented.

3.3.3   Basal Gas Exchange Between Snow and Soil

Soil microorganisms produce trace gases by a variety of processes, such as respiration, denitrification, and nitrification (Granli and Bøckman, 1994). The onset of snow cover usually results in a decrease in the overall rate of microbiological activity in the soil due to the decrease in temperature at the snow–soil interface. However, trace gas emissions under snow can still represent a significant fraction of annual emissions (CO_2, >20 percent [Winston et al., 1995]; N_2O, ~50 percent [Brooks, Williams, and Schmidt, 1996]). The extent to which seasonal snow cover influences trace gas emissions from soil varies with the duration and depth of snow during the cold, dry accumulation period and with meltwater discharge and chemistry during the melt season. Thus, wintertime emissions may vary considerably from year to year (van Bochove et al., 2000). Snow is a porous medium, so trace gases released from the soil permeate the snow cover and give rise to either consistent gaseous concentration profiles (Sommerfeld, Mosier, and Musselman, 1993; Jones, van Bochove, and Bertrand, 1999) or ephemeral localised gas-rich pockets of air within the snow cover (Zimov et al., 1993), depending on soil processes and the physical characteristics and dynamics of the snow (see Figure 3.4).

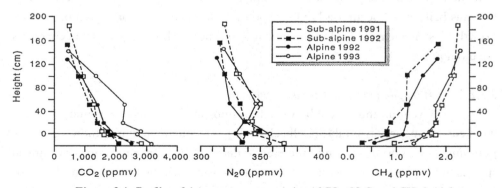

Figure 3.4. Profiles of the gaseous concentration of CO_2, N_2O, and CH_4 in alpine snow cover (after Sommerfeld, Mosier, and Musselman, 1993).

142

The soil beneath deep alpine snow cover rarely freezes because of the relatively thick (2–3 m) snow cover. Microbiological activity in the soil is thus maintained throughout the winter (Massman et al., 1995). The continual production of trace gases and a relatively homogeneous snow structure result in consistent concentration profiles of gases with snow depth. Typical profiles are as follows: CO_2, 1,800–2,500 ppmv m^{-1} (Solomon and Cerling, 1987; Massman et al., 1995); CH_4, 0.5 ppmv m^{-1} (Sommerfeld et al., 1993); N_2O, 15 ppmv m^{-1} (Sommerfeld et al., 1993). In the case of CH_4, the flux is from the atmosphere to the soil, showing that the soil was a sink for CH_4 (Figure 3.4). Windpumping can affect the gaseous fluxes through snow by modifying concentration gradients (Massman et al., 1997); the pressure pumping effects are more pronounced when the gradients are weak.

The combination of shallow snow cover and extreme cold air temperatures in arctic and subarctic environments causes the surface soils to freeze, in contrast to deep alpine snows. Microbiological activity ceases when the temperature of the frozen soil drops below −8°C (Coxson and Parkinson, 1987). However, the soil still can remain a source of trace gases even if biological activity in the upper layers of soil becomes dormant. Gases may be released from deeper soil horizons; for example, Zimov et al. (1993) showed that respiration in deep subsurface layers is a source of CO_2 in Siberian soils even in the coldest midwinter period. However, the emissions of gas at the snow–soil interface was episodic and spatially variable, in contrast to the case of deep alpine snow. This is due to the random nature of the physical stresses in frozen soil that mediate the release of the gases to the snow cover. Gases can be released from frozen soils even in the absence of microbiological activity. Coyne and Kelley (1971) believed that the source of CO_2 in snow-covered soil in the Arctic under conditions of extreme cold was due to the release of pockets of CO_2 from within or below the frozen soil layer formed by exclusion of the gas from the soil-water matrix during freezing of the soil in early winter. Thermal stress in mid- and late winter periodically occurred and the resulting episodic and highly variable release of CO_2 was similar to that observed by Zimov et al. (1993).

Finally, pockets of gas liberated from soil may remain trapped locally within the snow structure if the permeability of snow cover in an area is low because of the presence of wind-packed snow or ice layers (Hardy, Davis, and Winston, 1995; Jones et al., 1999). Winston et al. (1995) believed that the low values of CO_2 emissions measured over snow in open areas of a boreal forest site were due to the trapping of the gas under impermeable ice lenses formed during melt-freeze cycles. In open exposed subarctic snow covers, subsurface trapping of gases and windpumping can mediate gaseous fluxes between the soil and the atmosphere (Jones et al., 1999).

143

3.4   Chemistry of Wet Snow and Snow–Meltwater Systems

The cold ($<0°C$) accumulation winter period is followed by the spring melt in many temperate and higher latitude environments. The physical aspects of melt-water production are covered in Chapter 2 by Pomeroy and Brun. Snow meltwaters penetrate dry snow cover and produce a wetting front that separates an upper layer of wet isothermal snow at $0°C$ from a lower layer of dry snow with a temperature $<0°C$. Flow fingers develop at the leading edge of wetting fronts due to flow instabilities and structural discontinuities in the snow (crusts and ice lenses). The permeability of the flow fingers increases because of grain growth in wet snow; hence, these flow fingers become areas of higher flow discharge rates called macropores (Kattleman, 1985, 1989; Marsh, 1990, 1991). Rain-on-snow events generally accelerate the growth of macropores.

Percolation of meltwaters through the snow cover causes the chemical composition of both the snow matrix and the meltwaters to change. The concentration and distribution of solutes in the snow–meltwater system is controlled by various physical and biological processes (see Figure 3.5). These processes are the leaching of solute from

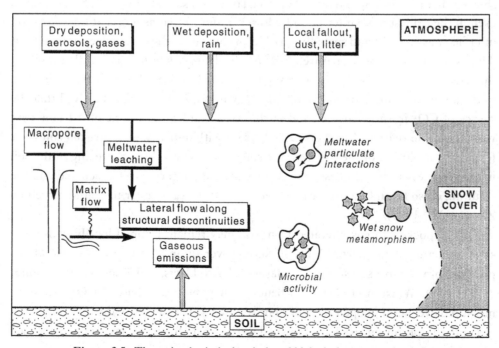

Figure 3.5. The main physical, chemical, and biological processes that influence the chemical composition of snow cover during the thaw.

snow grains and crystals, meltwater–particulate interactions, and microbiological activity. In addition, snow–atmosphere exchange is another factor, as dry deposition rates of certain species (e.g., SO_2, HNO_3, HCl) to wet snow crystals increase significantly because of their solubility in water (Cadle, 1991). Rain will also influence the chemistry of meltwaters (Tranter et al., 1992).

3.4.1   Leaching of Solute from Snow Crystals

The leaching of snow grains by meltwater causes fractionation of solute species between the grains and the liquid medium, and the meltwater front becomes progressively more concentrated as it moves through the pack (Tranter, 1991). The degree of fractionation of any solute species, x, between snow and meltwater is expressed by a nondimensional concentration factor, CF_x,

$$CF_x = {}^{x}M_m / {}^{x}M_p \qquad (3.4)$$

where ${}^{x}M_m$ is the concentration of species x in any meltwater fraction and ${}^{x}M_p$ is the concentration of x in the parent snow *prior* to melt. Values of CF_x during the initial stages of meltwater discharge may range from 1 to 50, but a more typical range is 2 to 7 (Tranter, 1991). CF_x decreases as melt progresses and the snowpack is leached to values of <0.1 in the final meltwaters. The efficiency of meltwater leaching (i.e., high values of CF in initial meltwater discharges) depends on the micro- and macroscale distribution of solute in snow grains and meltwater hydrology.

The microscale distribution of solute in snowpacks before melt is influenced by snow metamorphism (Davis, 1991; Pomeroy and Brun, Chapter 2). Highly concentrated solute is believed to be present initially as discrete droplets of liquid residing in the boundaries between associated snow grains and as quasi-liquid films on grain surfaces. (Mulvaney, Wolff, and Oates, 1988). These liquid forms contain the bulk of solute and consequently are more concentrated than the parent ice crystal by at least an order of magnitude. In the melt period, wet snow metamorphism occurs. This type of metamorphism is more rapid than dry snow metamorphism. Large grains grow at the expense of small grains in intimate contact with liquid water. Grain clusters dissociate and solute release from grain boundaries is favoured (Davis, 1991). The amount of solute able to diffuse from the grain structure into any discrete meltwater volume depends on the diffusion coefficient for the solute, the time of solute–meltwater contact (determined by flow rate), and the amount of snow leached.

Deeper snow increases the duration of snow–meltwater interaction and gives rise to higher snowmelt concentrations. For example, Marsh and Webb (1979) report

145

the approximate doubling of initial snowmelt concentrations with the doubling of snow depth. Low melt rates promote the more uniform flow of meltwater through the whole snow matrix and solute scavenging is maximised. Colbeck (1981) showed that slow-moving meltwater picks up solute by molecular diffusion at flow rates $<10^{-7}$ m s^{-1}. At high melt rates, contact time is less. In addition, preferential flow through vertical macropores also occurs (Kattleman, 1985, 1989), resulting in both a decrease in the snow–meltwater contact time and the mass of snow leached per unit volume of meltwater. Hence, solute scavenging is minimised. The composition of snowmelt thus may vary on a diurnal basis, because solute scavenging is related to the rate of melt (Tsiouris et al., 1985; Tranter et al., 1988b; Bales et al., 1993). Higher concentrations of solute in meltwater are found in the morning and evening, or during periods of shading, when melt rates are lowest.

Generally, meltwater flowing through macropores or flow fingers is more dilute than contemporaneous melt flowing through the snow matrix. The effects of heterogenous flow on fractionation are illustrated in Figure 3.6, where CF is shown for two flow paths, one with the lowest (matrix) and one with the highest (macropore) measured flow. Note that the maximum CF is 10 for the flow path with the lowest flow but only 8.0 for the path with the highest flow. Over the following 6 days, the CF of both flow paths gradually converged until all flow paths had similar values. However, the formation of horizontal macropores in some snow covers due to discontinuities between snow strata, or the impermeability of ice lenses, increases meltwater concentrations because the duration of snow–meltwater contact is increased by horizontal flow (Marsh and Pomeroy, 1993).

The mesoscale distribution of solute in snow cover will also affect the concentration of meltwaters. Discrete snowfalls give rise to snow strata of different compositions. Solute-rich bands also arise from the exclusion of solute from ice lenses formed by the refreezing of meltwater fronts or rain in cold, dry snow. The effect of diurnal melt–freeze cycles is to translocate solute from the surface to the interior or the base of the snowpack. Meltwater in contact with frozen ground refreezes to form basal ice layers, often overlaid with saturated wet snow layers (Marsh, 1990). Thus the result of several diurnal melt–freeze cycles is often to increase the concentration of ions in the first meltwaters issuing from the snowpack (Bales, Davis, and Stanley, 1989; Williams and Melack, 1993). Both laboratory and field experiments have shown that solute-rich layers give rise to more concentrated meltwaters (Colbeck, 1981; Tranter et al., 1986). However, this may be attenuated by macropore flow that causes meltwater to bypass areas of the solute-rich layers.

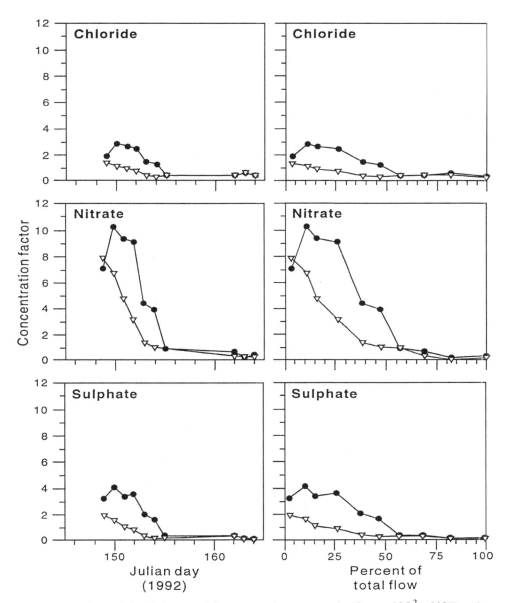

Figure 3.6. The impact of flow rate on the concentration factor of SO_4^{2-}, NO_3^-, and Cl^- in snowmelt. Open symbols denote high rates; closed symbols denote low flow rates (after Marsh and Pomeroy, 1993).

147

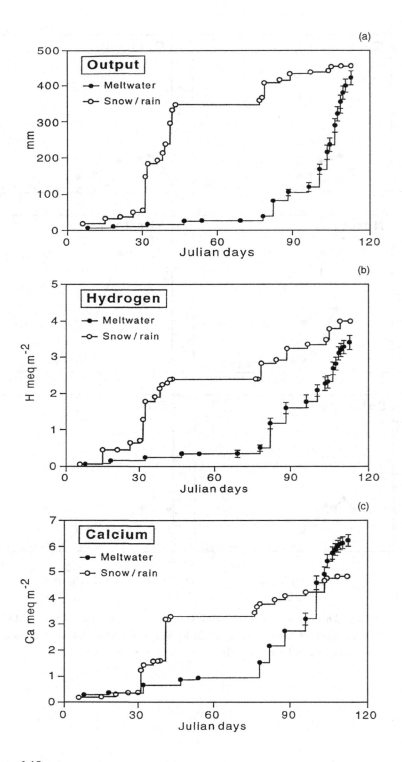

The distribution of solutes in snow and the dynamics of meltwater hydrology increases the difficulty of modeling solute leaching and meltwater composition. However, models for snow meltwater quality have been developed that generate CF as a function of meltwater discharge. These models ignore preferential meltwater flow paths and solute distribution (Goodison, Louie, and Metcalfe, 1986; Stein et al., 1986). They represent bulk models that combine total snow depth and a bulk leaching coefficient, k. The leaching coefficient is based on a first-order removal of the solute from snow by meltwaters (Hibberd, 1984). High values of k give rise to greater values of CF in the initial meltwaters. Values of k vary (e.g., from 0.002 to 0.01 μeq mm^{-1} for NO$_3^-$), depending on crystal habit and rate of flow.

3.4.2   Snowmelt–Particulate Interactions

Chemical reactions between meltwater and inorganic-organic particles can affect the concentration of solute in meltwater. Chemical weathering of dust scavenged by snow can take place in the atmosphere (Sequeira, 1991; Colin et al., 1987) or within the snow cover (Psenner and Nickus, 1986; Maupetit and Davies, 1991; Puxbaum, Kovar, and Kalina, 1991). Many studies have observed the neutralisation of snow acidity by carbonaceous dusts from a variety of sources, either of local (Colin et al., 1987; DeWalle et al., 1983; Hart et al., 1985; Delmas et al., 1996) or remote origin (Psenner and Nickus, 1986; Loye-Pilot, Martin, and Morelli, 1986).

Delmas et al. (1996) determined that the rate of chemical weathering of dusts in meltwaters depended on the location of the dust in the snow cover. Laboratory experiments showed that dust located in the lower strata of snow covers showed more rapid rates of weathering than dust in the upper strata because of increased partial pressures of CO_2 that arise during dust–meltwater interaction (see Figure 3.7). Dust in the lower regions of the snow cover, therefore, was more efficient than dust in the upper strata at neutralizing acidic meltwaters. However, the extent to which this weathering of dusts plays an effective role in the chemical composition of many regional meltwaters has not been truly evaluated or modeled. Delmas et al. (1996) showed that

Figure 3.7. The neutralisation of meltwater by calcareous dust in the French Alps (after Delmas et al., 1996). (a) The cumulative input of snow water equivalents and rain to the snow cover and the cumulative output of meltwater. Water balance is approximately equal. (b) The cumulative input of H$^+$ to the snow cover versus the cumulative output. There is a net loss of H$^+$ during the thaw. (c) The cumulative input of Ca^{2+} to the snow cover versus the cumulative output. There is a net increase in Ca^{2+} during the thaw.

149

meltwaters from an alpine site in the French Alps with aeolian dust inputs had far lower acidities than those predicted from the acidity of either snowfall or the snow cover before the melt period. Meltwater models that take into account only the bulk ionic composition of snow to compute the acidity of meltwaters thus may overestimate the acidity of meltwaters from snow covers containing mineral dust.

Along with aerosols and dusts, organic debris may be found in considerable quantities in some snow covers (e.g., canopy litter in boreal forest snow covers) (Jones, 1991; Taylor and Jones, 1990). Litter that falls onto snow is relatively unaffected by physical and microbiological activity during the cold accumulation period. However, the leaching of litter by meltwaters removes soluble organics and other chemical species (Fahey, 1979; Jones and Sochanska, 1985; Stottlemeyer, 1987). Surficial ionic exchange also may take place between meltwater and the organic debris (Cronan and Reiner, 1983). Leaching experiments in the field (Barry and Price, 1987; Jones and DeBlois, 1987; Courchesne and Hendershot, 1988) or in the laboratory (Moloney, Stratton, and Klein, 1983; Jones and DeBlois, 1987) with snow lysimeters show, in general, that large amounts of PO_4^{3-}, K^+, Mn^{2+}, Ca^{2+}, and Mg^{2+} are discharged from litter-laden snow covers and a decrease in the acidity of meltwaters may arise from cation exchange. In contrast, interaction between meltwaters and litter often leads to losses of NO_3^- and NH_4^+ in snow cover because of microbiological activity.

3.4.3   Microbial Activity

During springmelt, the presence of liquid water and the increase in solar radiation stimulates microbiological and invertebrate activity in snow cover. Jones and DeBlois (1987) showed that meltwater production increased microbiological activity on canopy fallout in forest snow cover. The presence of meltwater also results in photosynthetic activity of truly motile algal populations within the snow cover (Hoham, 1987; Hoham and Duval, Chapter 4). Photosynthesis results in an increase in algal biomass at the expense of nutrient concentrations in the meltwaters. Decreases in the concentrations of NH_4^+ and NO_3^- are particularly noted during the growth of algal populations and may be of the order of 0.67 eq[N] ha^{-1} day^{-1} and 1.05 eq[N] ha^{-1} day^{-1}, respectively (Gamache, 1992; Hoham et al., 1989). The decreases sometimes may be masked by dry deposition of the same species from the atmosphere. Jones and Sochanska (1985) showed that the losses (photosynthesis) or gains (dry deposition) of NH_4^+ or NO_3^- in the snowpack were related to the water content of the snow cover. High amounts of free water in snow cover lead to losses of NH_4^+ and NO_3^-, whereas NH_4^+ and NO_3^- concentrations increased when low amounts of free water were present. Low amounts of free water lower algal activity and the rate

150

of dry deposition of the N species to snow may then become higher than the rate of assimilation by algae. The loss of nutrients in snow meltwaters over the whole melt season may be appreciable, approaching 20–30 percent in some years (Jones, 1991). However, there is considerable annual variation due to annual variations in snowmelt climatology (Jones and DeBlois, 1987).

Laboratory studies on melting snow containing litter fallout from forest canopy (Jones and DeBlois, 1987; Jones and Tranter, 1989) and cultivated snow algae (Gamache, 1992) confirm the losses of nutrients observed in the field. NH_4^+ and NO_3^- depletion rates in meltwater simulators varied between 0.05 eq[N] ha^{-1} day^{-1} (Gamache, 1992) and 0.42 eq[N] ha^{-1} day^{-1} (Jones and Tranter, 1989) for NH_4^+ and 0.86 eq[N] ha^{-1} day^{-1} (Gamache, 1992) and 1.25 eq[N] ha^{-1} day^{-1} (Jones and Tranter, 1989) for NO_3^-. The total nutrient loss ($NH_4^+ + NO_3^-$) may represent up to 36 percent of the nitrogen available in the snow before melt (Gamache, 1992). The addition of coniferous forest litter to snow containing cultivated algae leads to increased depletion rates of 0.62 eq[N] ha^{-1} day^{-1} for NH_4^+ and 2.22 eq[N] ha^{-1} day^{-1} for NO_3^- in similar laboratory experiments. The total N loss, 2.8 eq[N] ha^{-1} day^{-1}, was equivalent to 62 percent of the total N available in the snow before melt (Gamache, 1992).

These measured depletions of total N in both laboratory and field studies (see Figure 3.8) are consistent with reported rates of algal biomass production in snow covers. Mosser, Mosser, and Bruck (1977) measured fixation of ^{14}C in Montana snow fields

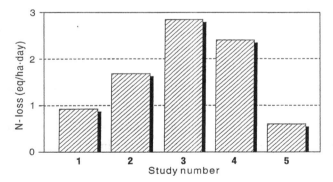

Figure 3.8. Nutrient (N) losses for field and laboratory experiments on snow/meltwater systems. Bars: 1, laboratory study (snow with added algae; Gamache, 1992); 2, laboratory study (snow with added coniferous litter; Jones and Tranter, 1989); 3, laboratory study (snow with added coniferous litter and algae; Gamache, 1992); 4, field study (Mosser, Mosser, and Brock 1977); 5, field study (Thomas, 1994).

and calculated an approximate depletion rate of 2.4 eq[N] ha^{-1} day^{-1} based on C:N ratios for algal biomass. A similar study by Thomas (1994) on fixation of ^{14}C by snow algae in the Sierra Nevada gave values that are equivalent to depletion rates of 0.6 eq[N] ha^{-1} day^{-1}. Thus rates of N losses measured directly in the field, simulated in the laboratory, and calculated from C-fixation rates in situ are consistent. They show that algal productivity can deplete up to two-thirds of the total inorganic N load in snow covers before the snow cover is completely melted.

In contrast to these snow covers, which have a significant input of organic matter, microbial activity may be significantly reduced in snow cover far from inputs of degradable litter. Jones, Duchesneau, and Handfield (1994) determined that the numbers of microorganisms within the snow cover of a High Arctic ice cap that underwent intermittent melting during the summer were not sufficient to significantly affect the measured concentrations of inorganic N in the snow.

3.4.4   Basal Processes: Soil–Meltwater Interactions

The release of liquid water and soluble nutrients from melting snow cover leads to significant changes in the rate and type of microbiological activity in the soil (Brooks, Williams, and Schmidt, 1997) and greater fluxes of gases to the snow cover (Brooks, Williams, and Schmidt, 1996). However, the subsequent diffusion of the gas through the snow may be counteracted by the dissolution of gas in the percolating meltwaters and consequent transport back into the soil water and groundwater systems (van Bochove et al., 2000).

Soil may be frozen at the end of winter, greatly affecting the volume and timing of meltwater runoff and its chemistry (Pomeroy and Brun, Chapter 2). Infiltration of meltwater into frozen soil is highly variable, ranging from near zero (i.e., the soil is impermeable) to a volume greater than the entire snowpack water equivalent (i.e., all meltwater enters directly into the soil) (Marsh, 1990, 1991). Infiltrating meltwater refreezes if the soil is sufficiently cold, thus delaying runoff of some snowmelt. As a result, the availability of N to both the soil and stream runoff may be limited during the initial runoff period when N concentrations are highest.

3.5   Snow Cover Nutrient Fluxes and Ecosystem Budgets

For most terrestrial ecosystems, nutrient input from the atmosphere is the major source for total pools of certain nutrients (e.g., N, S), while export by gaseous

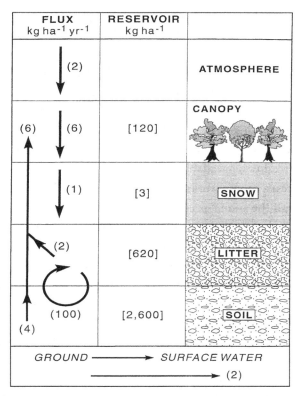

Figure 3.9. Simplified N pools and fluxes in a boreal forest (after Jones, 1991).

emissions back to the atmosphere and aquatic transport away from the system consti-tute the major outputs in ecosystems that are snow covered for much of the year. The major input arises in snow meltwaters in spring. The major output also may occur during the same period, when hydrological export by streams is the greatest because of the greater meltwater runoff (Williams, 1993).

On an annual basis, the input and/or output of certain key nutrients are relatively small with regard to the amount of nutrient that is being recycled within the system itself. This is particularly true in the case of nitrogen (see Figure 3.9), which is a limiting factor for growth in many terrestrial ecosystems. For example, an atmospheric input to a typical boreal forest of 0.15 keq N (2 kg) ha^{-1} year^{-1} (1.5 kg ha^{-1}, wet deposition; 0.5 kg ha^{-1}, dry deposition) represents only 0.2 percent of the N pool in the above-ground vegetation and forest floor (733 kg ha^{-1}; 116 kg ha^{-1}, above ground vegetation; 617 kg ha^{-1}, forest floor) and only 0.001 percent of the total pool of 3,350 kg ha^{-1}, including mineral soil horizons. As N fixation is relatively low in boreal forests, N cycling involves mostly intrasystem transformations. However, as only about 10–20 percent of the vegetative and forest floor N pool is involved in the annual turnover of the nutrient, precipitation inputs can contribute up to 1 percent of the annual turnover. This may not seem to be a major factor, but precipitation is one of the few long-term sources of new N to the system. In addition, the impact of the precipitation input on plant productivity may well outweigh its quantitative contribution to the N pool. This is because the ionic N species (NO_3^-, NH_4^+) in precipitation are the only major forms of N actively absorbed by plants (Haynes and Goh, 1978) and thus are readily bioavailable.

In the case of boreal forests, snow covers the ground for up to 6–7 months of the year and can represent up to half the annual precipitation (Pomeroy and Brun, Chapter 2). During the spring melt period, meltwaters deliver nutrients to the forest floor and, in spite of some nutrient losses of N from snow during both winter and spring, meltwaters can discharge up to 2–4 kg ha^{-1} of N within a period of a few weeks. However, there is little knowledge of the impact of such a relatively massive discharge of readily bioavailable N over a short time period on the N cycle of the forest ecosystem. It is known, however, that the export of inorganic N (NO_3^-) is greatest during the runoff period (Jones and Roberge, 1992; Stottlemeyer and Toczydlowski, 1990). This has also been found to be true for other ecosystems such as hardwood forests (Rascher, Driscoll, and Peters, 1987) and high–altitude alpine sites (Williams, Brown, and Melack, 1993). Typical values for the export of N (NO_3^-) by streamwaters during melt in a boreal forest are 0.65 kg N ha^{-1} (Skartveit and Gjessing, 1979) and 1.7 kg N ha^{-1} (Jones and Bedard, 1987). In general, the export of N as NH_4^+ is very low as stream concentrations of NH_4^+ are negligible. However, a significant part (55 percent) of total N export in the form of NH_4^+ has been reported before snowmelt during low-flow regimes in a stream draining a snow–covered mountain watershed (Lewis and Grant, 1980). In all studies, the export of total N is small compared with the amount of N cycling in the system.

The export of N as NO_3^- originates both from the solute in the meltwaters (Williams, 1993) and/or from the leaching of the species from soil after overwintering nitrification of organic matter (Peters and Driscoll, 1987; Rascher et al., 1987). In a study of a Colorado watershed, Lewis and Grant (1980) found that hydrological export of N as NO_3^- increased significantly in winters when the soil was frozen. They suggested that soil frost may inhibit denitrification or any other biological sink for NO_3^- and/or increase nitrification. The hypothesis about blocking denitrification seems to be supported by the work of Christensen and Tiedje (1990), who found that thawing of soils during spring resulted in very large increases in emissions of N_2O to the

154

atmosphere. However, the suggestion that soil frost increases nitrification is not supported by the study of Stottlemeyer and Toczydlowski (1990) on an upper-Michigan watershed. In this field experiment, the data from soil lysimeters under snow cover indicated that nitrification occurs throughout the winter in the upper organic soil horizons because of above-freezing temperatures. Slow sustained meltwater discharge moved the mineralized nitrogen to lower inorganic horizons where it was not taken up because of low biological activity. Some of the NO_3^- was then removed during the main melt period by macropore flow into surface water channels.

The studies described in this section have addressed only certain aspects of N cycling under snow. In a recent study on the influence of snow on the nutrient dynamics of alpine ecosystems, Brooks et al. (1996) attempted to quantify N budgets by determining the input of N due to snowmelt, the export of N by stream waters and gaseous emission (N_2O), and overwinter N mineralisation in the thawed soil beneath the snowpack. They reported that the inorganic N input (NO_3^-, NH_4^+) from snow represented only 3–6 percent of the net N mineralisation, showing that intrasystem cycling was the main source of inorganic N available for the start of the growth season. They also estimated that the winter/spring N losses by denitrification (N_2, N_2O) were equal to the NO_3^--N input by snowmelt and represented 50 percent of the annual gaseous N loss. In this particular system, the loss of N via runoff was negligible during snowmelt. This latter result contrasts the hydrological losses measured by Williams (1993) at another alpine site, by Peters and Driscoll (1987) at a hardwood forest site, and by Jones and Roberge (1992) at a coniferous boreal site. These studies show the different responses of ecosystems to the input of N by snowmelt. The factors controlling the hydrological export of N as NO_3^- are poorly known and remain to be elucidated by future research.

3.6   Summary and Future Research Needs

Recent studies of the chemistry of snow covers show that snow is not a passive reservoir of the chemical species scavenged from the atmosphere during snowfall. Snow–atmosphere exchange by dry deposition and volatilisation, accompanied by various physical processes such as air movement and snow metamorphism within the pack, can increase or decrease the quantity of certain species and/or lead to a redistribution of species within the snow cover. Chemical reactions also take place and the presence of microorganisms influences nutrient concentrations during the melt period. Snow covers often also show strong concentration gradients of CO_2 and trace gases as snow mediates the gaseous emissions from the soil to the atmosphere.

155

Some of these processes are understood well enough for models of the chemical evolution of snow cover to be constructed. However, for most of the processes described we have very little knowledge or data on which to base any viable models. This is particularly true for the evolution of chemical species during wind-blown snow events and dry cold snow metamorphism. There is practically no knowledge at all of photochemical reactions – even though they are thought to play a significant role in the chemistry of some snow covers. Finally, the factors that control the microbiological activity in snow and the effect of vegetative cover on snow cover composition are only now being brought to light. We need, above all, well-designed laboratory and field experiments that will permit us to acquire the data and knowledge to gradually construct the relationships that exist between snow and snow-covered ecosystems.

Thus, the relationships between the chemistry of snow, terrestrial nutrient cycles, and ecology will continue to be elucidated. Much remains to be done. We hope that we have given the reader an insight into the processes that form the basis of the chemical dynamics of snow, and more importantly, we hope that we have stimulated interest about the fascinating study of life in snow and snow-covered ecosystems. These subjects are further developed in the following chapters.

3.7   References

Bales, R.C. 1991. Modeling in-pack chemical transformations. In *NATO ASI Series G: Ecol. Sci., Vol. 28, Seasonal Snowpacks, Processes of Compositional Change*, T.D. Davies, M. Tranter, and H.G. Jones (Eds.). Springer-Verlag, Berlin, pp. 139–163.

Bales, R.C., and Wolff, E.W. 1995. Interpreting natural climate signals in ice cores. *Eos*, **76**(47), 477, 482–483.

Bales, R.C., Valdez, M.P., Dawson, G.A., and Stanley, D.A. 1987. Physical and chemical factors controlling gaseous deposition to snow. In *NATO-ASI Series C: Mathemat. Physic. Sci., Vol. 211, Seasonal Snowcovers: Physics, Chemistry, Hydrology*, H.G. Jones and W.J. Orville-Thomas (Eds.). Reidel, Dordrecht, The Netherlands, pp. 289–298.

Bales, R.C., Davis, R.E., and Stanley, D.A. 1989. Ionic elution through shallow, homogeneous snow. *Water Resources*, **25**, 1869–1877.

Bales, R.C., Davis, R.E., and Williams, M.W. 1993. Tracer release in melting snow: diurnal and seasonal variations. *Hydrol. Processes*, **7**, 389–401.

Barrie, L.A. 1991. Snow formation and processes in the atmosphere that influence its chemical composition. In *NATO ASI Series G: Ecol. Sci., Vol. 28, Seasonal Snowpacks, Processes of Compositional Change*, T.D. Davies, M. Tranter, and H.G. Jones (Eds.). Springer-Verlag, Berlin, pp. 1–20.

156

Barrie, L.A., and Vet, R.J. 1984. The concentration and deposition of acidity, major ions and trace metals in the snowpack of the Eastern Canadian Shield during the winter of 1980–1981. *Atmosph. Environ.*, **18**, 1459–1469.

Barry, P.J., and Price, A.G. 1987. Short term changes in the fluxes of water and of dissolved solutes during snowmelt. In *Seasonal Snowcovers: Physics, Chemistry, Hydrology*, H.G. Jones and W.J. Orville-Thomas (Eds.). NATO-ASI Series C, Mathematical and Physical Sciences, Vol. 211, Reidel, Dordrecht, The Netherlands, pp. 501–530.

Blais, J.M., Schindler, D.W., Muir, D.C.G., Kimpe, L.E., Donald, D.B., and Rosenberg, B., 1998. Accumulation of persistent organochlorine compounds in mountains of western Canada. *Nature*, **395**, 585–588.

Borys, R.D., Demott, P.J., Hindman, E.E., and Feng, D. 1983. The significance of snow crystal and mountain-surface riming to the removal of atmospheric trace constituents from cold clouds. In *Precipitation Scavenging, Dry Deposition and Resuspension. Vol. 1, Precipitation Scavenging*. H.R. Pruppacher, R.G. Semonin, and W.G.N. Slinn (coordinators). Elsevier, New York, pp. 181–190.

Borys, R.D., Del Vecchio, D., Jaffrezo, J.-L., Davidson, C.I., and Mitchell, D.L. 1993. Assessment of ice particle growth processes at Dye-3, Greenland. *Atmosph. Environ.*, **27A**, 2815–2822.

Brimblecombe, P., and Shooter, D.S. 1991. Chemical change in snowpacks. In *NATO ASI Series G: Ecol. Sci.*, *Vol. 28, Seasonal Snowpacks, Processes of Compositional Change*, T.D. Davies, M. Tranter, and H.G. Jones (Eds.). Springer-Verlag, Berlin, pp. 165–172.

Brooks, P.D., Williams, M.W., and Schmidt, S.K. 1996. Microbial activity under alpine snowpacks, Niwot Ridge, Colorado. *Biogeochemistry*, **32**, 93–113.

Brooks, P.D., Williams, M.W., and Schmidt, S.K. 1997. Inorganic nitrogen and microbial biomass dynamics before and during spring snowmelt. *Biogeochemistry*, **33**, 1–15.

Cadle, S.H. 1991. Dry deposition to snowpacks. In *NATO ASI Series G: Ecol. Sci.*, *Vol. 28, Seasonal Snowpacks, Processes of Compositional Change*, T.D. Davies, M. Tranter, and H.G. Jones (Eds.). Springer-Verlag, Berlin, pp. 21–66.

Cadle, S.H., and Dasch, J.M. 1987. The contribution of dry deposition to snowpack acidity in Michigan. In *NATO-ASI Series C: Mathemat. Physic. Sci.*, *Vol. 211, Seasonal Snowcovers: Physics, Chemistry, Hydrology*, H.G. Jones and W.J. Orville-Thomas (Eds.). Reidel, Dordrecht, pp. 299–230.

Cadle, S.H., Dasch, J.M., and Mulawa, P.A. 1985. Atmospheric concentrations and deposition velocity to snow of nitric acid, sulfur dioxide and various particulate species. *Atmosph. Environ.*, **19**(11), 1819–1827.

Cerling, T.E., and Alexander, A.J. 1987. Chemical composition of hoarfrost, rime and snow during a winter inversion in Utah, U.S.A. *Water Air Soil Pollut.*, **35**, 373–379.

Christensen, S., and Tiedje, J. M. 1990. Brief and vigorous N_2O production by soil at spring thaw. *J. Soil Sci.*, **41**, 1–4.

Clow, D.W., and Ingersoll, G.P. 1994. Particulate carbonate matter in snow from selected sites in the south-central Rocky Mountains. *Atmosph. Environ.*, **28**(4), 575–584.

Colbeck, S.C. 1981. A simulation of the enrichment of atmospheric pollutants in snow cover runoff. *Water Resources. Res.*, **17**(5), 1383–1388.

Colbeck, S.C. 1987. Snow metamorphism and classification. In *NATO-ASI Series C: Mathemat. Physic. Sci., Vol. 211, Seasonal Snowcover: Physics, Chemistry, Hydrology*, H.G. Jones and W.J. Orville-Thomas (Eds.). Reidel, Dordrecht, The Netherlands, pp. 1–35.

Colbeck, S.C. 1989. Air movement of snow due to windpumping. *J. Glaciol.*, **35**(120), 209–213.

Colin, J.L., Jaffrezo, J.L., Pinart, J., and Roulette-Cadene, S. 1987. Sequential sampling of snow in a rural area. Experimentation and identification of the acidifying agents. *Atmosph. Environ.*, **21**, 1147–1157.

Colin, J.L., Renard, D., Lescoat, V., Jaffrezo, J.L., Gros, J.M., and Strauss, B. 1989. Relationship between rain and snow acidity and air mass trajectory in eastern France. *Atmosph. Environ.*, **23**, 1487–1498.

Conklin, M.H. 1991. Dry deposition to snowpacks. In *NATO ASI Series G: Ecol. Sci., Vol. 28, Seasonal Snowpacks, Processes of Compositional Change*, T.D. Davies, M. Tranter, and H.G. Jones (Eds.). Springer-Verlag, Berlin, pp. 67–70.

Conklin, M.H., and Bales, R.C. 1993. SO$_2$ uptake on ice spheres: liquid nature of the ice-air interface. *J. Geophys. Res.*, **98**(D9), 16851–16855.

Courchesne, F., and Hendershot, W.H. 1988. Cycle annuel des elements nutritifs dans un bassin-versant forestier: contribution de la litiere fraiche. *Can. J. Forest Res.*, **18**, 930–936.

Coxson, D.S., and Parkinson, D. 1987. Winter respiratory activity in aspen woodland forest floor litter and soils. *Soil Biol. Biochem.*, **19**, 49–59.

Coyne, P.I., and Kelley, J.J. 1971. The release of carbon dioxide from frozen soil to the Arctic atmosphere. *Nature*, **234**, 407–408.

Cragin, J.H., and McGilvary, R. 1995. Can inorganic species volatilize from snow. In *Biogeochemistry of Seasonally Snow-Covered Catchments*, K.A. Tonnessen, M.W. Williams, and M. Tranter (Eds.). IAHS Publication No. 228, IAHS Press, Wallingford, UK, pp. 11–16.

Cronan, C.S., and Reiners, W.A. 1983. Canopy processing of acidic precipitation by coniferous and hardwood forests in New England. *Oecologia*, **59**, 216–223.

Dasch, J.M. 1987. Measurement of dry deposition to surfaces in deciduous and pine canopies. *Environ. Pollut.*, **44**, 261–277.

Dasch, J.M., and Cadle, S.H. 1985. Wet and dry deposition monitoring in southeastern Michigan. *Atmosph. Environ.*, **19**, 789–796.

Davidson, C.I., Honrath, R.E., Kadane, J.B., Tsay, R.S., Mayewski, P.A., Lyons, W.B., and Heidam, N.Z. 1987. The scavenging of atmospheric sulfate by Arctic snow. *Atmosph. Environ.*, **21**, 871–882.

Davies, T.D., Abrahams, P.W., Tranter, M., Blackwood, I, Brimblecombe, P., and Vincent, C.E. 1984. Black acid snow in the remote Scottish Highlands. *Nature*, **312**, 58–61.

Davies, T.D., Tranter, M., Jickells, T.D., Abrahams, P.W., Landsberger, S., Jarvis, K., and Pierce, C.E. 1992. Heavily contaminated snowfalls in the remote Scottish Highlands: a consequence of regional-scale mixing and transport. *Atmosph. Environ.*, **26A**, 95–112.

158

Davis, R.E. 1991. Links between snowpack physics and snowpack chemistry. In *NATO ASI Series G: Ecol. Sci., Vol. 28, Seasonal Snowpacks, Processes of Compositional Change*, T.D. Davies, M. Tranter, and H.G. Jones (Eds.). Springer-Verlag, Berlin, pp. 115–138.

Delmas, R., Briat, M., and Legrand, M. 1982. Chemistry of south polar snow. *J. Geophys. Res.*, **87**, 4314–4318.

Delmas, V., Jones, H.G., Tranter, M., and Delmas, R. 1996. The chemical weathering of aeolian dusts in alpine snows. *Atmosph. Environ.*, **30**, 1317–1325.

DeWalle, D.R., Sharpe, W.E., Izbicki, J.A., and Wirries, D.L. 1983. Acid snowpack chemistry in Pennsylvania 1979–81. *Water Resources Bull.*, **19**(6), 993–1001.

Duchesneau, M. 1993. M.Sc. thesis, No. 347, *Echange Neige-Atmosphère et Profil Physico-Chimique de la Neige en Période Froide*. Institut National de la Recherche Scientifique, Ste Foy, Québec.

Edwards, J.S., and Banko, P.C., 1976. Arthropod fallout and nutrient transport: a quantitative study of Alaskan snowpatches. *Arctic Alpine Res.*, **8**, 237–245.

Fahey, T.J. 1979. Changes in nutrient content of snow water during outflow from Rocky Mountain coniferous forest. *Oikos*, **32**, 422–428.

Frazén, L.G., Mallsson, J.O., Mårtensson, U., Nihlén, T., and Rapp, A. 1994. Yellow snow over the Alps from a dust storm in Africa, March 1991. *Ambio.*, **23**, 233–235.

Galloway, J.N., Henrey, J.R., Schofield, C.L., Peters, N.E., and Johannes, H. 1987. Processes and causes of lake acidification during spring snowmelt in the west central Adirondack Mountains, New York. *Can. J. Fish. Aquatic Sci.*, **44**, 1595–1602.

Gamache, S. 1992. M.Sc. thesis, No. 229, *Influence des Algues Nivales sur la Physico-Chimie de la Neige lors de la Fonte Printanière*. Institut National de la Recherche Scientifique, Ste Foy, Québec.

Gjessing, Y.T. 1977. The Filtering Effect of Snow. In *Isotopes and Impurities in Snow and Ice*. IAHS Publication No. 118, IAHS Press, Wallingford, UK, pp. 199–203.

Goodison, B.E., Louie, P.Y.T., and Metcalfe, J.R. 1986. Investigations of snowmelt acidic shock potential in south central Ontario, Canada. In *Modelling Snowmelt - Induced Processes*, E.M. Morris (Ed.). IAHS Publication 155, IAHS Press, Wallingford, UK, pp. 297–309.

Granberg, H.B. 1985. Distribution of grain sizes and internal surface area and their role in snow chemistry in a sub-Arctic snow cover. *Ann. Glaciol*, **7**, 149–152.

Granli, T., and Bøckman, O.Ch. 1994. Nitrous oxide from agriculture. *Norwegian J. Agri. Sci. Suppl.*, **12**.

Gregor, D.J. 1991. Organic micropollutants in seasonal snow cover and firn. In *NATO ASI Series G: Ecol. Sci., Vol. 28, Seasonal Snowpacks, Processes of Compositional Change*, T.D. Davies, M. Tranter, and H.G. Jones (Eds.). Springer-Verlag, Berlin, pp. 323–358.

Gunz, D.W., and Hoffmann, M.R. 1990a. Field investigations on the snow chemistry in Central and Southern California – II Carbonyls and carboxylic acids. *Atmosph. Environ.*, **24A**(7), 1673–1684.

Gunz, D.W., and Hoffmann, M.R. 1990b. Atmospheric chemistry of peroxides: a review. *Atmosph. Environ.*, **24A**(7), 1601–1633.

Hagen, A., and Langeland, A. 1973. Polluted snow in southern Norway and the effect of the meltwater on freshwater and aquatic organisms. *Environ. Pollut.*, **5**, 45–57.

Hardy, J.P., Davis, R.E., and Winston, G.C. 1995. Evolution of factors affecting gas transmissivity of snow in the boreal forest. In *Biogeochemistry of Seasonally Snow-Covered Catchments*, K.A. Tonnessen, M.W. Williams, and M. Tranter (Eds.). IAHS Publication No. 228, IAHS Press, Wallingford, UK, pp. 41–50.

Hart, J., Lockett, G.P., Schneider, R.A., Michaels, H., and Blanchard, C. 1985. Acid precipitation and surface-water vulnerability on the western slope of the high Colorado Rockies. *Water Air Soil Pollut.*, **25**, 313–320.

Haynes, R.J., and Goh, K.M. 1978. Ammonium and nitrate nutrition in plants. *Biol. Rev. Cambridge Philos. Soc.*, **53**, 465–510.

Hendershot, W.H., Mendes, L., Lalande, H., Courchesne, F., and Savoie, S. 1992. Soil and stream water chemistry during spring snowmelt. *Nordic Hydrol.*, **23**(1), 13–26.

Heubert, B.J., Fehsenfeld, F.C., Norton, R.B., and Albritton, D. 1983. The scavenging of nitric and vapor by snow. In *Precipitation Scavenging, Dry Deposition and Resuspension*, Vol. 1, H.R. Pruppacher, R.G. Semonin, and W.G.N. Slinn (coordinators). Elsevier, New York, pp. 293–300.

Hewitt, A.D., and Cragin, J.H. 1994. Determination of ionic concentrations in individual snow crystals and snowflakes. *Atmosph. Environ.*, **28**(15), 2545–2547.

Hibberd, S. 1984. A model for pollutant concentrations during snow melt. *J. Glaciol.*, **30**, 58–65.

Hinkley, T.K. 1994. Composition and sources of atmospheric dusts in snow at 3200 m in the St. Elias Range, southeastern Alaska, U.S.A. *Geochim. Cosmochim. Acta*, **58**(15), 3245–3254.

Höfken, K.D., Meixner, F.X., and Ehhalt, D.H. 1983. Deposition of atmospheric trace constituents onto different natural surfaces. In *Precipitation Scavanging, Dry Deposition and Resuspension*, H.R. Pruppacher, R.G. Semonin, and W.G.N. Slinn (coordinators). Elsevier, New York, pp. 825–835.

Hodgkins, R., Tranter, M., and Dowdeswell, J.A. 1997. Solute provenance, transport and denudation in a high-Arctic glacierised catchment. *Hydrol. Processes*, **11**, 1813–1832.

Hogan, A., and Leggett, D. 1995. Soil-to-snow movement of synthetic organic compounds in natural snowpack. In *Biogeochemistry of Seasonally Snow-Covered Catchments*, K.A. Tonnessen, M.W. Williams, and M. Tranter (Eds.). IAHS Publication No. 228, IAHS Press, Wallingford, UK, pp. 107–114.

Hoham, R.W. 1987. Snow algae from high-elevation temperate latitudes and semipermanent snow: their interaction with the environment. In *Proc. Eastern Snow Conf., 44th Annual Meeting*, pp. 73–79.

Hoham, R.W., Yatsko, C., Germain, L., and Jones, H.G. 1989. Recent discoveries of snow algae in upstate New York and Québec Province and preliminary reports on related snow chemistry. In *Proc. Eastern, Snow Conf., 46th Annual Meeting*, pp. 196–200.

160

Ibrahim, M., Barrie, L.A., and Fanaki, F. 1983. An experimental and theoretical investigation of the dry deposition of particles to snow, pine trees and artificial collectors. *Atmosph. Environ.*, **17**, 781–788.

Jaffrezo, J.L., Clain, M.P., and Maschet, P. 1994. Polycyclic aromatic hydrocarbons in the polar ice of Greenland: geochemical use of these atmospheric tracers. *Atmosph. Environ.* **28**, 1131–1145.

Jones, H.G. 1991. Snow chemistry and biological activity: a particular perspective of nutrient cycling. In *NATO ASI Series G: Ecol. Sci., Vol. 28, Seasonal Snowpacks, Processes of Compositional Change*, T.D. Davies, M. Tranter, and H.G. Jones (Eds.). Springer-Verlag, Berlin, pp. 21–66.

Jones, H.G., 1999. The ecology of snow-covered systems: a brief overview of nutrient cycling and life in the cold. *Hydrol. Processes*, **13**, 2135–2147.

Jones, H.G., and Bedard, Y., 1987. The dynamics and mass balances of NO_3 and SO_4 in meltwater and surface runoff during springmelt in a boreal forest. In: *Forest Hydrology and Watershed Management*, R.H. Swanson, P.Y. Bernier, and P.D. Woodard (Eds.). IAHS Publication 167, IAHS Press, Wallingford, UK, pp. 19–13.

Jones, H.G., and DeBlois, C. 1987. Chemical dynamics of N–containing ionic species in a boreal forest snowcover during the spring melt period. *Hydrol. Processes*, **1**, 271–282.

Jones, H.G., and Roberge, J. 1992. Nitrogen dynamics and sub-ice meltwater patterns in a small boreal lake during snowmelt. In *Proc. 49th Annual Meeting of the Eastern Snow Conf.*, pp. 169–180.

Jones, H.G., and Sochanska, W. 1985. The chemical characteristics of snow cover in a northern boreal forest during the spring run-off. *Ann. Glaciol.*, **7**, 167–174.

Jones, H.G., and Stein, J. 1990. Hydrogeochemistry of snow and snowmelt in catchment hydrology. In *Process Studies in Hillslope Hydrology*. M.G. Anderson and T.P. Burt (Eds.). Wiley, Chichester, pp. 255–298.

Jones, H.G., and Tranter, M. 1989. Interactions between meltwater and organic-rich particulate material in boreal forest snowpacks: evidence for both physico-chemical and microbiological influences. In *Proc. Sixth International Symposium on Water-Rock Interaction*, Malvern, UK, August 3–8, 1989, D.E. Miles (Ed.). Balkama, Rotterdam, pp. 349–352.

Jones, H.G., Pomeroy, J.W., Davies, T.D., Marsh, P., and Tranter, M. 1993. Snow-atmosphere interactions in Arctic snowpacks: net fluxes of NO_3, SO_4 and influence of solar radiation. In *Proc. 50th Annual Meeting of the Eastern Snow Conf.*, Canada, June 8–10, pp. 255–264.

Jones, H.G., Duchesneau, M., and Handfield, M. 1994. Nutrient cycling on the surface of an Arctic ice cap: snow-atmospheric exchange of N-species and microbiological activity. In *Snow and Ice Covers: Interactions with the Atmosphere and Ecosystems*, H.G. Jones, T.D. Davies, A. Ohmura, and E.M. Morris (Eds.). IAHS Publication No. 223, IAHS Press, Wallingford, UK, pp. 331–339.

Jones, H.G., van Bochove, E., and Bertrand, N., 1999. The transmission of soil gases through seasonal snow cover: an experiment to determine the diffusivity of snow in situ. In: *Interactions between the Cryosphere, Climate and Greenhouse Gases*, M. Tranter, R. Armstrong, E. Brun, H.G. Jones, M. Sharp, and M. Williams (Eds.). IAHS Publication No. 256, IAHS Press, Wallingford, UK, pp. 237–244.

Jones, H.G., Pomeroy, J.W., Davies, T.D., Tranter, M., and Marsh, P. 1999. CO_2 in Arctic snow cover: landscape form, in-pack gas concentrations gradients, and the implications for the estimation of gaseous fluxes. *Hydrol. Processes*, 13, 2977–2989.

Jones, A.E., Weller, R., Wolff, E.W., and Jacobi, H.-W., 2000. Speciation and rate of photochemical NO and NO_2 production in Antarctic snow. *Geophys. Res. Lett.*, 27, 345–348.

Kamai, M. 1976. Identification of nucleii and concentrations of chemical species in snow crystals sampled at the South Pole. *J. Atmos. Sci.*, 33, 833–841.

Kattleman, R.C. 1985. Macropores in snowpacks of Sierra Nevada. *Ann. Glaciol.*, 6, 272–273.

Kattleman, R.C. 1989. Spatial variability of snow-pack outlet flow at a site in the Sierra Nevada. *Ann. Glaciol.*, 13, 124–128.

Laberge, C., and Jones, H.G. 1991. A statistical approach to field measurements of the chemical evolution of cold ($<0°C$) snow cover. *Environ. Monitor. Assess.*, 17, 211–216.

Laird, L.B., Taylor, H.E., and Kennedy, V.C. 1986. Snow chemistry of the Cascade-Sierra Nevada Mountains. *Environ. Sci. Technol.*, 20, 275–290.

Laird, S.K., Buttry, D.A., and Sommerfeld, R.A., 1999. Nitric acid adsorption on ice: surface diffusion. *Geophys. Res. Lett.*, 26, 699–701.

Lamb, D.S., Mitchell, D., and Blumernstein, R. 1986. Snow chemistry in relation to precipitation growth forms. In *Proc. 23rd Conference on Radar Meteorology and the Conference on Cloud Physics*. Snowmass, CO, September 23–26, 1986. American Meteorological Society, Boston, pp. 77–80.

Landsberger, S., Davies, T.D., Tranter, M., Abrahams, P.W., and Drake, J.J. 1989. The solute and particulate chemistry of background snowfall on the Cairngorm Mountains, Scotland: a comparison with a black acid snowfall. *Atmosph. Environ.*, 23, 395–401.

Legrand, M.R., and Delmas, R.J. 1984. The ionic balance of Antarctic snow: a 10-year detailed record. *Atmosph. Environ.*, 18, 1867–1874.

Leuenberger, C., Czuczwa, J., Heyerdahl, E., and Giger, W. 1988. Aliphatic and polycyclic aromatic hydrocarbons in urban rain, snow and fog. *Atmosph. Environ.*, 22, 695–705.

Lewis, M.J., and Grant, M.C. 1980. Relationships between snow cover and winter losses of dissolved substances from a mountain watershed. *Arctic Alpine Res.*, 12, 11–17.

Likens, G.E., Edgerton, E.S., and Galloway, J.N. 1983. The composition and deposition of organic carbon in precipitation. *Tellus*, 35B, 16–24.

Loye-Pilot, M.D., Martin, J.M., and Morelli, J. 1986. Influence of Saharan dust on the rain acidity and atmospheric input to the Mediterranean. *Nature*, 321, 427–428.

Lyons, W.B., Wake, C., and Mayewski, P.A. 1991. Chemistry of snow at high altitude, mid/low latitude glaciers. In *NATO ASI Series G: Ecol. Sci., Vol. 28, Seasonal Snowpacks, Processes of Compositional Change*, T.D. Davies, M. Tranter, and H.G. Jones (Eds.). Springer-Verlag, Berlin, pp. 359–384.

Magono, C., and Lee, C.W. 1966. Meteorological classification of natural snow crystals. *J. Faculty Sci., Hokkaido Univ. Ser. VII Geophys.*, **II**(4), 321–335.

Marsh, P. 1990. Snow hydrology. In *Northern Hydrology: Canadian Perspectives*, T.D. Prowse, and C.S.L. Ommanney (Eds.). NHRI Science Report No. 1, Minister of Supply and Services, Ottawa, Canada, pp. 37–62.

Marsh, P. 1991. Water flux in melting snow covers. In *Advances in Porous Media*, M.Y. Corapcioglu (Ed.). Vol. 1, pp. 61–124. Elsevier Science, New York.

Marsh, P., and Pomeroy, J.W. 1993. The impact of heterogeneous flowpaths on snowmelt runoff chemistry. In *Proc. 50th Annual Meeting of the Eastern Snow Conf.*, pp. 231–238.

Marsh, A.R.W., and Webb, A.H. 1979. *Physico-Chemical Aspects of Snow-Melt*. CEGB Reports, RD/L/N, 60/79. Central Electricity Generating Board. London.

Massman, W., Sommerfeld, R., Zeller, K., Hehn, T., Hudnell, L., and Rochelle, S. 1995. CO_2 flux through a Wyoming seasonal snowpack: diffusional and pressure pumping effects. In *Biogeochemistry of Seasonally Snow-Covered Catchments*. K.A. Tonnessen, M.W. Williams, and M. Tranter (Eds.). IAHS Publication No. 228, IAHS Press, Wallingford, UK, pp. 71–80.

Massman, W.J., Sommerfeld, R.A., Mosier, A.R., Zeller, K.F., Hehn, T.J., and Rochelle, S.G., 1997. A model investigation of turbulence-driven pressure-pumping effects on the rate of diffusion CO_2, N_2O and CH_4 through layered snowpacks. *J. Geophys. Res.*, **102**, 18851–18863.

Maupetit, F., and Davies, T.D. 1991. Chemical composition and fluxes of wet deposition at elevated sites in the eastern Alps. In *NATO ASI Series G: Ecol. Sci., Vol. 28, Seasonal Snowpacks, Processes of Compositional Change*, T.D. Davies, M. Tranter, and H.G. Jones (Eds.). Springer-Verlag, Berlin, pp. 299–302.

Maupetit, F., and Delmas, R.J. 1994. Carboxylic acids in high-elevation snow. *J. Geophys. Res.*, **99**(D8), 16491–16500.

Misra, P.K., Chan, W.H., Chung, D., and Tang, A.J.S. 1985. Scavenging ratios of acidic pollutants and their use in long-range transport models. *Atmosph. Environ.*, **19**, 1471–1475.

Mitra, S.K., Barth, S., and Pruppacher, H.R. 1990. A laboratory study on the scavenging of SO_2 by snow crystals. *Atmosph. Environ.*, **24A**, 2307–2313.

Molina, M.J., Tso, T.-L., Molina, L.T., and Wang, F.C.-Y. 1987. Antarctic stratospheric chemistry of chlorine nitrate, hydrogen chloride and ice: Release of active chlorine. *Science*, **238**, 1253–1257.

Moloney, K.A., Stratton, L.J., and Klein, R.M. 1983. Effects of simulated acidic, metal-containing precipitation on coniferous litter decomposition. *Can. J. Bot.*, **61**, 3337–3342.

Mooney, H.A., Vitousek, P.M., and Matson, P.A. 1987. Exchange of materials between terrestrial ecosystems and the atmosphere. *Science*, **238**, 926–932.

Mosser, J.L., Mosser, A.G., and Brock, T.D. 1977. Photosynthesis in the snow: The alga *Chlamydomonas nivalis* (Chlorophycae). *J. Phycol.*, **13**, 22–27.

Mulvaney, R., Wolff, E.W., and Oates, K., 1988. Sulphuric acid at grain boundaries in ice. *Nature*, **331**, 247–249.

163

Muniz, I.P. 1991. Freshwater acidification: its effects on species and communities of fresh-water microbes, plants and animals. *Proc. R. Soc. Edinburgh*, **97B**, 227–254.

Neftel, A. 1991. Use of snow and firn analysis to reconstruct past atmospheric composition. In *NATO ASI Series G: Ecol. Sci., Vol. 28, Seasonal Snowpacks, Processes of Compositional Change*, T.D. Davies, M. Tranter, and H.G. Jones (Eds.). Springer-Verlag, Berlin, pp. 385–416.

Neubauer, J., and Heumann, K.G. 1988. Nitrate trace determinations in snow and firn core samples of ice shelves at the Weddel Sea, Antarctica. *Atmosph. Environ.*, **22**, 537–545.

Nicholson, K.W., Branson, J.R., and Giess, P. 1991. Field measurements of the below-cloud scavenging of particulate material. *Atmosph. Environ.*, **25A**, 771–777.

Palmer, T.Y., Smith, L.R., and Neirink, J. 1975. *Ozone in the Alleys of the High Sierra Nevada*, Paper 32-7. 1975, Las Vegas, IEEE, New York.

Peters, N.E., and Driscoll, C.T. 1987. Sources of acidity during snowmelt in the west-central Adirondack Mountains, New York. In *Forest Hydrology and Watershed Management*. R.H. Swanson, P.Y. Bernier, and P.D. Woodward (Eds.). IAHS Publication No. 167, IAHS Press, Wallingford, UK, pp. 99–108.

Pomeroy, J.W., and Jones, H.G. 1996. Wind-blown snow: sublimation and changes to polar snow. In *Processes of Chemical Exchange Between the Atmosphere and Polar Snow*, E. Wolff and R.C. Bales (Eds.). NATO-ASI Series I, Vol. 43, Springer Verlag, New York, pp. 453–490.

Pomeroy, J.W., Davies, T.D., and Tranter, M. 1991. The impact of blowing snow on snow chemistry. In *NATO ASI Series G: Ecol. Sci., Vol. 28, Seasonal Snowpacks, Processes of Compositional Change*, T.D. Davies, M. Tranter, and H.G. Jones (Eds.). Springer-Verlag, Berlin, pp. 71–114.

Pomeroy, J.W., Marsh, P., and Lesack, L. 1993. Relocation of major ions in snow along the tundra-taiga ecotone. *Nordic Hydrol.*, **24**, 151–168.

Psenner, R., and Nickus, U. 1986. Snow chemistry of a glacier in the central Eastern Alps (Hintereisferner, Tyrol, Austria). *Z. Gletscherkunde Glazialgeol.*, **22**, 1–18.

Puxbaum, H., Kovar, A., and Kalina, M. 1991. Chemical composition and fluxes at elevated sites (700-3105 m a.s.l.) in the eastern Alps (Austria). In *NATO ASI Series G: Ecol. Sci., Vol. 28, Seasonal Snowpacks, Processes of Compositional Change*, T.D. Davies, M. Tranter, and H.G. Jones (Eds.). Springer-Verlag, Berlin, pp. 273–298.

Rahn, K.A., and McCaffrey, R.J. 1979. Compositional differences between Arctic aerosol and snow. *Nature*, **280**, 479–480.

Rango, A., Wergin, W.P., and Erbe, E.F. 1996. Snow crystal imaging using scanning electron microscopy. I: Precipitated snow. *Hydrol. Sci. J.*, **41**(2), 219–234.

Rascher, C.M, Driscoll, C.T., and Peters, N.E. 1987. Concentration and flux of solutes from snow and forest floor during snowmelt in the west-central Adirondack region of New York. *Biogeochemistry*, **3**, 209–224.

Raynor, G.S., and Haynes, J.V. 1983. Differential rain and snow scavenging efficiency implied by ionic concentration differences in winter precipitation. In *Precipitation*

Scavenging, Dry Deposition, and Resuspension, Vol. 1, H.R. Pruppacher, R.G. Semonin, and W.G.N. Slinn (Eds.). Elsevier, New York, pp. 249–264.

Reynolds, B. 1983. The chemical composition of snow at a rural upland site in mid-Wales. *Atmosph. Environ.*, **17**, 1849–1851.

Schöndorf, T., and Herrmann, R. 1987. Transport and chemodynamics of organic micro-pollutants and Ions during snowmelt. *Nordic Hydrol.*, **18**, 259–278.

Scott, B.C. 1981. Sulfate washout ratios in winter storms. *J. Appl. Meteorol.*, **20**, 619–625.

Semkins, R.G., and Jeffries, D.S. 1988. Chemistry of atmospheric deposition, the snowpack, and snowmelt in the Turkey lakes Watershed. *Can. J. Fish. Aquat. Sci.*, **45**(suppl. 1), 38–46.

Sequeira, R. 1991. A note on the consumption of acid through cation exchange with clay minerals in atmospheric precipitation. *Atmosph. Environ.*, **25A**, 487–490.

Sigg, A., Neftel, A., and Zircher, F. 1987. Chemical transformation in a snowcover at Weissfluhjoch, Switzerland. Situated at 2500 m.a.s.l. In *NATO-ASI Series C: Mathemat. Physic. Sci.*, *Vol. 211, Seasonal Snowcovers: Physics, Chemistry, Hydrology*, H.G. Jones and W.J. Orville-Thomas (Eds.). Reidel, Dordrecht, The Netherlands, pp. 269–280.

Skartveit, A., and Gjessing, Y.T. 1979. Chemical budgets and chemical quality of snow and runoff during spring snowmelt. *Nordic Hydrol.*, **10**, 141–154.

Solomon, D.K., and Cerling, T.E. 1987. The annual carbon dioxide cycle in a montane soil: Observations, modeling, and implications for weathering. *Water Resour. Res.*, **23**, 2257–2265.

Sommerfeld, R.A., Mosier, A.R., and Musselman, R.C. 1993. CO_2, CH_4 and N_2O flux through a Wyoming snowpack and implications for global budgets. *Nature*, **361**, 140–142.

Stein, J., and Amundsen, C.C., 1967. Studies on snow algae and snow fungi from the Front Range of Colorado. *Can. J. Bot.*, **45**, 2033–2045.

Stein, J., Jones, H.G., Roberge, J., and Sochanska, W. 1986. The prediction of both runoff quality and quantity by the use of an integrated snowmelt model. In *Modelling Snowmelt-Induced Processes*, E.M. Morris (Ed.). IAHS Publication No. 155, IAHS Press, Wallingford, UK, pp. 347–358.

Stottlemyer, R. 1987. Snowpack ion accumulation and loss in a basin draining to Lake Superior. *Can. J. Fish. Aquat. Sci.*, **44**(11), 1812–1819.

Stottlemyer, R., and Toczydlowski, D. 1990. Pattern of solute movement from snow into an Upper Michigan stream. *Can. J. Fish. Aquat. Sci.*, **47**, 290–300.

Sturm, M. 1991. The role of thermal convection in the heat and mass transport in the Subarctic snow cover. Cold Regions Research and Engineering Laboratory (CRREL) Report 91-19 October 1991, U.S. Army Corps of Engineers, Hanover.

Suzuki, K. 1987. Spatial distribution of chloride and sulfate in the snow cover in Sapporo, Japan. *Atmosph. Environ.*, **21**(8), 1773–1778.

Taylor, B.R., and Jones, H.G. 1990. Litter decomposition under snow cover in a balsam fir forest. *Can. J. Bot.*, **68**, 112–120.

Thomas, W.H. 1994. Tioga Pass revisited: Interrelationships between snow algae and bacteria. *Proc. 62nd Western Snow Conf.*, New Mexico, May 1994, pp. 56–66.

Thornton, J.D., and Eisenreich, S.J. 1982. Impact of land-use on the acid and trace element composition of precipitation in the North central U.S. *Atmosph. Environ.*, **16**, 1945–1955.

Tranter, M. 1991. Controls on the chemical composition of snowmelt. In *NATO ASI Series G: Ecol. Sci., Vol. 28, Seasonal Snowpacks, Processes of Compositional Change*, T.D. Davies, M. Tranter, and H.G. Jones (Eds.). Springer-Verlag, Berlin, pp. 241–270.

Tranter, M., Brimblecombe, P., Davies, T.D., Vincent, C.E., Abrahams, P.W., and Blackwood, I. 1986. The composition of snowfall, snowpack and meltwater in the Scottish Highlands – evidence for preferential elution. *Atmosph. Environ.*, **20**, 517–525.

Tranter, M., Davies, T.D., Brimblecombe, P., Abrahams, P.W., Blackwood, I., and Vincent, C.E. 1987. Spatial variability of the chemical composition of snow cover in a small, remote Scottish catchment. *Atmosph. Environ.*, **21**, 853–862.

Tranter, M., Abrahams, P.W., Blackwood, I.L., Brimblecombe, P., and Davies, T.D. 1988a. The impact of a single, black, snowfall on streamwater chemistry in upland Britain. *Nature*, **332**, 826–829.

Tranter, M., Davies, T.D., Brimblecombe, P., and Vincent, C.E. 1988b. The composition of acidic meltwater during snowmelt in the Scottish Highlands. *Water Air Soil Pollut.*, **36**, 75–90.

Tranter, M., Tsiouris, S., Davies, T.D., and Jones, H.G. 1992. A laboratory investigation of the leaching of solute from snowpack by rainfall. *Hydrol. Processes*, **6**, 169–179.

Tsiouris, S., Vincent, C.E., Davies, T.D., and Brimblecombe, P. 1985. The elution of ions through field and laboratory Snowpacks. *Ann. Glaciol.*, **7**, 196–201.

Ushakova, L.A., and Troshkina, E.S. 1974. Role of the Liquid-Like Layer in Snow Metamorphism. In *Snow Mechanics*. IAHS Publication No. 114, IAHS Press, Wallingford, UK, pp. 62–65.

Valdez, M.P., Bales, R.C., Stanley, D.A., and Dawson, G.A. 1987. Gaseous deposition to snow: I. Experimental study of SO_2 and NO_2 deposition. *J. Geophys. Res.*, **92**, 9779–9789.

van Bochove, E., Jones, H.G., Bertrand, N., and Prévost, D., 2000. Winter fluxes of greenhouse gases from snow-covered agricultural soil: intra- and interannual variations. *Global Biogeochem. Cycles*, **14**, 113–125.

White, P.J., Garrolf, R.A., and Heisey, D.M., 1997. An evaluation of snow-urine ratios as indices of ungulate nutritional status, *Can. J. Zool.*, **75**, 1687–1694.

Williams, M.W. 1993. Snowpack storage and release of nitrogen in the Emerald Lake watershed, Sierra Nevada. In *Proc. 50th Annual Meeting of the Eastern Snow Conf.*, pp. 239–246.

Williams, M.W., and Melack, J.M. 1991. Solute chemistry of snowmelt and runoff in an alpine basin, Sierra Nevada. *Water Resources Res.*, **27**, 1575–1588.

Williams, M.W., Brown, A., and Melack, J.M. 1993. Geochemical and hydrological controls on the composition of surface waters in a high-elevation basin, Sierra Nevada. *Limnol. Oceanogr.*, **38**, 775–797.

166

Winston, G.C., Stephens, B.B., Sundquist, E.T., Hardy, J.P., and Davis, R.E. 1995. Seasonal variability of CO_2 transport through snow in a boreal forest. In *Biogeochemistry of Seasonally Snow-Covered Catchments*, K.A. Tonnessen, M.W. Williams, and M. Tranter (Eds.). IAHS Publication No. 228, IAHS Press, Wallingford, UK, pp. 61–70.

Zimov, S.A., Zimova, G.M., Daviodov, S.P., Davidova, A.I., Voropaev, Y.V., Voropaeva, Z.V., Prosiannikov, S.F., Prosianikova, O.V., Semiletova, I.V., and Semiletov, I.P. 1993. Winter biotic activity and production of CO_2 in Siberian soils: a factor in the Greenhouse Effect. *J. Geophys. Res.*, **98**(D3), 5017–5023.

4 Microbial Ecology of Snow and Freshwater Ice with Emphasis on Snow Algae

RONALD W. HOHAM AND BRIAN DUVAL

4.1 Overview

The physical and chemical properties of snow and ice result in a habitat that can support microbiological activity. Microorganisms such as bacteria, algae, and fungi are commonly found thriving in snow, glacial ice, lake ice, ice shelves, and sea ice (Kol, 1968a; Horner, 1985). Microbes found living in these habitats may encounter extreme conditions of temperature, acidity, irradiation levels, and minimal nutrients, and they are subjected to desiccation when liquid water is no longer available (Hoham, 1992). These environments are suitable habitats for microorganisms only when liquid water is present for at least part of the year, allowing for growth and reproduction of the microbes (Pollock, 1970). Even though viable microbial cells and spores are present, the surface of the vast Antarctic polar plateau and the Greenland ice sheet is virtually lifeless because of the lack of water in the liquid phase (Catranis and Starmar, 1991). In contrast, the bottom surfaces of the Antarctic glaciers are wet because of basal melting and may serve as an important habitat for microbes (Scherer, 1991). The sea ice of the Arctic Ocean often contains blooms of snow algae during the summer when meltwater is present on the ice surface (Melnikov, 1989; Gradinger and Nürnberg, 1996). Microbes are abundant in snow fields and in depressions and ponds found on glaciers (Kol, 1968a; Wharton et al., 1985); those found in snow and ice have special adaptations and mechanisms for living in cold temperatures and tolerating the long period of desiccation during the time of nonactive growth and the high irradiance levels characteristic of summer (Bidigare et al., 1993). Microorganisms play a fundamental role in the biogeochemistry of snow and ice (Jones, 1991) and are closely involved in the primary production, respiration, nutrient cycling, decomposition, metal accumulation, and food webs associated with these habitats (Fjerdingstad et al., 1978; Hoham, 1980; Jones, 1991; Hoham et al., 1993; Hoham and Ling, 2000). Microbes also contribute to melting of snowpacks and formation of cryoconite holes (Kol, 1968a; Wharton et al., 1985).

4.2   Introduction

This chapter focuses on the microbial ecology of snow fields and glaciers with some discussion of lake ice and the relatively freshwater surfaces of ice shelves. It does not include sea ice that is predominantly a marine environment (see reviews on this subject by Horner, 1985; Palmisano and Garrison, 1993; Kirst and Wiencke, 1995; Horner, 1996) or biological ice nucleation (see the comprehensive bibliography on this subject by Warren, 1994).

4.2.1   Historical Perspective

Snow and ice microorganisms have been recorded since the time of Aristotle but were first brought to the attention of the scientific community in 1819 (Bauer, 1819). Most of the work in snow and freshwater ice microbiology has focused on the snow algae (Kol, 1968a; Hoham, 1980) and ice algae and cyanobacteria from dry valley lakes in Antarctica (Parker et al., 1982a, 1982b). However, there is an increasing awareness of algae from permanent glaciers (Ling and Seppelt, 1990, 1993; Yoshimura, Kohshima, and Ohtani, 1997). Much less is known about fungi from snow (Hoham et al., 1993) and glacial ice (Abyzov, 1993), and, until recently, eubacteria from snow (Margesin and Schinner, 1994) and glacial ice have been scarcely studied (Schinner, Margesin, and Pümpel, 1992). The effects of grazers, such as ciliates, rotifers, and collembola, on snow microbe populations have been included in some studies (Pollock, 1970; Aitchison, 1989; Hoham et al., 1993).

Between 1819 and the mid–1960s, research emphasized the systematics, taxonomy, and distribution of snow and ice algae, with most investigations coming from Europe (Wille, 1903; Kol, 1968a), North America (Kol, 1968a), Japan (Fukushima, 1963), and one of significance from the South Orkney Islands (Fritsch, 1912). In the past three decades, however, the ecology and physiological ecology, life histories, taxonomy, distribution, ultrastructure, biochemistry, physiology, and interrelationships with snow chemistry of snow microorganisms have been examined. This chapter emphasizes this more recent research.

4.2.2   Locations of Snow and Ice Microorganisms

Snow and ice algae and other microbes are well known from alpine and high latitude regions of Europe (Kol, 1968a; Schinner et al., 1992). Their distributions in western North America have been reported regionally (Kol, 1942; Stein and Brooke, 1964; Garric, 1965; Stein and Amundsen, 1967; Hoham and Blinn, 1979; Wharton and

169

Vinyard, 1983) and more recently in the eastern parts of the continent (Handfield et al., 1992, Duval, 1993; Hoham et al., 1993, Duval and Hoham, 2000). Other notable reports include those from Japan (Kobayashi and Fukushima, 1952; Fukushima, 1953, 1963), Australia (Marchant, 1982, 1998), New Zealand (Kol, 1968b; Thomas and Broady, 1997), New Guinea (Kol and Peterson, 1976), the Himalayas (Kohshima, 1984b, 1987a, 1987b; Yoshimura et al., 1997), and some parts of South America (Lagerheim, 1892; Kol, 1968a), the Arctic (Kol, 1942, 1963; Sinclair and Stokes, 1965; Kobayashi, 1967; Kol, 1968a; Kol and Eurola, 1973; Melnikov, 1989; Gradinger and Nürnberg, 1996), and Antarctica and surrounding islands (Fritsch, 1912; Llano, 1962; Hirano, 1965; Fogg, 1967; Kol, 1971; Akiyama, 1979; Ishikawa, Matsuda, and Kawaguchi, 1986; Bidigare et al., 1993; Broady, 1996; Ling, 1996; Mataloni and Tesolin, 1997; Ling and Seppelt, 1998). Little is known about snow and ice microbes from Africa and many alpine areas of Asia and South America, but Duval, Duval, and Hoham (1999) recently discovered a snow alga from the Atlas Mountains of Morocco. Because similar extreme environments may exist or may have existed in the past on the planet Mars, NASA has included these microbes as one of four life systems on Earth as possible analogs for life on early and present day Mars (Wharton et al., 1989a, 1989b; Rothschild, 1990).

4.3   Populations

4.3.1   Population Diversity and Density

This section emphasizes the bacteria, fungi, algae, protozoa, and some invertebrates that are found in snow and ice. Of these microbes, the algae have received the most attention. Algae found in snow and ice include Chlorophyta (green algae), Euglenophyta (euglenoids), Chrysophyta (Xanthophyceae [yellow-green algae], Chrysophyceae [golden algae], Bacillariophyceae [diatoms]), Pyrrhophyta (dinoflagellates), and Cryptophyta (cryptomonads). The Cyanophyta (blue-green algae) are discussed as cyanobacteria. Specific overviews on these groups in snow and ice have been published for the green algae (Kol, 1942, 1968a; Hoham, 1980; Hoham et al., 1993), euglenoids (Kiener, 1944; Hoham and Blinn, 1979; Hoham et al., 1993), yellow-green algae (Kol, 1968a; Hoham et al., 1993), golden algae (Fukushima, 1963; Stein, 1963; Hoham and Blinn, 1979, Hoham et al., 1993), diatoms (Wharton, Parker, and Simmons, 1983; Wharton and Vinyard, 1983), dinoflagellates (Kol, 1968a; Gerrath and Nicholls, 1974), cryptomonads (Javornický and Hindák, 1970), and cyanobacteria

Table 4.1. *Maximum populations of snow and ice algae recorded for green algae (Division Chlorophyta).*

Species	Snow color	Location	Cells mL^{-1} of snow water equivalent (SWE)	Reference
Chlamydomonas nivalis and Trochiscia americana	Red	California, USA	6.3×10^4	Thomas (1994)
Chlamydomonas nivalis	Red	Oregon, USA	2.3×10^5	Sutton (1972)
Chlamydomonas sp.	Green	Czech Republic	1.4×10^6	Lukavský (1993)
Chloromonas brevispina	Green	Washington, USA	5.0×10^5	Hoham et al. (1979)
Chloromonas pichinchae	Green	Washington, USA	1.0×10^6	Hoham (1989a)
Chloromonas rubroleosa	Red	Antarctica	2×10^5	Ling and Seppelt (1993)
Chloromonas sp.-B	Salmon-orange	Massachusetts, USA	8.6×10^5	Hoham et al. (1993)
Chloromonas sp.-C	Salmon-orange	Massachusetts, USA	3.0×10^5	Hoham et al. (1993)
Mesotaenium berggrenii	Grey-pink	Antarctica	1.0×10^5	Ling and Seppelt (1990)

(Wharton et al., 1981, 1983). Several of these microbes are illustrated in Figures 4.1 to 4.30 (Hoham and Blinn, 1979).

Population sizes of snow and ice algae are best known for the green algae that color snow green, red, and orange in North America, but other reports include those from Europe and Antarctica (Table 4.1) (also see Müller et al., 1998a). In a snow algal study on a Himalayan glacier, algal biomass estimated by total cell volume rapidly decreased as the altitude increased (Yoshimura et al., 1997). Populations of eubacteria in snow and ice have been studied only recently, and relatively little is known about their species composition in this habitat. Moiroud and Gounot (1969) confirmed the presence of psychrophilic bacteria from glacier ice. A relatively low eubacterial recovery was made from the Agassiz ice cap, Ellesmere Island, Canada (Handfield et al., 1992). Of the 17 microorganisms that grew at 1°C and 4°C and were defined as psychrophilic, there were 4 Gram–positive bacteria, 9 algae (after publication they were determined to be

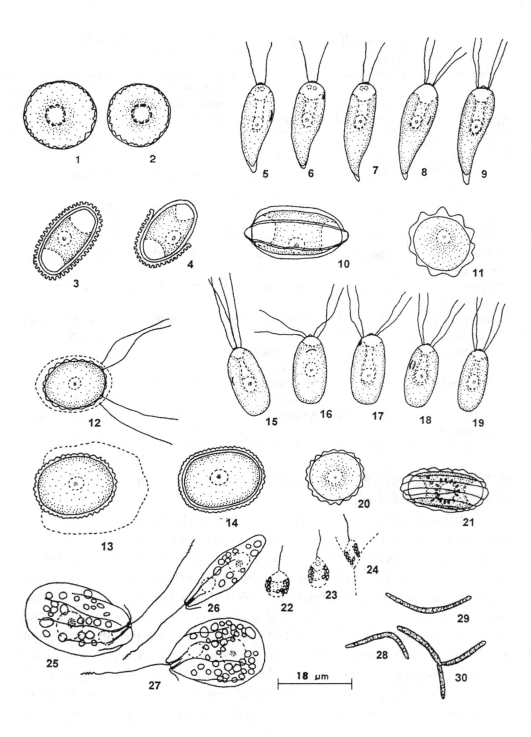

Figures 4.1 to 4.30. Cryophilic organisms found in southwestern United States (bar = 18 μm) (from Phycologia [Hoham and Blinn, 1979], 18, 141). Figures 4.1 and 4.2: *Chlamydomonas nivalis*, a green alga. Resting spores with smooth cell wall, red protoplast, and central pyrenoid surrounded by starch plates. Figures 4.3 and 4.4: *Chloromonas brevispina* (formerly *Cryocystis brevispina*), a green alga. Zygospores with central nucleus surrounded by parietal chloroplast. Note two large lipid bodies, one at each pole of the cell. Note shedding of primary wall with spines and exposure of inner smooth secondary wall (Figure 4.4). Figures 4.5 to 4.11: *Chloromonas nivalis*, a green alga. Figures 4.5 to 4.7: Vegetative cells with two flagella, papilla, two apical contractile vacuoles, centrally located nucleus, and parietal cup-shaped chloroplast. Note median, lens-shaped eyespot (Figure 4.5), anterior oval eyespot (Figure 4.6), and cell without eyespot (Figure 4.7). Figures 4.8 and 4.9: Planozygotes with four flagella otherwise similar to vegetative cells. Note median, lens-shaped eyespot (Figure 4.8) and cell without eyespot (Figure 4.9). Figures 4.10 and 4.11: Zygospores with central nucleus surrounded by parietal chloroplast. Note two large lipid bodies, one at each pole of cell (Figure 4.10). Note wall with flanges extending full length of cell (Figure 4.10) and flanges in cross section (Figure 4.11). Figures 4.12 to 4.14: *Cryocystis granulosa* cell type (the zygote of a *Chloromonas* [unpublished]), a green alga. Figure 4.12: Quadriflagellate cell giving rise to *Cryocystis granulosa*. Note central nucleus surrounded by protoplast, cell with developing wart-like protuberances, and membrane surrounding cell. Note expansion of membrane around cell (Figure 4.13) and shedding of membrane and development of primary and secondary walls (Figure 4.14). Figures 4.15 to 4.21: *Chloromonas polyptera*, a green alga. Figures 4.15 to 4.19. Planozygote (formerly *Carteria nivale*) with four flagella, papilla, centrally located nucleus, and parietal, cup-shaped chloroplast. Note median, lens-shaped eyespot parallel to longitudinal axis of cell (Figure 4.15), anterior, lens-shaped eyespot perpendicular to longitudinal axis of cell (Figure 4.16), anterior oval eyespot (Figure 4.17), two median, lens-shaped eyespots (Figure 4.18), and cell without eyespot (Figure 4.19). Figures 4.20 and 4.21: Zygotes (formerly *Scotiella polyptera*) with central nucleus surrounded by central chloroplast. Note small lipid granules (Sudan IV test) surrounding nucleus, and two large lipid bodies, one at each pole of the cell (Figure 4.21). Note wall with numerous flanges in cross section (Figure 4.20) and extending full length of cell, sometimes branching (Figure 4.21). Figures 4.22 to 4.24: *Chromulina chionophilia*, a golden alga. Cells with a parietal, band chloroplast, one contractile vacuole, a single flagellum, and 8–25 chrysolaminarin bodies. Note three rhizopodia extending from the plasmalemma (Figure 4.24). Figures 4.25 to 4.27: *Notosolenus* sp., a colorless euglenoid. Cells with ridges on pellicle, central to posterior nucleus, numerous paramylon bodies, one contractile vacuole at anterior end of cell, and two flagella, the longer projected anteriorly, the shorter posteriorly. Cells in narrow diameter view (Figure 4.26) and wide diameter view (Figures 4.25 and 4.27). Figures 4.28 to 4.30: *Selenotila nivalis*, a fungus. Cells with radiating arms possessing numerous vacuoles alternating with granular cytoplasm.

yeasts; M. Handfield, personal communication, 1994), 3 yeasts, and 1 uncertain (this was also a bacterium; M. Handfield, personal communication, 1994). Of the 220 microorganisms that were isolated at 25°C, 104 were eubacterial isolates and most were fungi. The eubacteria were categorized as Gram-positive rods (37), Gram-negative rods (30), cocci (20), streptomyces (5), and uncertain forms (12). Of these 104, 25 (24 percent) grew at 4°C and were defined as psychrotrophic. The origin of these microbes was considered terrestrial because they grew poorly in seawater. The low recovery values suggested that environmental conditions on the ice cap permit only minimal microbial activity.

Thomas (1994) reported a direct correlation between eubacterial and algal populations in snow from Tioga Pass, California. The highest populations of eubacteria (3.2×10^5 cells mL^{-1}) were found with large populations of algae (4.9×10^4 cells mL^{-1}) in red-colored snow, and snow not colored by algae contained 11 to 33 percent of the eubacterial populations and 0.1 to 5 percent of the algae found in red snow. Three separate eubacterial species were recognized but not identified from these samples, and the green algae identified were red-colored spores of *Chlamydomonas nivalis* and *Trochiscia americana* (*T. americana* is probably a resting spore of a green algal biflagellate). Baross and Morita (1978) suggested that large populations of psychrophilic bacteria exist in association with snow algae.

4.3.2 Snow Microbe Associations

Weiss (1983) found encapsulated Gram-negative bacteria in the loose fibrous network surrounding cells of the snow alga, *Chlamydomonas nivalis*, collected from Yellowstone National Park, Wyoming, and the Sierra Nevada Mountains, California. Bacteria were not found in control samples taken from white snow. Weiss concluded that the association between bacteria and snow algae was characteristic of red snow populations and was not a chance association. Paerl (1980) indicated that such interactions can be a mutual benefit in the form of increased nutrients, but Weiss (1983) did not have the experimental evidence to support this possibility. Interestingly, Thomas (1994) located bacteria in controls (white snow samples) from similar study sites in the Sierra Nevada Mountains. Hoham et al. (1993) reported a similar association between bacteria and fungi with *Chloromonas* snow algae from eastern North America. They found bacteria and fungi adhering to the outer gelatinous matrix surrounding resting spores in this *Chloromonas*, and it was not known if a symbiotic association occurred between these microbes.

Kol (1968a) reported that, of 466 species of microorganisms found in snow, 77 were fungi and 35 were bacteria; the remainder were algae. She indicated, however, that none of the bacteria and only 6 fungal species were cryobionts. Three of the fungi were parasites on *Chlamydomonas* spp. and the desmid *Ancylonema nordenskioldii*. Kol (1968a) illustrated several developmental stages of the remaining 3 fungi, *Chionaster nivalis*, *Chionaster bicornis*, and *Selenotila nivalis*. These fungi are found most commonly with large populations of snow algae (Hoham and Blinn, 1979; Hoham, Mullet, and Roemer, 1983; Hoham et al., 1993) and do not appear to enter into any type of symbiotic association with the algae, but they probably receive their required carbon for growth from them (Hoham et al., 1989). Schinner et al. (1992) reported 430 strains of microbial bacteria and fungi from alpine soils and cryoconite of glaciers from the European Alps. Of these isolates, 77 percent were bacteria, 5 percent were actinomycetes, 20 percent were yeasts, and 3 percent were hyphomycetes. Active microbial communities were found in the slush layers of the winter cover of lakes in the Pyrenean and Tyrolean Alps in spite of the low temperature and seasonal occurrence of the habitat (Felip et al., 1995). The microbes included bacteria, flagellates (algae and heterotrophs), and ciliates. The snow alga *Chlamydomonas nivalis* was restricted to surface pools.

Sinclair and Stokes (1965) reported psychrophilic yeasts that were isolated from polar ice, snow, and soil. It is now apparent that many fungi found in snow are yeasts. Yeasts were reported from snow bacterial studies (Handfield et al., 1992; Thomas, 1994), and Hoham and Clive (unpublished data) found a direct correlation of large yeast populations in snow with large snow algal populations from Killington Ski Area, Vermont, and Whiteface Mountain, New York. Thus, the data collected on populations of snow bacteria and snow fungi indicate a direct correlation between large populations of these microbes with large populations of snow algae.

In some parts of the Pacific Northwest, the snow fungus *Phacidium infestans* grows throughout melting snowpacks (Hoham, 1975b; Hoham and Mullet, 1977; Hoham, Roemer, and Mullet, 1979; Hoham, 1987). Snow algae passively adhere to this fungus, but physical connections between them do not occur as in lichen symbiosis. However, exchange of metabolites may take place between these microbes. Upon complete melting of the snowpacks, *Phacidium* with its adhering algae becomes dried threads or strands draping over soil, rocks, tree branches, shrubs, and forest cover such as mosses and lichens. *Phacidium* also grows during winter under a heavy layer of snow over unfrozen soil when temperatures remain near 0°C at the soil surface and under snow cover where air humidity is sufficient (Vuorinen and Kurkela, 1993). Under these circumstances, the blight may parasitize coniferous seedlings in winter as it does during the time of spring snowmelt. It is not known if *Phacidium* exchanges metabolites with other microbes such as bacteria, algae, and fungi during the winter.

175

In Antarctic dry valley lakes, the algal mats located under the ice were dominated by the cyanobacteria *Phormidium frigidum* and *Lyngbya martensiana* (Wharton et al., 1983). Heterocystis cyanobacteria (*Anabaena cylindrica* and *Nostoc commune*) were restricted to moat, pond, ice surface, and cryoconite areas (Wharton et al., 1981). Except for moat and anaerobic prostrate mats, pennate diatoms were abundant in all samples, particularly in surface layers of liftoff and aerobic prostrate mats. These diatoms were often seen gliding between filaments of the cyanobacteria and included species of *Navicula*, *Nitzschia*, and *Pinnularia*. Green algae were rare in all below-ice samples. Heterotrophic bacteria and yeasts were observed in the surface layers of mats, but these microbes were not identified. Viruses may regulate heterotrophic and autotrophic protozoan, algal, and bacterial community successions in ice-covered Antarctic lakes (Kepner, Wharton, and Galchenko, 1997).

4.3.3   Food Chains and Food Webs

Primary consumers, protozoa and rotifers, are part of the food web that occurs in snow (Pollock, 1970; Hardy and Curl, 1972; Hoham et al., 1979, 1983; Hoham, 1987; Bidigare et al., 1993; Hoham et al., 1993). Some of these protozoa and rotifers from more temperate latitudes were reported to select and digest green cells over the more brightly colored orange and red cells (Pollock, 1970). It was suggested that the latter may be spores that were hard to digest. Ciliated protozoa from Antarctic snow, however, consumed both red- and green-colored algal cells (Bidigare et al., 1993). A population of ciliated protozoa was reported from Stowe ski area, Vermont, where there were large numbers of the snow alga *Chloromonas* and the snow fungus *Chionaster* (Hoham et al., 1993). However, there were no reports of consumption of the snow fungus by the protozoa as there were for the snow alga. Some protists called chytrids parasitize snow and ice algae (*Chlamydomonas* spp. and *Ancylonema*) (Kol, 1968a).

Larger Oligochaete invertebrates known as ice worms are common to certain Alaskan glaciers (Moore, 1899) and on snow fields at Mt. Rainier, Washington (Welch, 1916; Wailes, 1935). Welch (1916) reported two species of the ice worm *Mesenchytraeus* from Mt. Rainier and that certain snow algae were probably their chief sources of food. Kohshima (1984a, 1984b, 1987a) reported cold-tolerant insects and copepods that lived on a Himalayan glacier by feeding on the algae and bacteria. Aitchison (1983, 1989) reported that subnivean, winter-active collembola in Manitoba prefer feeding on species of fungi from the family Dematiaceae (see Aitchison, Chapter 5). During winter, the snow cover ranged from 30 to 60 cm and the subnivean temperature varied from $-3°C$ to $-6°C$. Even though the collembola were more active at $+5°C$, they continued to feed at temperatures of $-2°C$ and $-5°C$ (Aitchison, 1989). These fungi are subnivean and therefore unlike the fungal species that are found in and on snow at the time of snowmelt. In addition, Antor (1995) discussed the importance of arthropod fallout on alpine snowpatches for the foraging of alpine birds in the Spanish Pyrenees.

Hoham et al. (1993) discussed food webs in snow involving bacteria, algae, fungi, protozoa, rotifers, tardigrades, insects, arachnids, subnivean mammals, and higher plants. Trophic categories included autotrophs, heterotrophs, decomposers, detritivores, and contaminants.

4.4 Cell Structure and Cell Physiology

4.4.1 Cell Structure

Snow- and ice-microbe cell structure was reviewed by Kol (1968a) and Hoham (1980). Having flagella, for example, allows cells to move in the meltwater that surrounds the snow crystals at the time of snowmelt (see Pomeroy and Brun, Chapter 2). The dominant snow algae are flagellates that belong to the order Volvocales of the green algae including *Chlamydomonas* and *Chloromonas* (Kol, 1968a; Hoham, 1980) and occasionally *Smithsonimonas* and *Chlainomonas* (Kol, 1968a; Hoham, 1974a, 1974b). Even though *Carteria* with four flagella has been identified from snow (Fukushima, 1963; Stein and Amundsen, 1967; Kol, 1968a), it is probable that all *Carteria* from snow are planozygotes (resting phases) of other biflagellate green algae (Hoham et al., 1983). Other flagellates in snow include the golden algae (Chrysophyceae) *Chromulina* and *Ochromonas* (Fukushima, 1963; Stein, 1963; Hardy and Curl, 1972; Hoham and Blinn, 1979; Hoham et al., 1993), euglenoids *Euglena* and *Notosolenus* (Kiener, 1944; Hoham and Blinn, 1979; Hoham et al., 1993), cryptomonad *Cryptomonas* (Javornický and Hindák, 1970), and dinoflagellates *Gyrodinium* and *Gymnodinium* (Kol, 1968a; Gerrath and Nicholls, 1974). However, other snow microbes that are not flagellates (algae, fungi, bacteria) are passively moved in the snow meltwater. These diverse microbes have different nutritional requirements that affect the dynamics of snow chemistry in the snowpacks (see Tranter and Jones, Chapter 3).

Cell wall structures reduce desiccation, penetration of ultraviolet (UV) light, and grazing by protists and animals, and they adhere cells to substrates. Although the cell wall of red-colored cells of *Chlamydomonas nivalis* includes both protein and cellulose components (Tazaki et al., 1994b), the chemical nature of coverings surrounding other snow algal cells has not been identified. However, the wall composition may be similar to that described for the different algal groups to which snow algae belong

(Bold and Wynne, 1985; Sze, 1993). The cell wall ultrastructure of some Volvocalean snow algae has been examined by scanning electron microscopy (SEM). *Chlainomonas* has two external envelopes in *Chlainomonas kolii* (Hoham, 1974b, 1980), and the one in *Chlainomonas rubra* (Stein and Brooke, 1964; Hoham, 1974a, 1980), which may be proteinaceous (Hoham, 1980), has not been examined by SEM. The SEM of the zygote wall in species of *Chloromonas* shows an outer primary wall composed of flanges or ridges (*Chloromonas nivalis*) (Hoham and Mullet, 1977, 1978) (*Chloromonas polyptera*) (Hoham et al., 1983) or short protuberances or spines (*Chloromonas brevispina*) (Hoham et al., 1979) and a secondary inner wall that is smooth in these species (Hoham, 1980). Similarly, the zygote wall of *Chlamydomonas nivalis* has both a primary wall of truncate extensions and a smooth inner wall (Kol, 1968a; Kawecka and Drake, 1978). The combination of a secondary and primary wall may reduce desiccation and UV penetrance. In a separate study of *Chlamydomonas nivalis* (probably asexual resting spores), SEM shows an outer smooth wall surface associated with bacteria (Weiss, 1983) and a possible symbiotic association between these microbes was suggested. Transmission electron microscopy (TEM) reveals that this wall has three layers: a fibrous inner layer, a dense central core, and a smooth compact outer layer (probably asexual resting spores). These cells were surrounded by a loose fibrous network in which encapsulated Gram-negative bacteria and unidentified surface debris were seen, thus showing a close physical association between the alga, bacteria, and debris. Using single cell dielectric spectroscopy, the cell wall of *Chlamydomonas nivalis* resting spores was found to have an extremely low permittivity of 3 to 5 (Müller, Schnelle, and Fuhr, 1998b). In *Chloromonas* and *Chlamydomonas*, the flanges, spines, and truncate processes may reduce grazing and help anchor these cells to substrates after snowmelt. SEM also has helped to resolve difficult taxonomic questions associated with these species.

TEM reveals information about internal cell structure that may give insight on how cells are more suited for the environment where they live, and it has been used to examine two species of snow algae, *Chlamydomonas nivalis* and a *Chloromonas* sp. Weiss (1983) found abundant clear granules in the cytoplasm and several thylakoid membrane stacks and occasional starch grains in the chloroplast of *Chlamydomonas nivalis*. Marchant (1982) reported massive accumulations of carotenoid vesicles in the cytoplasm of *Chlamydomonas nivalis* from Australian snow samples visible by TEM, and Weiss (1983) suggested that these pigment vacuoles were the same as the granules (which contained lipids) found in his samples of *Chlamydomonas nivalis* from Wyoming and California. Lipids are more common in mature zygotes and zygospores that are

178

usually orange or red or in vegetative cells that have been exposed to subfreezing conditions (Hoham, 1975a, 1975b). Large accumulations of red-colored astaxanthin esters that occur in extrachloroplastic lipid globules in *Chlamydomonas nivalis* reduce photoinhibition and photodamage (Bidigare et al., 1993). Andreis and Rodondi (1979) used TEM to infer that the red stage of *Chlamydomonas nivalis* was in a condition of high metabolic activity because of the large number of mitochondria present. They also found intracytoplasmic globules of pigments that joined together and flowed in the vacuoles after local tonoplast dissolution and active cell wall synthesis in asexual reproductive stages. TEM sections of *Chlamydomonas nivalis* from Svalbard revealed ribosomal-rich cytoplasm in green biflagellate zoospores and a domination of oil droplets and starch grains in red asexual cysts (Müller et al., 1998a). Fjerdingstad et al. (1974) examined TEM sections prepared from red snow samples but published micrographs of bacteria labelled as *Chlamydomonas nivalis*.

Food reserves stored in snow algal cells are potential sources of nutrition for grazers. The common food reserves in the snow and ice algae that belong to the Chlorophyta or green algae are carbohydrates and lipids (Kol, 1968a; Hoham, 1980). Starch is found in rapidly dividing cells or in newly formed zygotes that are usually green (Hoham, 1980). Stein and Bisalputra (1969) reported crystalline bodies in the chloroplast of a *Chloromonas* (published as *Chlamydomonas* sp.). These inclusions were hexagonal, 0.2–0.9 μm in diameter, and it was suggested that they were proteinaceous. A structural relationship between these inclusions and the algal chloroplast was not apparent. It was also suggested that these crystalline bodies were involved in storage of photosynthates (carbohydrates). Food reserves in algae other than green algae are typical for their respective groups (Bold and Wynne, 1985; Sze, 1993), but little is known about developmental stages or growth conditions in snow algae associated with these reserves.

Pigments in the snow algae have not been studied critically for most species. However, the pigments in the green alga *Chlamydomonas nivalis* have been identified (Viala, 1966; Czygan, 1970; Bidigare et al., 1993). Viala (1966) first reported that the red pigment was the xanthophyll astaxanthin, and Czygan (1970) indicated that the red pigments were ketocarotenoids and their synthesis paralleled chlorophyll decomposition. (For more information about pigment shifts in snow algae, see the next section on cell physiology and special adaptations.) In the laboratory, Czygan reported a disappearance of red carotenoids that coincided with increasing concentrations of carotenes, other xanthophylls, and chlorophylls a and b, when cells were placed in a nutrient solution for 4 weeks with a light regime of 6,000 lux. However, Mosser,

179

Mosser, and Brock (1977) recorded a shortwave radiation flux of 86,000 lux in the field for *Chlamydomonas nivalis*, or fifteen times greater than that used by Czygan in the laboratory. Thus, it was not clear if the pigment shift observed by Czygan in the laboratory was due to added nutrients (nitrogen) or to reduced light intensity. Hoham (unpublished data) observed similar pigment shifts in the laboratory with *Chlamydomonas nivalis* cultures where nutrients were not added to the original snow meltwater, but cells were maintained in low level light regimes. In those cultures, however, bacterial populations may have been responsible for the nitrogen levels that caused the pigment shift from carotenoids to chlorophylls in the algae.

4.4.2 Cell Physiology and Special Adaptations

For Antarctic snow algae collected from open exposures, it was hypothesized that species of *Chlamydomonas* remain green when growth is rapid (abundant nutrients and high level of light) and cells maintain a rapid turnover of the Q_b protein that reduces the cells' susceptibility to photoinhibition from damaging wavelengths of light (Bidigare et al., 1993). In the same study, at low growth rates (few nutrients and high level of light), cells synthesize and accumulate red astaxanthin esters to protect against photoinhibition of photosynthesis. Astaxanthin accumulation was linked directly to a decline in nitrogen (Czygan, 1970; Bidigare et al., 1993), and a similar correlation was found for the green algal flagellate *Haematococcus lacustris* (Lee and Soh, 1991). However, depleted P, S, K, or Fe can mimic the effects of N depletion (Czygan, 1968). Ratios of chlorophyll a : astaxanthin esters are highest in green vegetative cells and lowest in red cysts (asexual spores and zygotes) in species of *Chlamydomonas* found in snow in polar regions (Bidigare et al., 1993; Müller et al., 1998a). Red astaxanthin pigments probably accumulate in other genera of green snow algal flagellates – i.e., *Chlainomonas*, *Chloromonas*, and *Smithsonimonas* – through a similar correlation of rapid growth and low nitrogen supply, but this needs verification. However, not all snow algae respond this way. *Chloromonas* collected from snow in Québec and Whiteface Mountain, New York, reduce NH_4^+-N and NO_3^--N nitrogen levels to near zero but display no visual evidence of secondary carotenoid synthesis and remain green (Hoham et al., 1989). These samples, however, were collected from beneath forest canopies where maximum irradiances were not achieved or were minimal (thus few nutrients and low level of light). Perhaps the forest canopy screens out the damaging UV light in this ecosystem, and thus these algal cells do not need Q_b proteins or red astaxanthin esters for protection against photoinhibition.

Orange to yellow–orange colored carotenoids develop in the zygotes of *Chloromonas nivalis* (Hoham and Mullet, 1977, 1978), *Chloromonas brevispina* (Hoham et al., 1979), and *Chloromonas polyptera* (Hoham et al., 1983). In these species, the brightly colored pigments are located in large lipid bodies of which there is usually one at each pole of the zygote. These orange to yellow-orange zygotes are located in the upper snow layers where the irradiation levels are highest, and zygotes of these three species collected from lower snow layers are usually green. Another snow alga, *Cryocystis granulosa* cell type (a zygote in the life cycle of an unnamed species of *Chloromonas*) (Hoham, unpublished data), accumulates secondary carotenoids progressively from yellow-green to orange to a sometimes red-orange color in small lipid vesicles dispersed throughout the cell (Hoham and Blinn, 1979). This secondary carotenoid development appears to be more related to aging of cells than to irradiation levels (Hoham, 1980). However, it is not known whether the pigment change in this species is correlated with nutrient (nitrogen) depletion as reported for *Chlamydomonas nivalis* by Czygan (1970) and Bidigare et al. (1993).

The fatty acid (FA) composition of *Chlamydomonas* spp. was determined from Antarctic snow samples collected at Hermit Island (Bidigare et al., 1993) (Table 4.2). Green cells contained more saturated FA (mostly 16:0 and 18:0) and fewer mono-unsaturated FA, and red cells contained fewer saturated FA and more monounsaturated FA (mostly 16:1 and 18:1). Red cells also contained fewer saturated FA and more monounsaturated FA in their astaxanthin esters, and these fractions were dominated by palmitic acid (16:0) and oleic acid (18:1). In the same study, 91 percent of the astaxanthin present in Antarctic red vegetative cells were diesters, whereas 91 percent of the asta-xanthin present in *Chlamydomonas nivalis* red cysts from Wyoming were monoesters. In the red snow samples collected from Hermit Island, approximately 5 percent of the total pool of FAs was associated with the astaxanthin esters. In a separate study with red-colored *Chlamydomonas nivalis* collected near Resolute, Cornwallis Island, Canada, Tazaki et al. (1994a, 1994b) found palmitic, stearic, oleic, and behenic acids in these cells. High concentrations of n-alkanes with n-C_{24} was characteristic of their red-colored snow algae, suggesting the presence of hydrocarbons that could be derived from the Arctic cold desert and/or organic debris from wind-transported bacteria.

The shift in FA composition of red-pigmented snow algae may serve as a cry-oprotective function (Bidigare et al., 1993). Samples of red and green *Chlamydomonas* were transported from Antarctica to Texas A&M University in 1989 at $-70°C$ (the 1990 samples were transported at $0°C$); after thawing, the red cells were intact but the green cells had lysed from the freezing process at $-70°C$. The green cells were primarily vegetative; however, the red cells were in transition between vegetative cells and cyst formation. Survival of the red cells may be explained partly by the predomi-nance of unsaturated FAs, which increases membrane fluidity at this low temperature

181

Table 4.2. *Fatty acid composition (relative %) and content of whole cells and the astaxanthin esters of* Chlamydomonas *spp. sampled from Hermit Island, Antarctica.*

Fatty acid (FA)*	Total fatty acids[†]		Astaxanthin esters[†]
	Green cells	Red cells	Red cells
14:0	2.3	3.9	4.1
15:0i	0.4	ND*	ND
15:0a	0.8	ND	ND
15:0	1.3	ND	TR[†]
$16:3\Delta^{7,9,12}$	ND	2.0	ND
$16:2\Delta^{9,12}$	1.8	1.5	ND
$16:1\Delta^{7}$	0.4	0.4	ND
$16:1\Delta^{9}$	4.0	17.9	1.8
16:0	26.5	0.5	26.8
17:0	0.8	0.3	0.3
$18:3\Delta^{9,12,15}$	ND	1.4	ND
$18:2\Delta^{7,10}$	ND	ND	1.6
$18:2\Delta^{9,12}$	2.3	2.4	0.7
$18:1\Delta^{9}$	11.1	59.2	51.5
$18:1\Delta^{11}$	3.1	2.0	ND
18:0	34.5	0.2	5.1
$20:5\Delta^{5,8,11,14,17}$	0.5	0.9	0.7
20:0	3.0	0.7	0.8
$20:6\Delta^{4,7,10,13,16,19}$	0.6	0.3	4.3
$22:2\Delta^{10,13}$	ND	ND	0.7
$22:1\Delta^{13}$	2.0	0.5	0.7
22:0	1.6	5.4	0.4
$24:1\Delta^{15}$	1.6	ND	ND
24:0	1.3	0.4	ND
Total %	100.0	100.0	100.0
% saturated FA	72.5	11.4	37.5
% monounsaturated FA	22.2	80.0	54.0
Total FA (μg mg^{-1} dry weight)	47.9	45.4	2.5

*Superscript numbers stand for references.
[†]ND; not detectable, TR; trace levels.
Note: From Bidigare et al. (1993).

182

(Roessler, 1990). In their review of bacteria, Margesin and Schinner (1994) stated that psychrotrophs synthesize neutral lipids and phospholipids containing more unsaturated FAs when grown at low temperatures to a greater extent than do mesophiles. An increase of unsaturated FAs leads to a decrease in the lipid melting point and maintains the lipid in a liquid and mobile state, which is necessary for survival. It was proposed that the growth temperature range of an organism depends on the ability of the organism to regulate its lipid fluidity. The lower growth temperature limits on cold-adapted organisms is fixed by the freezing properties of dilute aqueous solutions inside and outside the cell and not by chemical properties of cellular macromolecules.

The snow alga *Chlamydomonas nivalis* was more tolerant to stress (freezing injury, shrinkage, and rehydration) than other more temperate species of *Chlamydomonas* (i.e., *Chlamydomonas reinhardii*, *Chlamydomonas moewusii*, and *Chlamydomonas eugametos*) (Morris, Coulson, and Clarke, 1979). The mechanism suggested for this resistance was that more unsaturated FAs in *Chlamydomonas nivalis* allows for greater membrane fluidity. They found that the average number of double bonds per fatty acid was higher in *Chlamydomonas nivalis* than in *Chlamydomonas reinhardii*. This was mainly due to the differences in the 16:3 and 18:4 content, which made up more of the total FAs in *Chlamydomonas nivalis* than in *Chlamydomonas reinhardii*. Viable cells of *Chlamydomonas nivalis* were recovered from $-196°C$ following cooling rates between $0.3°C$ and $8°C$ min^{-1} with optimal survival at a cooling rate of $2.4°C$ min^{-1}. *Chlamydomonas nivalis* was also more resistant to the stresses of shrinkage and rehydration; the median lethal concentration of NaCl was 1.2 M for *Chlamydomonas nivalis* and 0.4 M for *Chlamydomonas reinhardii*. A higher unsaturated to saturated FA ratio was reported in *Chlamydomonas nivalis* than in *Chlamydomonas reinhardii* (Morris et al., 1981). A reduction of viable cells in the mesophilic species *Chlamydomonas reinhardii* was found by cooling to $-2.5°C$ and below, and this was correlated with changes in structure and function of the cell membrane (Grout, Morris, and Clarke, 1980). A significant difference in ^{86}Rb uptake was indicated for *Chlamydomonas nivalis* versus *Chlamydomonas reinhardii* in media undercooled to $-5°C$ (Clarke, Leeson, and Morris, 1986). It was suggested that this uptake difference, which was greater in *Chlamydomonas nivalis*, may relate to the more highly unsaturated FAs in the cell membrane of this snow alga and that an evolutionary adjustment of membranes in *Chlamydomonas nivalis* of living at near $0°C$ involved an increase in the unsaturation of membrane phospholipids. (The taxon, *Chlamydomonas reinhardtii* is the correct spelling, but we have followed the spellings from individual publications that use *C. reinhardtii*, *C. reinhardii*, and *C. reinhardi*.)

183

At culture temperatures of 10°C and 20°C, the Antarctic green alga, *Chlorella vulgaris* strain SO-26, was compared with the mesophilic *Chlorella sorokiniana* strain C-133 for FA composition (Nagashima et al., 1995). Even though these microbes were not reported from snow or ice, they are representative when cold-tolerant and temperate strains are being compared. When culture temperatures were changed from 20°C to 10°C, the increase of the ratio of unsaturated to total FAs was considerably greater in strain SO-26 than in strain C-133. The results indicated that, in photosynthesis, the properties of the Antarctic *Chlorella* SO-26 were more psychrophilic than those of the mesophilic *Chlorella* C-133; both strains could be acclimated by culture temperature, at least partly because of FA unsaturation. In both strains, the major fatty acids were palmitic (16:0), linoleic (18:2), and linolenic (18:3).

Green cells of *Chlamydomonas nivalis* from the Canadian Arctic rich in Ca were involved in active photosynthesis and red cells low in Ca were in a resting stage (Tazaki et al., 1994b). Protamine, stearic acid, and decanoic acid were found in Ca-rich green cells and carminic acid and nopalcol BR-13 were found in Ca-poor red cells.

The properties of cold-adapted bacteria and fungi were reviewed by Margesin and Schinner (1994). Even though snow and ice are just two of the several cold-adapted habitats in their review, the information is included here because of its applicability to snow and ice habitats. They reported that proteins from cold-adapted species are not prone to cold denaturation and their enzymes have higher catalytic efficiencies than those from warm-adapted species. These efficiencies are associated with formation of loose, more flexible structures that allow catalytic conformational changes to occur with less energy input. Activation free energies of metabolic reactions are proportional to adaptation temperature; thus, these values are lower in cold-adapted homologs of an enzyme than in warm-adapted homologs. Enzymes from cold-adapted microorganisms have a shift of the optimum activity toward lower temperatures, which reflects adaptation to their natural habitat. Greatest enzyme formation occurs at temperatures much lower than the optimum temperature for growth; this may compensate for the slow rate of enzymatic activity and probably ensures a high utilization of substrate in the cold environment (Devos et al., 1998). Optimal enzyme activity in the psychrophilic alga *Chloromonas* ANT1 from Antarctica was compared with the mesophilic *Chlamydomonas reinhardtii*; optimal activity was 20°C lower in ANT1 for the two enzymes studied (Loppes et al., 1996).

Margesin and Schinner (1994) also discussed a rapid protein turnover in psychrotrophic bacteria that could be an energy-saving mechanism that provides amino acids during synthesis of new proteins for adaptation, especially when these organisms live in nutrient-poor environments. Protein synthesis at low temperatures might be the result of specialized proteins that allow for reinitiation of protein synthesis. The ability of an organism to adapt to low temperatures depends not only on the ability to synthesize protein but also may involve regulation and interaction of a number of cellular components and processes. Five psychrophilic bacteria collected near the Antarctic station Dumont d'Urville (the habitat was not given) secreted exoenzymes maximally in culture at temperatures close to that of their environment ($-2°C$ to $4°C$) (Feller et al., 1994). Higher temperatures ($17°C$ to $25°C$) induced faster growth rates but reduced cell development and enzyme secretion. The suggested optimal activity temperature for these enzymes was $30°C$ to $40°C$, about $20°C$ lower than that of mesophiles.

In the same review, Margesin and Schinner (1994) indicated that synthesis of polysaccharides is greater in cold-adapted microorganisms than in mesophiles, and some psychrotrophs show a different attack on metabolizing substrates at different temperatures. These same organisms display greater and more efficient transport of solutes across cell membranes than do mesophiles. Membrane transport at low temperature is facilitated by the high lipid content in the membranes of psychrophiles and psychrotrophs. The increased lipid content could reflect an increased cell size in some psychrophiles at lower temperatures and be part of a mechanism to increase the cell membrane surface area and the ability of the cell to take up nutrients at $0°C$.

Using gas liquid chromatography, Roser et al. (1992) found high levels of carbohydrates in the snow alga *Mesotaenium berggrenii* from Antarctica. Sucrose and glucose made up 61 percent of the total carbohydrate, glycerol accounted for 11 percent, and 26 percent was unknown. The other snow algae in their study showed much lower levels of intracellular carbohydrates. These algae, including species of *Chloromonas*, *Chlorosarcina*, and *Chlamydomonas*, caused algal blooms in low nutrient areas away from penguin rookeries. The dominance of inositol in one sample of *Chlorosarcina* was not considered major because of the very low sugar/polyol levels detected on a chlorophyll a basis. The lack of detectable polyols or sugars in most of these snow algae raised questions about how these cells tolerated freezing during the winter months. Using ^{13}C nuclear magnetic resonance analysis, Chapman, Roser, and Seppelt (1994) confirmed the earlier findings of Roser et al. (1992) from Antarctica that *Mesotaenium berggrenii* contained sucrose and glucose, and no significant quantities of sugars, polyols, or amino acids were found in extracts of a *Chloromonas* that caused red snow. In an earlier study, Tearle (1987) detected high levels of polyols in two species of Antarctic snow algae but did not indicate which compounds were detected.

4.5   Life Cycles, Laboratory Mating Experiments, and Cultures

4.5.1   Life Cycles and Speciation

Life cycles of snow algae have emphasized green algal flagellates (Hoham, 1980), and the phases of these life cycles correlate with physical and chemical factors at the time of snowmelt (Hoham, 1980; Hoham et al., 1989; Jones, 1991; Hoham and Ling, 2000) (see Chapters 2 and 3). Active metabolic phases occur in spring or summer when the snow melts, nutrients and gases are available, and light penetrates through the snowpack (Hoham, 1980). The process begins with germination of resting spores at the snow–soil interface (old snow–new snow interface in persistent snow fields) producing biflagellate zoospores (Hoham, 1980). These cells swim in the liquid meltwater surrounding the snow crystals toward the upper part of the snowpack, and their position in the snowpack is determined by irradiance levels and spectral composition. Visible blooms of snow algae occur a few days after germination. Both asexual and sexual biflagellates develop in some species. The sexual cells (gametes) fuse to form resting zygotes; in other species, asexual resting spores develop directly from asexual biflagellate vegetative cells (Hoham, 1980; Hoham et al., 1993; Ling, 1996). The resting spores eventually adhere to the soil or debris over the soil when the snowpack has melted or remain on old snow in persistent snow fields. From year to year, populations of snow algae stay in approximately the same localities. The resting spores remain dormant during summer; may form daughter cells through cell division after the first freezes in autumn; are covered with new snow in fall, winter, and spring; and do not germinate again until the factors repeat themselves (Hoham, 1980). The entire life cycle process is illustrated in Figure 4.31. It is not known how long resting spores remain viable, but some retained in their original meltwater have remained viable for more than 25 years (R. Hoham, personal observation).

Since snow algal flagellates typically are restricted to the snow environment and are not found in lakes or other bodies of water, the species selected for in snow are those that require minimal nutrients. It is important that the snow alga produces some type of resistant spore before complete snowmelt, and nutrient depletion in snow appears to correlate with the transition from vegetative phase to resistant spore (Czygan, 1970; Hoham et al., 1989; Jones, 1991; Hoham and Ling, 2000) (Table 4.3). The selection

186

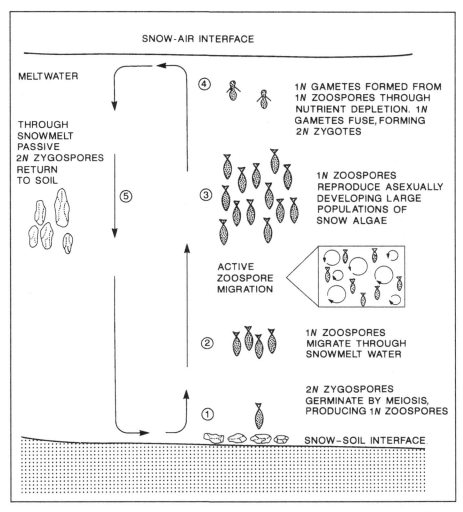

Figure 4.31. Life cycle of snow algal flagellate (*Chloromonas*) with sexual life history (modified from Gamache, 1990; Jones, 1991).

for phototactic species allows populations to migrate for optimal photosynthesis and to locate new nutrient sources.

Extensive studies of snow algal life histories have been done with the green algae (order Volvocales), and these life histories are complex (Fukushima, 1963; Hoham, 1975b; Hoham and Mullet, 1977, 1978; Kawecka and Drake, 1978; Hoham et al., 1979; Hoham, 1980; Kawecka, 1981; Hoham et al., 1983; Kawecka, 1983/1984; Ling, 1996), and one of these life histories is illustrated in Figure 4.32. A number of snow algae that were assigned to the order Chlorococcales and other volvocalean algae are zygotic

Table 4.3. *Snow algal populations (cells ml⁻¹) of* Chloromonas
(Chlorophyta) from Lac Laflamme, Laurentian Mountains, Québec.

9 May 1988 samples*	Vegetative cells	Resting spores	Total
1 Algae	73,300	0	73,300
1 Control	4,200	0	4,200
2 Algae	123,600	0	123,600
2 Control	2,200	0	2,200
17 May 1988 samples†	Vegetative cells	Resting spores	Total
1 Algae	107,800	800	108,600
2 Algae	193,900	20,000	213,900
3 Algae	39,300	26,500	65,800

*Control samples without visible algae were taken adjacent to samples with
  visible algae.
†Note shift from vegetative cells to resting spores when compared with
  9 May samples; controls were not tabulated for 17 May.
Note: From Hoham et al. (1989).

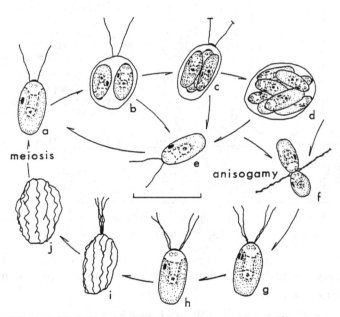

Figure 4.32. Life cycle of *Chloromonas polyptera* (bar = 18 μm).
(a) Vegetative cell; (b–d) cleavage of vegetative cells; (e) zoospore;
(f) anisogamy; (g) planozygotes (formerly *Carteria nivale*) with two
eyespots; (h) planozygote with one eyespot; (i) zygote (formerly
Scotiella polyptera) with four flagella; (j) zygote without flagella
(from Hoham et al., 1983).

188

stages of the snow alga *Chloromonas*. In this list are *Scotiella* (Hoham, 1975b; Hoham and Mullet, 1977, 1978; probably Kawecka, 1983/1984), *Cryocystis*, *Cryodactylon*, *Oocystis*, and *Trochiscia* (Hoham et al., 1979; Ling, 1996), and *Carteria* (Hoham et al., 1983). In one study (Hoham et al., 1979), the zygotes of *Chloromonas brevispina* were recognized as seven previously described taxa assigned to the Chlorococcales. Ling (1996) reported that a species of *Trochiscia* was the zygote of the green algal flagellate *Desmotetra* in the family Chlorosarcinaceae. *Trochiscia rubra* and *T. cryophila* are stages of the zygotic phase of the snow alga *Chloromonas brevispina* (Hoham et al., 1979).

Zygotes of *Chloromonas* and the closely related *Chlamydomonas* have not been studied in most species of these genera (Huber-Pestalozzi, 1961; Ettl, 1970, 1976; Ettl and Schlösser, 1992). A number of life histories have been suggested for the snow alga *Chlamydomonas nivalis* (Fukushima, 1963; Kol, 1968a; Sutton, 1972; Hoham, 1980), with more specific studies on asexual aplanospore formation (Kawecka, 1981) and sexual reproduction (Kawecka, 1978; Kawecka and Drake, 1978). Bridge formations form between gametes in *Chlamydomonas nivalis* (Kawecka and Drake, 1978) and probably in *Chlainomonas rubra* (Hoham, 1974a). Abnormal sexual mating configurations occur in *Chlamydomonas nivalis* (Kawecka and Drake, 1978), *Chloromonas rostafiński* (Kawecka, 1983/1984), and other species of *Chloromonas* (Hoham, 1992; Hoham et al., 1993, 1997, 1998).

Asexual reproduction would be the fastest route to produce resting spores because more time is needed to complete the sexual phase of the snow algal life cycle (Hoham, 1992). *Chloromonas* sp.-A (Hoham et al., 1993) produces asexual resting spores that resemble *Scotiella cryophila* in field material. Ling (1996) reported that *Scotiella polyptera* from Windmill Islands, Antarctica, appears to be asexually produced from a *Chloromonas* and differs from *Scotiella polyptera* reported in North America, which are the sexual zygotes of *Chloromonas polyptera*. Thus, in localities where the snowpack is inconsistent from year to year, asexual reproduction may be favored by natural selection. However, some of these populations of snow algae that lack in genetic diversity and that live in an inconsistent habitat may be headed for evolutionary extinction.

Species definition within *Chlamydomonas* and other related genera in the order Volvocales (this includes the snow algae *Chloromonas* and *Chlainomonas*) has been difficult. Ettl and Schlösser (1992) stated that is not possible to recognize species of *Chlamydomonas* as populations of interfertile individuals because reproduction is unknown in more than 80 percent of the described species. Relationships within these genera were suggested on the basis of cell wall structure and/or the specificity of cell

189

wall autolysins. Nucleic acid base sequence data may permit more reliable lineages at the species and generic levels. Using restriction fragment length polymorphism analysis of the region of DNA that codes for the 18S ribosomal RNA subunit (rDNA) and the adjacent internal transcribed spacer (ITS) amplified by polymerase chain reaction, Hoopes (1995, personal communication), however, indicated that three strains of the snow alga *Chloromonas* sp.-A (Hoham et al., 1993) from the same locality (White Mountains, Arizona) have dissimilar nuclear rDNA sequences, suggesting that genetic differences may have evolved in the same species from the same location. This is in contrast to the findings of Judge, Scholin, and Anderson (1993) where strains of an algal dinoflagellate species from a single locality had similar rDNA sequences, suggesting that these sequences could be used to delineate regional populations for morphologically similar species.

Using nuclear 18S rDNA gene sequence data, it was suggested that cold-tolerant species of *Chlamydomonas* and *Chloromonas* have a common ancestry because they align into a single group or clade (Buchheim, Buchheim, and Chapman, 1997). However, when additional species of *Chloromonas* from snow were added to the study (*Chloromonas brevispina*, *Chloromonas pichinchae* and *Chloromonas* sp.-D), they formed a second clade implying that cold tolerant species of *Chloromonas* may have evolved or originated in snow habitats at least twice (Bonome, Leebens-Mack, and Hoham, 2000). These three species of *Chloromonas* have unique cell divisions where parental cells retain their flagella when they form cell packs in *Chloromonas brevispina* (Hoham et al., 1979) and *Chloromonas pichinchae* (Hoham, 1975b) or show cell transformations from oblong to spherical cells in *Chloromonas* sp.-D (Marcarelli et al., 2000). The alignment of the cold tolerant taxa presented by Buchheim et al. (1997) is also supported by chloroplast rbcL gene sequence data (Morita et al., 1999).

4.5.2 Laboratory Mating Experiments

Using modifications of Hoshaw (1961), laboratory mating experiments revealed that strains 593A and 593C were normal + and − mating strains from Québec that produced normal mating pairs (Figure 4.33 [see plate section]) and zygospores (Hoham et al., 1993). These were the first normal mating strains of *Chloromonas* isolated from snow in North America, and zygospores with markings similar to those of *Chloromonas brevispina* developed in the laboratory in these strains (Hoham et al., 1979; Hoham and Clive, unpublished data). Additional strains of *Chloromonas brevispina* from Québec and Vermont have been crossed with strains 593A and 593C, some of which are self-mating and others appear to have no mating potential (Yoder, Knapp,

190

and Martin, 1995, personal communication). Similar variation in genetic compatibility within a species complex was also reported for the green algal desmid *Micrasterias thomasiana* (Blackburn and Tyler, 1987). In the normal sexual mating *Chloromonas* sp.-D from the Tughill Plateau, New York, three types of mating pairs were observed (oblong–oblong, oblong–sphere, and sphere–sphere). The oblong-oblong mating pairs diminished through the 8-hour time duration of the experiments and the spherical-type mating pairs peaked 6–7 hours after the experiments began (Hoham et al., 1997, 1998; Hoham and Ling, 2000). Another *Chloromonas* from widely separate geographical areas (Adirondack Mountains, New York, White Mountains, Arizona and Jay Peak, Vermont) is asexual and belongs to *Chloromonas* sp.-A (Hoham et al., 1993) and not to *Chloromonas polyptera*, as suggested earlier by Hoham (1992). Even though *Chloromonas* sp.-A produces rare abnormal mating configurations, resting zygotes did not develop. The abnormal mating pairs (Figure 4.34 [see plate section]) observed in the laboratory for *Chloromonas* sp.-A were similar to those observed in the field for *Chlamydomonas nivalis* (Kawecka and Drake, 1978) and *Chloromonas rostafiński* (Kawecka, 1983/1984). In very rare cases, triple fusions were observed (Hoham, 1992), and these were mostly between geographically isolated strains of *Chloromonas* sp.-A. Similar triple fusions were noted in *Chlamydomonas nivalis* from red snow in Poland (Kawecka and Drake, 1978) and in *Chloromonas* sp.-D from the Tughill Plateau, New York (Hoham et al., 1997, 1998). Populations of snow algae that reproduce asexually instead of sexually result in later generations that lack in genetic diversity. Abnormal mating pairs would promote asexual reproduction if nonviable zygotes are the end product. In another volvocalean green alga, *Gonium viridistellatum*, crosses between geographically isolated strains resulted in deterioration of 90 percent of the zygotes (Nozaki, 1989). In a fusion-arrested strain, zym-26-3 of *Chlamydomonas monoica* obtained after UV irradiation, genetic analysis revealed two unlinked mutations, one (cf-1) responsible for the failure to complete cell fusion and the other (ger-8) interfering with late stages of spore maturation and germination (VanWinkle-Swift, Aliaga, and Pommerville, 1987).

4.5.3 Culture Collections and Growth Media

Hoham (1989a, 1992) and Hoham et al. (1993) indicated that a culture collection of about 100 strains of snow algae was maintained at Colgate University, Hamilton, New York. This collection now includes about 200 strains representing approximately three dozen species, many of which are identified and bacteria free. However, there are several other strains not identified with certainty or not bacteria free. These algae are grown on M-1 medium, which was developed for snow algae

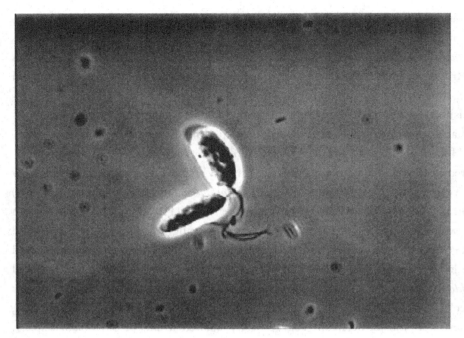

Figure 4.33

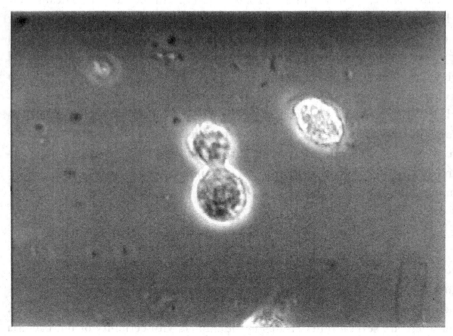

Figure 4.34

(Hoham et al., 1979), and this medium, which has been slightly modified (Hoham, unpublished data), has been most successful for growing a variety of snow algae.

Strains of algae isolated from snow are available from the UTEX collection, Texas (Starr and Zeikus, 1993); the SAG collection, Germany (Schlösser, 1994); and other worldwide collections (Miyachi et al., 1989). Sugawara et al. (1993) published a world directory of cultures of bacteria, fungi, and yeasts that includes snow microorganisms.

Historically, different media have been used to grow snow algae. These include Chu No 10, Benecke's and Bristol's in Japan (Fukushima, 1963), BMT and Tris in Canada (Stein, 1964; Stein and Brooke, 1964), mineral medium supplemented with vitamins in Colorado (Stein and Amundsen, 1967), Holm-Hansen's and Pringsheim's in Oregon (Sutton, 1972), PFW in Washington (modified Provasoli's [1966] enriched sea water medium) (Hoham, 1973), M-1 in New York (Hoham et al., 1979), and Bold's Basal with and without soil extract in Antarctica (Ling and Seppelt, 1993). Handfield et al. (1992) used brain heart infusion (BHI), trypticase soy broth (TSB), plate count agar (PCA), and sea water agar (SWA) media for growing snow and ice bacteria in Canada.

4.6   Evolution and Origins

4.6.1   Snow and Ice Microorganisms

Hypotheses of microbial evolution indicate that thermophiles were the first to evolve (Brock, 1967), followed by mesophiles and psychrophiles. The evolution of psychrophiles and psychrotrophs is probably due to many genetic events (Morita, 1975) because they are distinguished by more than one parameter from mesophiles. There are no reports that indicate how the code or sequence of codons for mesophilic enzymes differs from the code for cold-adapted enzyme homologs (Margesin and Schinner, 1994). Elucidation of the three-dimensional structure of proteins from cold-adapted organisms and a comparison with closely homologous mesophilic proteins may contribute to a better understanding of protein folding and dynamics (Feller, Thiry, and Gerday, 1991).

There are no known fossils of snow algae, but fossils may be difficult to locate because the cells are not typically in lake beds or in rock layers (Hoham, 1980). It was

Figures 4.33 and 4.34. Normal and abnormal mating in *Chloromonas* sp.-D
(see Hoham et al., 1993). Figure 4.33: Normal mating pair of gametes showing two pairs of intertwined flagella. Figure 4.34: Abnormal mating pair of gametes without flagella showing a connecting bridge.

A colour version of these plates is available for download from
www.cambridge.org/9780521188890

suggested that present day populations of snow algae may be remnants of much larger populations that existed through the ice ages (Kol, 1968a). If this is the case, were they derived during recent ice ages during the Quarternary or do they go back to the Carboniferous or before? Snow algae probably were derived from species of soil or aquatic algae by natural selection when glaciers and semipermanent snow fields were more prevalent during the ice ages (Hoham, 1980).

Snow algae maintain populations from year to year in approximately the same microhabitats (Hoham, 1980). Hundreds to thousands of snow algal resting spores adhere to dried fungal strands of *Phacidium infestans*, and these strands could be broken up and distributed by wind or animals to new locations (Hoham, 1987). These potential inoculae would have better chances of producing populations of snow algae simply because of their numbers compared with individual resting spores that might be distributed by wind to unsuitable habitats.

Chloromonas sp.–B, found exclusively on artificial snow, may be distributed from one ski slope to another on skis (Hoham et al., 1993; Dybas, 1998). Resting spores of this species have a gelatinous matrix to which other microbes such as bacteria and fungi adhere. This same material would permit these cells to stick to skis and survive indoors at room temperature in storage until the next season, when they could inoculate a new ski slope. This suggestion is supported by the alga's occurrence at four disjunct ski slopes in three states in New England; it has yet to be found in natural alpine snow areas in New England or elsewhere. Possibilities for the origin of this species include the mutation of an existing species found in snow or a lotic species (from rivers or streams) that adapted to snow when artificial snow was processed from flowing water near ski resorts.

A stable habitat under extreme conditions was observed in perennially ice-covered Antarctic lakes (Wharton et al., 1989c). This ecosystem harbors a diverse assemblage of microbes including those in Antarctic lake ice and most of the life in the Ross Desert (Wharton et al., 1983, 1993). Cells of the desmid *Mesotaenium* were found in the surface slush of an Antarctic lake (Ling and Seppelt, 1990); however, these cells were washed in from the meltwater of surrounding snow. Barsdate and Alexander (1970) reported ice bubbles containing purple sulfur bacteria and algae. Benthic microbial communities were torn from an Antarctic lake bottom and transiently found in lake ice (Parker et al., 1982a) and may be later removed and dispersed by wind (Simmons, Vestal, and Wharton, 1993). In southern Victoria Land, Antarctica, the macroalga *Prasiola calophylla* was described living on wet ablating glacial ice walls (Vincent, Howard-Williams, and Broady, 1993). This alga remained unattached around small ice projections and probably is dispersed as wind-borne inoculae. After snowmelt, propagules of unicellular snow algae become air borne in Antarctica when the soil surface dries (Marshall and Chalmers, 1997).

Microscopic observations of shallow ice samples (0–1.2 m) collected from the Agassiz Ice Cap in the Canadian Arctic showed extremely low numbers of bacteria (Handfield et al., 1992). In this study, bacterial distribution was consistent with wind deposition rather than the presence of an autochthonous or self-maintained population. Bacterial concentrations differed with summer and winter layers of snow, similar to pollen deposition patterns. Superficial meltpools that form during mid-summer periods may support the growth of psychrophilic strains (Handfield et al., 1992).

4.7   Interrelationships of Physical Factors with Snow and Ice Microorganisms

4.7.1   Temperature

Are the microbes found living in snow and ice optimally adapted to living at temperatures near 0°C? Temperature studies done in the laboratory for snow microbes were reviewed by Hoham (1975a, 1980), and it was suggested that true snow algae have optimal growth at temperatures below 10°C. The snow alga *Chloromonas pichinchae* was designated as an obligate cryophile (psychrophile) because the motile vegetative cells grew at temperatures between 1°C and 5°C, developed into abnormal clumps at 10°C, and did not survive at temperatures above 10°C (Hoham, 1975a, 1975b). Working primarily with bacteria, Morita (1975) defined a true psychrophile with an optimal temperature for growth below 16°C, an upper limit for growth at 20°C, and a minimal temperature for growth at 0°C or lower. Psychrotrophic bacteria were defined as organisms that can grow at 0°C but grow optimally at 20°C to 25°C (Morita, 1975). These concepts were discussed further by Baross and Morita (1978).

An optimum temperature was reported at 5°C for a *Chloromonas* sp. from snow (Stein and Bisalputra, 1969), 4°C for several strains of the snow alga, *Chlamydomonas nivalis* (Czygan, 1970), and 4°C for *Raphidonema tatrae* with a maximum temperature for growth below 10°C (Hindák and Komárek, 1968). *Cryptomonas frigoris* grew at temperatures between 2°C and 10°C (Javornický and Hindák, 1970) and *Chromulina chionophilia* grew below 10°C (Stein, 1963). Using a laboratory cooling stage, cells of *Chlainomonas rubra* and *Chlainomonas kolii* shed their flagella at temperatures above

195

4°C, and cells of *C. rubra* were damaged when frozen at −1°C (Hoham, 1975a). All snow algal cultures from the Windmill Islands, Antarctica, grew well at 3°C, *Chloromonas rubroleosa* and *Chloromonas polyptera* died at 10°C, and most *Chlorosarcina antarctica* were dead at 10°C (Ling, 1996). These all appear to be examples of obligate cryophiles. An optimum temperature range of 3.5°C to 10°C was reported for the first psychrophilic obligate methanotrophic bacterium isolated from the tundra soil in the polar Ural Mountains of Russia (Omelchenko, Vasilyeva, and Zavarzin, 1993).

Ling (1996) also reported that *Palmellopsis* sp. and *Desmotetra* sp. 1 grew at 10°C but died at 15°C, and *Desmotetra* sp. 2 and *Chloromonas* sp. 1 were barely living at 15°C. A temperature range for growth at 0°C to 15°C was reported for *Raphidonema nivale* (optimum temperature 5°C), 1°C to 20°C for *Cylindrocystis brébissonii* (optimum temperature 10°C), and 1°C to 15°C for *Chromulina chionophilia* (Hoham, 1975a). These appear to be examples of nonobligate cryophiles where growth stages should be able to live in environments other than snow such as soils or water. However, of these, only *Cylindrocystis brébissonii* was found living outside of snow. The extracellular protease-producing psychrotrophic bacteria from the cryoconite of glaciers and high elevation alpine soils from the Alps of Europe were studied by Schinner et al. (1992). Most bacterial strains isolated had an optimum temperature within the range of 10°C to 25°C, and almost half of the bacterial strains excreted protease into the medium at a cultivation temperature of 10°C. As reported earlier, Handfield et al. (1992) isolated 17 psychrophilic species of bacteria and yeasts that grew at 1°C and 4°C and 25 psychrotrophic species of bacteria and yeasts that grew at 4°C and 25°C from Agassiz Ice Cap, Ellesmere Island, Canada.

The changes in morphology and viability in 20 species of fungi belonging to 5 different groups were examined by Morris, Smith, and Coulson (1988) during freezing in relation to cooling rate and the presence of glycerol. They found that several species survived freezing and thawing in the absence of glycerol. The conventionally used cooling rate of 1°C min^{-1} was found not to be optimal for all the fungal strains studied. The resistance to freezing in the red-colored snow alga *Chlamydomonas nivalis* may be explained by the presence of astaxanthin ester-rich lipid vacuoles that occupy a large volume of the cell (Marchant, 1982; Bidigare et al., 1993), thus reducing the water content that otherwise would increase the potential for lysing during crystalline ice formation (Giese, 1973). The halite chlorine, present in aerosols and soot materials (Cornwallis Island, Canada), may be critical in depressing the freezing point of the liquid films associated with the snow alga *Chlamydomonas nivalis* and may inhibit evaporation in the low-humidity environment (Tazaki et al., 1994a, 1994b). However, they did not indicate the concentrations of NaCl or soot particles present in this film,

196

which is an important factor to consider when determining the lowering of the freezing point (J. Raymond, personal communication).

4.7.2   Meltwater Flow and Water Content

Meltwater flow, horizontal ice layers, and vertical ice fingers affect the position of the algae within the snowpacks (Hoham, 1975b; Hoham et al., 1979, 1983) (see Pomeroy and Brun, Chapter 2). Dyes were used to illustrate that meltwater travels laterally several meters in shallow snowpacks (Hardy, 1993, personal communication). Food color dyes were applied at different points from upper surface to ground level in snowpacks under 60 cm in depth to measure rate and directional flow of liquid meltwater (Hoham, 1975b), and water flowed either gravitationally or laterally and often followed the horizontal ice layers and vertical ice fingers. The range of water flow varied between 3 and 150 cm hr^{-1}, and maximum flow occurred during midafternoon when air temperatures were highest, there was a direct sun exposure, and snowbanks were connected. Minimum water flow occurred during late evening and early morning hours when air temperatures were lowest, there was no sun exposure, and in isolated snowbanks. This meltwater flow was correlated with developmental stages and potential sources of nutrients for the snow alga *Chloromonas pichinchae* (Hoham, 1975b). Nutrients in the meltwater would be available at the snow–soil interface for algal growth at the time of resting spore germination and throughout the snowpack for growth of vegetative cells (Hoham et al., 1983).

Significant negative correlations were found between albedo and snow water content and between albedo and snow algal cell numbers (Thomas and Duval, 1995). Water melt, however, in general was not affected by the presence of algae because of their patchiness.

Percent water content (mL of H_2O melted from 100-mL core samples of snow) was reviewed by Hoham et al. (1983) from snowbanks in Washington (44 to 69 percent), Arizona (43 to 63 percent), and Montana (52 to 72 percent), and these percentages were correlated with algal life cycle phases in different species of *Chloromonas*. The highest percentages (57 to 63 percent) associated with the snow alga *Chloromonas pichinchae* occurred in early to midafternoon in Washington snow when air temperatures and light intensities were highest (Hoham, 1975b). At this time, the alga was in the swimming asexual vegetative phase. Gamete production, release, and fusions (sexual phase) occurred during lower percentage readings (47 to 54 percent). Water content percentages of 42 to 52 percent and 60 to 96 percent were recorded in association with the snow alga *Mesotaenium berggrenii* and the pink ice alga *Chloromonas rubroleosa* respectively, from Antarctica (Ling and Seppelt, 1990, 1993).

197

4.7.3   Light, Cryoconite Holes, and Suncups

Snow readily absorbs long-wave radiation from its warmer surroundings as indicated by "tree well" basins that form around tree trunks (Marchand, 1991). Dust and other particulates on snow reflect little light (Clow, 1987; Dozier, 1987), such that pure snow reflects more light than does snow with dust and particulates (Warren, 1982; Thomas and Duval, 1995) (see Pomeroy and Brun, Chapter 2). Physical and biological forces act together to form cryoconite holes, suncups, ice fingers, and channels (Gerdel and Drouet, 1960; McIntyre, 1984). The absorption of solar radiation by particulates and organic material and subsequent warming above 0°C cause melting and formation of cryoconite holes in glacial ice (McIntyre, 1984; Wharton et al., 1985). Wharton et al. (1985) reviewed cryoconite holes on glaciers and reported several life forms including algae, rotifers, pollen, insects, and cyanobacteria. They also discussed how cryoconite holes may increase the rate of glacial ice wastage. Cryoconite holes or depressions yielded seven pennate diatoms and three green algae on Mt. Athabasca's north glacier, Alberta, Canada (Wharton and Vinyard, 1983); two green algal desmids, *Ancylonema nordenskioldii* and *Mesotaenium berggrenii*, from Columbia Glacier, Alaska (Kol, 1942); the desmid *Cylindrocystis brébissonii* from the Thule icefields, Greenland (Gerdel and Drouet, 1960); three green algae, *C. brébissonii*, *M. berggrenii*, and *Trochiscia* sp., that dominated Yala Glacier, Nepal, from lower to higher elevations, respectively (Yoshimura et al., 1997); two cyanobacteria, *Gloeocapsa* and *Nostoc*, from glaciers in Marie Byrd Land, Antarctica (Broady, 1989); and psychrotrophic bacteria from glaciers in the European Alps (Schinner et al., 1992). Prescott (1978) reported surface algal blooms of *Chlamydomonas* and *Ancylonema*, the latter forming black snow or black ice that developed meltpores in ice by light absorption. Drifting Arctic ice was also reported to contain autochthonous microbial populations (Kawecka, 1986). Red-colored snow algae and associated cryophilic microbes darkening snow are often associated with suncups (Hardy and Curl, 1972). From field data collected in the Sierra Nevada Mountains, California (Thomas and Duval, 1995), red-colored snow algae enhanced shortwave radiation absorption in snow on the order of 7 to 12 percent. However, red snow caused by algal blooms did not decrease mean albedos in representative snow fields because of algal patchiness, and much of the albedo was reduced by dirt and debris over the snow. Snow algae on a Himalayan glacier accelerated glacial melting, affected the mass balance of the glacier by forming dark-colored areas on the glacier, and reduced surface albedo (Kohshima, Seko, and Yoshimura, 1993).

The amount of light that penetrates snow is a function of wavelength, snow density, and depth (Pomeroy and Brun, Chapter 2; Richardson and Salisbury, 1977; Warren,

1982). Of importance to subnivean phototrophs is the amount of photosynthetically active radiation (PAR) received at a particular depth. This region of the light spectrum includes visible or white light (400–700 nm) with blue light penetrating the deepest (Sze, 1993). Curl, Hardy, and Ellermeier (1972) reported that the major portion of spectral energy in snow is between 450 and 600 nm, with a peak at 475 nm. Also, blue and green light at 455 and 526 nm (Hoham et al., 1983) and at 454 and 530 nm, respectively (Hoham, 1975b) penetrated the deepest in snow for most time periods sampled. Although most light is attenuated by snow, 1 percent of incident PAR was measured at a depth of 1 m in wet snow, and this promoted photosynthesis and algal germination (Curl et al., 1972). Light was measured penetrating through 2 m of wet snow, suggesting that photoactive responses can occur under minimal light conditions (Richardson and Salisbury, 1977). In laboratory mating experiments with *Chloromonas* sp.-D isolated from Tughill Plateau snow, New York State, it was found that maximum mating occurred at a photon irradiance of 95 μmol m^{-2} s^{-1} (PAR 400–700 nm) under both wide-spectrum and cool-white regimes, and blue light regimes produced more matings than green or red light regimes (Hoham et al., 1997, 1998).

In polar regions (Antarctica), red snow is characteristic of low nutrient areas, whereas green snow is most often associated with higher nutrients near seabird rookeries (Bidigare et al., 1993; Broady, 1996). In temperate regions, however, the distribution of red, green, and orange snow, colored by populations of algae (Figure 4.35a–f [see plate section]), is related to the amount of sunlight received by the populations (Kol, 1968a; Pollock, 1970; Hoham, 1971). It was noted by Fukushima (1963) that green snow algae occur under shaded tree canopies with light regimes of <50 percent sunlight, whereas red snow generated by the alga *Chlamydomonas nivalis* generally is found in areas of higher light intensity. Photosynthesis in *Chlamydomonas nivalis* is not inhibited at high light intensity ($\leq$86,000 lux) in exposed alpine snow and is well adapted to a high light regime (Mosser et al., 1977). Snow algae are often found concentrated in horizontal ice layers several centimeters below the snowpack surface (Hoham, 1975b, 1980, Hoham et al., 1983). The algae adjust their vertical photoposition in the snowpack and support populations of up to 1×10^6 cells mL^{-1} in these horizontal layers (Hoham, 1987). Snow algal populations are often associated with vertical ice fingers as well (Hoham et al., 1979). Although not phototrophic, concentrated bands of bacteria were noted to occur approximately 20 cm above the soil in Rocky Mountain spring snowpacks (Brooks et al., 1993).

The effects of solar UV-B irradiation on photomovement, motility, and velocity in the snow alga *Chlamydomonas nivalis* and in the colorless euglenoid *Astasia* were studied by Häder and Häder (1989). They found that motility and velocity were

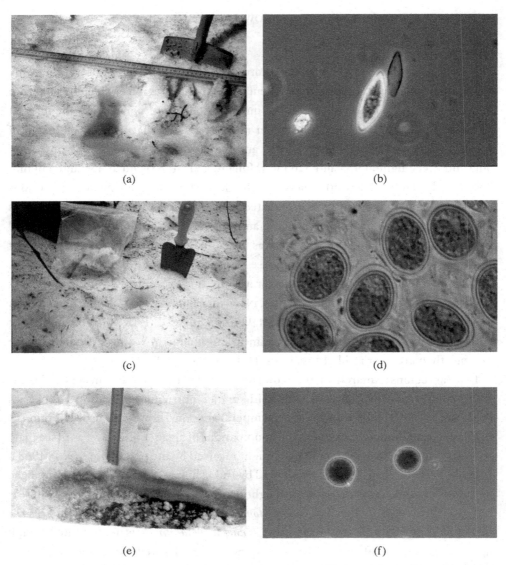

(a)   (b)

(c)   (d)

(e)   (f)

Figure 4.35. (a) Green and orange snow caused by *Chloromonas* sp.-D, Whetstone Gulf State Park, Tughill Plateau, New York [380 M]. (b) Dormant asexual green resting spore of *Cloromonas* sp.-A, Whiteface Mountain, New York [1,340 M]. The resting spore was formerly known as *Scotiella cryophila* (Hoham, unpublished data). (c) Orange snow caused by *Chloromonas* sp.-B, Killington Ski Area, Vermont [1,070 M]. This species is known only from manmade ski slopes. (d) Dormant orange resting spores of *Cloromonas* sp.-B, Wachusett Mountain Ski Area, Massachusetts [525 M]. (e) Red–burgundy-colored snow caused by *Chlamydomonas nivalis*, near Ingalls Pass, Stuart Range, Washington [1,890 M]. (f) Dormant red–burgundy asexual resting spores known as cysts of *Chlamydomonas nivalis*, Raspberry Springs, Coconino National Forest, Arizona [2,957 M].

A colour version of these plates is available for download from
www.cambridge.org/9780521188890

impaired in the snow alga within 90 min of exposure compared with 3 hours before impairment occurred in *Astasia*. They also found that neither alga demonstrated an obvious phototaxis, but they stated that the test organisms used in their laboratory studies were not adapted to natural conditions. This is interesting because the temperature used in the laboratory was 23°C for *Chlamydomonas nivalis* instead of near 0°C, which is the natural temperature for metabolizing, and personal observations of this alga in North American snow indicate that it is phototactic. The origin of the culture of *Chlamydomonas nivalis* used in their study was not given. Using field experiments in Norwegian snow, the vertical distribution of *Chlamydomonas nivalis* associated with the water surface surrounding the snow crystals and its appearance in the snow layers was related to the melting process instead of to phototaxis (Grinde, 1983). It also appears that Grinde used resting spores stages of *Chlamydomonas nivalis* in the experiments, and these cells would not reform flagella even if wetted. Thus, it is difficult to conclude that *Chlamydomonas nivalis* is not phototactic from this study. Kessler, Hill, and Häder (1992) suggested that biflagellate green cells of *Chlamydomonas nivalis* were strongly oriented by gravity, but direction of gravitaxis was degraded by collisions between cells. For the effects of UV light on photosynthesis, see the next section on primary productivity and respiration.

4.8   Productivity and Biogeochemical Cycles in Snow and Ice

4.8.1   Primary Productivity and Respiration

Snow microbes are often subjected to overnight freezes or repeated freeze-thaw events, and photosynthesis was reported from frozen samples of snow algae that were later thawed (Hoham, 1975a, 1975b; Mosser et al., 1977). Optimum temperatures for photosynthesis in certain strains of snow algae were reported at −3°C to 4°C (Hoham, 1975a; Mosser et al., 1977). Studies of Antarctic lichens also showed near-freezing photosynthetic optima at low PAR levels (Kappen, 1993).

Chlorophyll a concentrations that were very heterogeneous in snow were attributed to populations of snow algae (Thomas, 1972). Using ^{14}C ($H^{14}CO_3^-$ and $^{14}CO_2$), Mosser et al. (1977) reported that *Chlamydomonas nivalis* photosynthesized optimally in the field at 10°C or 20°C but retained substantial activity at temperatures as low as 0°C or −3°C. Using ^{14}C, similar quantities of μg C fixed mm^{-3} cell volume hr^{-1} were recorded for *Chlamydomonas nivalis* (0.05–0.97) (Mosser et al., 1977), mixed populations of snow algae including *Chlamydomonas* sp. (0.04–1.85) (Komárek,

Hindák, and Javornický, 1973), and *Chlamydomonas nivalis* (0.002–0.86) (Fogg, 1967). Larger amounts of fixed carbon (5.7–34.2) were reported for *Chlamydomonas nivalis* (Thomas, 1972), but he questioned these values because of a high assay of CO_2 concentration in the snow meltwater. The carbon concentrating mechanism (CCM) is an inducible mechanism that concentrates CO_2 at the fixation site, which allows for acclimation to a wide range of CO_2 concentrations (Kaplan and Reinhold, 1999). CCM was reported for first time in algae without pyrenoids (*Chloromonas*) (Morita et al., 1998). In cold tolerant species of *Chloromonas*, a biological relationship exists between the absence of pyrenoids and the inability to form a large pool of inorganic carbon in the CCM (Morita et al., 1999).

Carbon production in algal photosynthesis ranged from 1.2×10^{-2} to $12.3 \times 10^{-2} \mu g$ of C (ml snow)$^{-1}$ hr^{-1} in red snow containing *Chlamydomonas nivalis* and *Trochiscia americana* (Thomas, 1994). In adjacent white snow samples, photosynthesis values ranged from 0.00 to $0.16 \times 10^{-2} \mu g$ of C (ML snow)$^{-1}$ hr^{-1}. Even though algal photosynthesis occurred in white snow, ratios of photosynthesis in red snow to white snow ranged from 27 to 79. Thomas (1994) also converted the algal photosynthesis values to units used by Mosser et al. (1977) for red snow comparisons and found that the values reported from both studies fell into the same range. A bacterial production of 2 to $9 \times 10^{-4} \mu g$ of C (ml snow)$^{-1}$ hr^{-1} was recorded in red snow, but only 0.35 to $4.47 \times 10^{-5} \mu g$ of C (ml snow)$^{-1}$ hr^{-1} was recorded in white snow, and ratios of red snow to white snow ranged from 1.2 to 56.4 (Thomas, 1994). In red snow at one site, algal production was 141 to 180 times higher than bacterial production, and it was suggested that the bacteria were utilizing organic matter produced photosynthetically by the algae (Thomas, 1994). These same snow algal and bacterial production interrelationships were discussed further by Thomas and Duval (1995).

The snow alga *Chromulina chionophilia* may possess a photoreactivation enzyme that repairs damage done by UV irradiation to the chlorophyll and other photosynthetic pigments (Hardy and Curl, 1972). The effects of total UV on photosynthetic uptake of radioactive carbon in green and red snow from the Sierra Nevada Mountains, California, were studied by Thomas and Duval (1995). They found UV inhibited uptake by 85 percent in green snow containing *Chloromonas* but inhibited uptake by only 25 percent in red snow containing spores of *Chlamydomonas nivalis* (1994 field data). They concluded that red snow found in open, sunlit areas was better adapted to UV than green snow found in forested, shaded locations. Thomas (1995, personal communication) found that *Chlamydomonas nivalis* photosynthesis was not inhibited by UV as was the case in 1994 (Thomas and Duval, 1995). Thomas attributed these differences in UV inhibition to a snowpack four times greater in 1995 than in 1994 that allowed the

202

cells in 1995 to become better adapted to UV because of the longer growth season. As mentioned earlier, Häder and Häder (1989) found that UV-B impaired motion in the snow alga *Chlamydomonas nivalis* greater than it did in the colorless euglenoid *Astasia*. However, nothing was mentioned in their study about impairment of photosynthesis.

4.8.2 Dissolved Gases and pH

The dissolved gases CO_2 and O_2 are important criteria concerning microbial populations in the snowpack. The concentration of dissolved CO_2 in snow ranged from 2.5 to 5.0 mg L^{-1}, and it was 9 to 13 mg L^{-1} for dissolved O_2 in snow containing populations of snow algae (Hoham, 1975b; Hoham and Mullet, 1977). Brooks et al. (1993) reported a CO_2 flux at the snow–air interface of 320–360 mg of $C\,m^{-2}\,day^{-1}$ and suggested a minor source within the snowpack. They also indicated that most snow is probably oxygenated enough to support metabolic activity for aerobic microbes. In their review of psychrophilic and psychrotrophic bacteria, Margesin and Schinner (1994) indicated that, at low temperatures, solubility and availability of oxygen are increased. Therefore, these organisms are more favorably affected by aeration than are mesophiles.

Snow algae affect pH values in snow (Hoham et al., 1989; Hoham and Ling, 2000). Green snow samples with algae from Whiteface Mountain, New York, had higher pH values than snow without algae (5.87 versus 5.63 at 1341 m and 5.17 versus 4.98 at 1265 m), reputedly the result of CO_2 consumption during photosynthesis (Hoham et al., 1989). Similar increases in pH values were reported in green snow from Svalbard where biflagellate cells were photosynthetically active (Müller et al., 1998a). Newton (1982), however, reported a lower pH (6.2) in regions of Svalbard snow colonized by *Chlamydomonas nivalis* compared with areas that were not colonized (pH 7.0–7.6), suggesting that the alga excreted organic materials (acids and polysaccharides) into the snow, lowering the pH. However, Newton's study was conducted when the asexual spore (cyst) stage was present, a phase that would have reduced metabolic activity. Interestingly, Müller et al. (1998a) found a similar situation in Svalbard where snow with red-orange resting spores of algae were found in lower pH (up to 0.4–0.7 pH unit lower) than in control samples without algae. It appears that the relationship between pH and algae in snow depends on the metabolic state and phase of the snow algal life cycle (Hoham et al., 1989, 1993).

Snow microbes are subjected to high acidity. Observations of pH in snow ranged from 4.0 to 6.3 in western North America (Hoham et al., 1983) to as low as 3.4 for meltwater in the Adirondacks, New York (Schofield and Trojnar, 1980). The pH in snow ranged from 3.5 to 5.4 in south central Ontario, with the lowest pH during the

203

initial snowmelt (Goodison, Louie, and Metcalf, 1986). In Antarctic snow populated by the green algal desmid *Mesotaenium*, the pH was between 4.5 and 5.7 (Ling and Seppelt, 1990); for the green alga *Chloromonas rubroleosa*, the pH was between 4.6 and 6.2 (Ling and Seppelt, 1993). Lukavský (1993) reported a pH range of 4.6–4.9 from snow fields in the High Tatra Mountains, Europe, and a higher pH in old snow from the Bohemian Forest Mountains (no values given), and both habitats were associated with snow algae. A pH range of 4.4–6.2 was reported from Svalbard snow algal samples (Müller et al., 1998a).

The effects of acidic snow may result in the natural selection of snow microbes with greater tolerance to acidity. Hoham and Mohn (1985) reported that strains of the snow alga *Chloromonas* (currently thought to belong to *Chloromonas* sp.-A) from the Adirondack Mountains, New York, had growth optima between pH 4.0 and 5.0 compared with other strains isolated from the White Mountains, Arizona, with pH optima of 4.5–5.0 (Figures 4.36 to 4.39). The significant difference in growth between these geographically isolated strains at pH 4.0 ($P < 0.05$) suggests that snow algae were adapting to the more acidic precipitation found in eastern North America.

4.8.3 Nutrients, Nutrient Cycling, and Conductivity

Snow is an environment that is limiting in nutrients (Hoham et al., 1989; Jones, 1991; Hoham and Ling, 2000), and nutrient loads in the snowpack are spatially distributed (Tranter et al., 1987; Davies et al., 1989). This spatial variability may correlate with the spatial distributions of microbes such as the snow algae (Hoham, 1980; Tranter, 1993, personal communication). Nutrient depletion (particularly NO_3^-) coincided with shifts in phases of the life cycle of snow algae such as *Chloromonas* (Hoham et al., 1989). The importance of NO_3^- on the forest floor, such as in the Adirondacks, New York (Rascher, Driscoll, and Peters, 1987), may play an important role in the life cycles of some snow algal species at the time of their germination. The snowmelt waters concentrated in sulfuric acid and nitrogenous anions probably affect the snow microbiota (Hoham and Mohn, 1985; Bartuma et al., 1990; Williams, 1993), and microbial processes in surface soils beneath the snow may also contribute to this acidity (Arthur and Fahey, 1993). Many of the processes and features occurring in snow such as sublimation, melt, and meltwater channels determine the concentration of nutrients that affect the snow community (Goodison et al., 1986) (see Chapters 2 and 3). Snow ecosystem models have neglected the microbial processes that remove CO_2 and nutrients that help buffer acidic conditions in snow (Hornbeck, 1986).

Even though snow is considered oligotrophic, high nutrient loads of nitrogen (≤ 4.76 mg L^{-1} N-NH_4^+, ≤ 3.0 mg L^{-1} N-NO_3^-) and phosphorus (≤ 0.93 mg L^{-1})

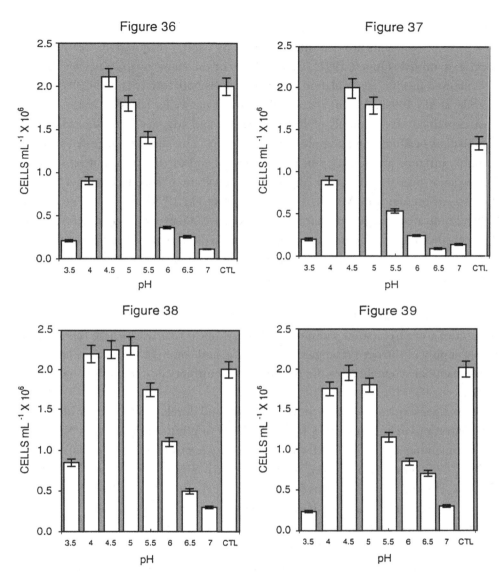

Figures 4.36 to 4.39. Final growth of *Chloromonas* sp.-A (see Hoham et al., 1993) at stationary phase in M-1 medium at pH 3.5–7.0. Ninety-five percent confidence intervals at top of each bar ($N = 3$ with an average of 12 counts each). Figures 4.36 and 4.37: White Mountain, Arizona, strains C381F (Figure 4.36) and C381G (Figure 4.37). Figures 4.38 and 4.39: Adirondack Mountain, New York, strains C204 (Figure 4.38) and C479A (Figure 4.39) (modified from Hoham and Mohn, 1985).

were reported in eastern European snow comparable to eutrophic waters (Komárek et al., 1973). Nutrients are deposited on snow by wind, precipitation, weathering of rock, and animals (Jones, 1991). Cell surfaces of the snow alga *Chlamydomonas nivalis* showed prolific accumulation of aerosol debris both locally and globally derived (Tazaki et al., 1994a, 1994b). In samples from Cornwallis Island, Canada, aerosol dust contained clay minerals of P, S, K, Si, Ca and Mg, sea salts, soot and other combustion products, and complex organic debris. The authors suggested that this complex mixture provided a thin film of aerosol soil for the algal nutrient supply. Yellow snow over the European Alps and the Subarctic was derived from a Saharan dust storm in Africa in March, 1991 (Franzén et al., 1994a, 1994b). The authors (1994a) indicated that pollen that fell over Fennoscandian snow originated from the Alps, northern parts of central Europe, and the Mediterranean. Weathering of parent rock may also add to the nutrient composition of snow (Kawecka, 1986). The desmid *Mesotaenium* was observed in snowmelt downslope from moraine and rock aggregations, and this observation was probably a result of enhanced melting or mineral leachate from the upgradient rock (Ling and Seppelt, 1990). Most snow algae from Antarctica were from snow samples located at seabird rookeries where there was an ample supply of nitrogen (Bidigare et al., 1993), and common vegetation, birds, and small mammals contribute to the patchiness of nutrients described in snow (Jones, 1991).

Snow algae deplete nutrients for their growth and development in their life cycles. Concentrations of nutrients (μeq L^{-1}) were lower in Whiteface Mountain, New York, snow containing algae compared with surrounding control samples without algae for N-NO$_3^-$ (7.7 versus 9.5), N-NH$_4^+$ (5.3 versus 6.5), S-SO$_4^{2-}$ (9.4 versus 11.4), Ca^{2+} (7 versus 11), and K$^+$ (31.3 versus 58.2) (Hoham et al., 1989; Jones, 1991). Similar data from Lac Laflamme, Québec, showed lower concentrations of nutrients (μeq L^{-1}) in snow samples containing algae versus surrounding samples without algae for N-NO$_3^-$ (2.3 versus 13.4), N-NH$_4^+$ (1.5 versus 6), S-SO$_4^{2-}$ (5.9 versus 10.8), Ca^{2+} (1.9 versus 3.5), and Mg^{2+} (1.5 versus 1.6) (Gamache, 1991; Germain, 1991; Jones, 1991) (Figure 4.40). Lower levels of SO$_4^{2-}$, NO$_3^-$ and NH$_4^+$ were also found in snow samples with algae in the Himalayas compared with control samples lacking algae (Yoshimura et al., 1997). Lower levels of N (2.6 μeq L^{-1}) and higher levels of Ca (1000 μeq L^{-1}) and P (0.4 μeq L^{-1}) were found in snowpacks where the dinoflagellate *Gymnodinium pascheri* caused snow coloration in Ontario, Canada, compared with other snowpacks in the region lacking the alga (Gerrath and Nichols, 1974). Their nutrient data, however, suggested that surface water may have contaminated their snowpacks containing the dinoflagellate (Jones, 1991).

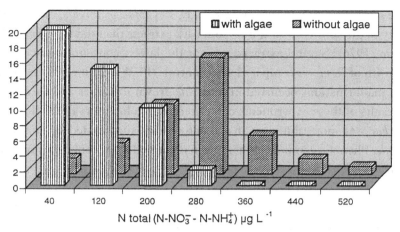

Figure 4.40. Distribution frequency of total N (N-NO$_3^-$ and N-NH$_4^+$) concentrations (μg L^{-1}) in a boreal forest snowpack during the spring melt. Snow classed as algal snow is snow with populations of >4,000 cells mL^{-1} (modified from Gamache, 1990; Jones, 1991).

Coniferous litter, dust, and debris are important sources of nutrients for microbes living in snow. Hoham (1976) investigated the effects that extracts from coniferous litter and different snow meltwaters had on growth of the snow algae *Raphidonema nivale* and *Chloromonas pichinchae*. The conifers used in the leaf litter and bark experiments in the laboratory were the same as those that grew adjacent to snowbanks containing these algae in the Stuart Range of western Washington. The pine pollen used, however, was from a species not found in the Stuart Range. The results of these experiments indicated increased growth for *Chloromonas pichinchae* in all extract concentrations (five species of conifers were used) from the leaf litter and bark except for the highest concentrations where some growth inhibition occurred. A similar result of enhanced growth for *Chloromonas pichinchae* took place in the pine pollen experiment. Results of growth experiments of *Chloromonas pichinchae* in filtered snow meltwater collected from three different sites in the Stuart Range correlated with the coniferous leachate experiments. The best growth occurred in snow meltwater collected from beneath conifers compared with the two samples collected from open exposures. When the two open exposure experiments were compared, *Chloromonas pichinchae* grew best in the sample containing the most atmospheric dust. The tree canopy meltwaters contained more P-PO$_4$ (18 μg L^{-1}) than snow from open areas (2.2 μg L^{-1} from the site with more dust and 1.6 μg L^{-1} from the second site). Increases in P-PO$_4$ were reported in snow meltwater containing wetted forest litter (Moloney, Stratton, and Klein, 1983; Jones, 1987).

207

The second snow alga used in this study, *Raphidonema nivale*, responded very differently to the coniferous leaf litter and bark extracts than did *Chloromonas pichinchae* (Hoham, 1976). Several extracts, even at the lowest concentrations used, inhibited the growth of *R. nivale* compared with controls (PFW medium). Even some morphological malformations occurred in *R. nivale* grown in higher extract concentrations. The results of these experiments correlated with field observations that *Chloromonas pichinchae* was more abundant in the snow fields surrounded by conifers than was *R. nivale*. Phenolic compounds have been identified from coniferous spruce leachates at high altitudes in snow that are potentially responsible for allelopathic interferences (Gallet and Pellissier, 1997).

Hoham et al. (unpublished data) conducted additional coniferous leaf litter experiments in the laboratory with the snow alga *Chloromonas* sp.-A (Hoham et al., 1993) from the Adirondack Mountains, New York. This snow alga showed some enhanced growth in balsam fir leaf litter extracts, the same conifer that grows adjacent to the snowbanks containing this alga.

Vitamins such as B_1 and B_{12} are needed by certain microbes that live in snowpacks beneath coniferous tree canopies (Hoham et al., 1989). These vitamins are needed for growth by the golden alga *Chromulina chionophilia* and the colorless euglenoid *Notosolenus*. It is not known, however, if the vitamins are derived directly from the canopy or from other microbes in the snowpack such as bacteria, fungi, or lichen pieces. In laboratory experiments with axenic cultures, the filamentous green snow alga *Raphidonema nivale* required vitamin B_1 but later lost that requirement presumably because of mutation (Hoham, 1971). However, the green snow algae *Chloromonas pichinchae* and *Cylindrocystis brébissonii* demonstrated no vitamin requirements. In the Colgate University Culture Collection, all strains of *Chloromonas* grow without vitamins (Hoham, personal observation). Species of the green algal flagellate group (Volvocales) to which *Chloromonas* and *Chlamydomonas* belong are often found in enriched eutrophic water (Sze, 1993). However, in the case of the snow environment, it appears that volvocalean species have been selected for that require minimal nutrients and do not require specialized molecules such as vitamins for growth.

Tazaki et al. (1994a) found the major presence of Si, P, S, and organics in red and green cells of *Chlamydomonas nivalis* from Cornwallis Island, Canada, and high Ca content in the green cells only. They suggested that both P and S were of vital importance to the algae under conditions of such extreme low temperature. In Svalbard snow, high levels of Fe, Ca, Mg, K, P, and Al were found in algal cells despite the very low concentration of these ions in the extracellular meltwater (Müller et al., 1998a).

The interaction between snow algae and snow chemistry affect conductivity readings in snow (Hoham et al., 1989; Hoham and Ling, 2000). From Whiteface Mountain, New York, conductivity values were lower in snow samples containing algae than in snow samples without algae (13.1 versus 19.5 μS cm^{-1} at 1341 m and 9.6 versus 16.4 μS cm^{-1} at 1265 m; there were not enough samples for statistics). These differences were due to nutrient uptake and algal metabolism. However, conductivity in Svalbard snow was higher in regions of algal colonization (12 μS cm^{-1}) than in regions without algae (4–7 μS cm^{-1}) (Newton, 1982), and it was suggested that algal activity results in an increase of ionic concentrations as well as preferring regions that receive more windblown materials (the latter probably raised the conductivity values here). Other conductivity values reported from snow involving snow algal studies include 8–15 (Arizona; Hoham et al., 1983), 4–6 and 18–20 (Montana; Hoham et al., 1983), 6–33 (Antarctica; Ling and Seppelt, 1990), 25–85 (Antarctica; Ling and Seppelt, 1993), and 0.3–17 (Svalbard; Müller et al., 1998a) μS cm^{-1}. It was suggested that the snow alga *Chloromonas rubroleosa* required a slightly higher nutrient status inhabiting Antarctic snow with a conductivity of 25–85 μS cm^{-1} (Ling and Seppelt, 1993). The relationship between conductivity values and snow algae may depend on the species, the metabolic state and phase of the snow algal life cycle, the metabolic state of other microbes such as bacteria and fungi, and the degree to which the snow has been leached by meltwater (see Tranter and Jones, Chapter 3).

4.8.4   Bioaccumulation of Heavy Metals

Metals enter the nival food chain and accumulate in algal cells thousands of times their concentration in surrounding snow (Fjerdingstad, 1973; Fjerdingstad et al., 1974; Hoham et al., 1977; Fjerdingstad et al., 1978; Hoham, 1980). High levels of trace metals were recorded in red snow from Kulusuk, Greenland (Fjerdingstad, 1973; Fjerdingstad et al., 1974), and lesser amounts of the same trace metals were reported from red snow in eastern Greenland and Spitzbergen (Fjerdingstad et al., 1978). The concentrations of these metals accumulated hundreds to thousands of times higher in the cells of the snow alga *Chlamydomonas nivalis* compared with the values from the surrounding snow meltwater. Hoham et al. (1977) reported that concentrations of trace metals varied in green snow caused by the alga *Chloromonas pichinchae* from Washington, USA, and within the red snow samples caused by the alga *Chlamydomonas nivalis*, from eastern Greenland and Spitzbergen (Table 4.4). This study raised questions concerning whether different species and strains within species have different nutritional requirements. Other studies on heavy metals in snow from Greenland

Table 4.4. *Maximum concentration of elements in snow meltwater expressed in mg L^{-1}*.

Element	Washington, USA*	East Greenland[†]	Spitzbergen[‡]
As			0.006
Ba		0.13	
Br	0.002	0.03	0.025
Ca	2.25	16.50	0.73
Cl	0.72	2.13	9.31
Cr		0.005	0.007
Cu	0.01	0.01	0.02
Fe	0.04	0.80	2.83
K	0.21	0.72	2.42
Mn	0.05	0.03	0.04
Mo			0.52
Nb		0.002	0.003
Ni		0.58	0.004
P	0.29	0.25	0.27
Pb		0.003	0.05
Rb	0.002	0.015	0.014
S	0.83	0.13	1.00
Si		2.09	10.2
Sr	0.002	0.06	0.01
Ti		0.25	0.24
Zn	0.71	0.05	0.04
Zr		0.003	0.005

*Green snow, collected June 1976, near Mt. Stuart, elev. 1,387 m.
[†]Red snow, collected summer 1976, on westbank of Fiord Loch Fyne, Hudson Land, elev. near sea level.
[‡]Red snow, collected summer 1976, near Sveagruva Coal Mine, elev. 210 m.
From Hoham et al. (1977).

include those of Boutron et al. (1993), Boutron, Candelone, and Hong (1994) and Savarino, Boutron, and Jaffrezo (1994).

4.9   Human Aspects, Interests, and Considerations

4.9.1   Biotechnology
The properties of cold–adapted microorganisms (primarily bacteria) and their potential role in biotechnology were reviewed by Margesin and Schinner (1994)

210

and Feller et al. (1996). The response to high temperatures by these psychrophiles and psychrotrophs was disruptive in protein synthesis by the inability of RNA formation, alterations of the structure of nucleic acids, inactivation of thermolabile enzymes, activation of lytic enzymes, alteration of the cell morphology, inhibition of cell division, and induction of heat shock proteins. At low temperatures these organisms have slower metabolic rates and higher catalytic efficiencies than do mesophiles. The genetic basis of cold adaptation is not clear. Cold-adapted microorganisms have considerable potential in biotechnological applications of waste treatment at ambient temperatures, enzymology, the food industry, medicine, detergents, environmental bioremediations, biotransformations, and applications in molecular biology.

4.9.2 Human Food

Perhaps a more novel use of snow algae is as a source of human food. Microbio Resources, Inc. (1989), used the microalga *Dunaliella* to produce the 100 percent natural food, Provatene, a natural beta-carotene concentrate. Their 1989 brochure emphasized beta-carotene as an antioxidant. In the 1980s, Paul Bubrick, Microbio Resources, Inc., asked the senior author of this chapter how to grow the snow alga *Chlamydomonas nivalis* in very large volumes to mass produce beta-carotene. However, their "production farms" were located in the hot dry desert east of San Diego, California, an environment not suitable for growing this snow alga. Other difficulties with this project were perceived; for example, the beta-carotene was produced in the resting stages of *Chlamydomonas nivalis*, the phase of the life cycle not readily manipulated in laboratory culture. Even though this project was not undertaken, there may be a potential for using extracts from snow algae such as *Chlamydomonas nivalis* as a food supplement. Microbio Resources, Inc., in business between 1985 and 1995, is no longer in operation.

Another question frequently asked is whether eating snow algae directly may be harmful to humans. A study was undertaken to see if direct consumption of red snow by alpine mountain hikers may cause diarrhea (Fiore, McKee, and Janiga, 1997). Seven healthy volunteers aged 24–56 were given 500 g of red snow containing *Chlamydomonas nivalis*, and none of the volunteers developed diarrhea as suggested in previous unpublished communications from alpine mountain hikers. However, the Denver Medical Hospital reported an acute case of diarrhea and dehydration in September 1997 from a patient who consumed large quantities of red snow while hiking in the Colorado Rocky Mountains (Hoham; personal communication).

4.9.3   Exobiology

Snow and ice microbes were suggested as one of four Earth analogs for life on early and present-day Mars (Rothschild, 1990). The other three analogs were endoevaporites, endoliths, and chemoautotrophs. The search on Mars for signs of microbial life has recommended the polar cap regions as one possible site (McKay and Stoker, 1989; Wharton et al., 1989a, 1989b). The residual ice cap (cap left at the height of summer) at the south pole is composed of frozen carbon dioxide, but on the residual north polar cap the seasonal covering of carbon dioxide frost sublimes by the beginning of summer and exposes an underlying deposit of water ice (Haberle, 1995; Zent, 1996). Bacterial cells frozen in water ice are the only organisms to have survived for a geologically significant time on Earth, and it was suggested that bacterial-like cells may have left traces or remains in a Martian permafrost (Gilichinsky, 1993). Recently, it was hypothesized that 3.6-billion-year-old, bacterial-like fossils from a past Martian biota were found in Martian meteorite ALH84001 collected from Antarctica (McKay et al., 1996). Thus the question of whether Earth microbial life forms such as bacteria, fungi, algae, etc. exist elsewhere in our solar system has not been resolved. Future exploration for life on Mars and the Jovian satellite Europa may resolve this question.

4.10   Other Future Research

4.10.1   Genetics, Molecular Biology, and Ultrastructure

Margesin and Schinner (1994) discussed the transfer of genetic material from mesophilic bacteria to psychrotrophic strains and the expression of three lipase genes from an Antarctic psychrotrophic bacterium into a mesophilic one. In one study using an Antarctic psychrotrophic bacterium (Apigny, Feller, and Gerday, 1993), the molar ratio of stabilizing basic residues Arg/(Arg + Lys) was considerably lower than values obtained from mesophiles and thermophiles, suggesting that this ratio contributes to a more flexible tertiary structure at cold temperatures. Gene transfer and isolation studies have not been done with other snow and ice microbes, and this should be a targeted area for future research.

Speciation of snow microbes is complex (Hoham, 1980). Life cycles of some snow algae have revealed up to seven morphologies or forms in a single species of *Chloromonas* (Hoham et al., 1979). In this study, it was suggested that the different forms of the zygote were related to nutrient depletion. Laboratory mating experiments indicate that complex genetic patterns are emerging in species of *Chloromonas* as well (Hoham et al., 1997, 1998). Using restriction fragment length polymorphism analysis and polymerase chain reaction amplification, dissimilar rDNA sequences have been found in the same population of a *Chloromonas* from snow in Arizona (Hoopes et al., 1995, personal communication). The use of nuclear 18S rDNA gene sequence data in *Chlamydomonas* and *Chloromonas* has restructured our understanding of these genera and relationships between their cold tolerant taxa (Buchheim et al., 1990; Buchheim et al., 1997; Bonome et al., 2000). Chloroplast rbcL gene sequence data further supports these relationships (Morita et al., 1999). Future research using a combination of laboratory mating experiments, nuclear sequence data, and possibly ultrastructural studies (TEM) will be needed to give a more clear picture of speciation within the snow algae. Snow fungi such as *Chionaster* and *Selenotila* are poorly understood snow microbes. These organisms and other related snow fungi need to be cultured and reassessed. Nuclear sequence data for the snow fungi and snow bacteria would be most revealing.

4.10.2 Ecology and Physiology

More productivity studies are needed to better elucidate community interactions. The recent productivity inquiries on snow bacteria and algae (Thomas, 1994; Thomas and Duval, 1995) are a good beginning. Analysis of glacial ecology is in pioneer stages (Kohshima, 1994; Yoshimura et al., 1997). Productivity studies on snow fungi have not been undertaken. Symbiotic associations between snow and ice microbes need further research. It is not always clear whether symbioses actually occur between microbial populations or whether microbes live passively together.

Snow and freshwater ice ecosystems may serve as models to measure continuous changes in concentrations of UV irradiation on biological systems. Changes in UV concentrations in alpine and polar environments correlated with ozone depletion over the past two decades were discussed by Thomas and Duval (1995). There have not been enough studies with snow microbes to give us a clear picture of their sensitivity to UV wavelengths; however, Thomas and Duval (1995) reported that green-colored snow algae are much more sensitive to UV wavelengths than the red-colored cells. The effect of carotene composition on resistance to UV-C radiation in the green alga *Chlamydomonas reinhardtii* was studied by Ladygin and Shirshikova (1993). UV-C (200–280 nm) light has lethal effects and produces nuclear mutations. These injuries can be repaired by photoreactivating enzymes (photolyases), which have their greatest activity in the region 350–520 nm, or the main maxima of carotenoid absorption. The role that carotenoids may play in minimizing damage to snow algal cells from UV-C damage is not known. An increase in carotenoids was reported in aplanospores of *Chlamydomonas nivalis* collected from the Sierra Nevada Mountains, California, after short-term exposure (2 days) to UV-A (365 nm) (Duval, Shetty,

213

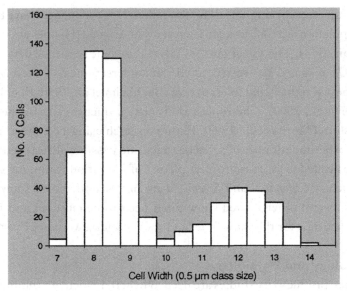

Figure 4.41. Frequency distribution of cell width classes for *Mesotaenium berggrenii* collected near Casey Station (from Ling and Seppelt, 1990).

and Thomas, 1999). In the same study, UV-C exposure (254 nm for 7 days) resulted in greater phenolic-antioxidant production (12–24 percent) when compared to 5–12 percent production under UV-A exposure for 5 days, and UV-C promoted an increase in free proline that did not occur with UV-A. UV-absorbing mycosporine-like compounds were not detected in resting spores of *Chlamydomonas nivalis* from the European Alps (Sommaruga and Garcia-Pichel, 1999).

Snow algal populations of *Mesotaenium* from Antarctica appear to fall into two distinct groups according to cell size (Ling and Seppelt, 1990) (Figure 4.41). Hoham and Blinn (1979) also reported cell size differences within snow algal species from different geographical locations in alpine areas of the American Southwest. The basis of these differences in cell sizes between populations needs to be investigated. Are these differences genetic or might they be caused by an environmental factor? Hoham and Mohn (1985) also suggested that snow algae are adapting to the more acidic precipitation recorded for eastern North America during this century. The potential of snow algae as ecological bioindicators needs further consideration.

It is not known to what degree expansive blooms of snow algae may contribute to the melting of snowpacks. By following the development of these blooms, their impact on alpine watersheds may be more clear. Experiments designed to alter environmental snowpack regimes, such as the one at Niwot Ridge, Colorado, may give insights into how microbial processes might respond to the disturbance of snow ecosystems (Walker et al., 1993) (see Walker, Billings, and de Molenaar, Chapter 6).

4.11   Acknowledgments

The authors thank Dean Blinn, Northern Arizona University, Flagstaff, USA; H.U. Ling, Antarctic Division, Kingston, Tasmania, Australia; Bob Wharton, University of Nevada, Reno, USA; Susan Brawley, University of Maine, Orono, USA; Skip Walker, University of Alaska, Fairbanks, USA; Gerry Jones, INRS, Québec City, Canada; and John Pomeroy, University of Wales, Aberystwyth, UK for their critical reviews of this manuscript. We also thank the Colgate University Research Council for supporting Jubel Caudill for assistance with the graphics and tables and the students at Colgate University who contributed to the laboratory mating experiment data in the phycology lab and rDNA analysis in Barbara Hoopes' molecular biology lab.

4.12   References

Abyzov, S.S. 1993. Microorganisms in the Antarctic ice. In *Antarctic Microbiology*, E.I. Friedmann (Ed.). Wiley-Liss, New York, pp. 265–295.

Aitchison, C.W. 1983. Low temperatures and preferred feeding by winter-active Collembola (Insecta, Apertygota). *Pedobiologia*, **25**, 27–36.

Aitchison, C.W. 1989. The ecology of winter-active collembolans and spiders. *Aquilo Ser. Zoologica*, **24**, 83–89.

Akiyama, M. 1979. Some ecological and taxonomic observations on the colored snow algae found in Rumpa and Skarvsnes, Antarctica. In *Proceedings of the Symposium on Terrestrial Ecosystem in the Syowa Station Area*, T. Matsuda and T. Hoshiai (Eds.). National Institute of Polar Research, Tokyo, pp. 27–34.

Andreis, C., and Rodondi, G. 1979. Ultrastructural aspects of cryoplankton. *Caryologia*, **32**, 119–120.

Antor, R.J. 1995. The importance of arthropod fallout on snow patches for the foraging of high-alpine birds. *J. Avian Biol.*, **26**, 81–85.

Apigny, J.L., Feller, G., and Gerday, C. 1993. Cloning, sequence and structural features of a lipase from the Antarctic facultative psychrophile *Psychrobacter immobilis* B10. *Biochim. Biophys. Acta*, **1171**, 331–333.

Arthur, M.A., and Fahey, T.J. 1993. Controls on soil solution chemistry in a subalpine forest in north-central Colorado. *Soil Sci. Soc. Am. J.*, **57**, 1122–1130.

Baross, J.A., and Morita, R.Y. 1978. Life at low temperatures: Ecological aspects. In *Microbial Life in Extreme Environments*, D.J. Kushner (Ed.). Academic Press, London, pp. 11–56.

Barsdate, R.J., and Alexander, V. 1970. Photosynthetic organisms in subarctic lake ice. *Arctic*, **23**, 201.

Bartuma, L.A., Cooper, S.D., Hamilton, S.K., Kratz, K.W., and Melack, J.M. 1990. Responses of zooplankton and zoobenthos to experimental acidification in a high-elevation lake (Sierra Nevada, California, U.S.A.). *Freshwat. Biol.*, **23**, 571–586.

Bauer, F. 1819. Microscopical observations on the red snow. *Q. J. Lit. Sci. Arts*, **7**, 222–229.

Bidigare, R.R., Ondrusek, M.E., Kennicutt II, M.C., Iturriaga, R., Harvey, H.R., Hoham, R.W., and Macko, S.A. 1993. Evidence for a photoprotective function for secondary carotenoids of snow algae. *J. Phycol.*, **29**, 427–434.

Blackburn, S.I., and Tyler, P.A. 1987. On the nature of eclectic species – a tiered approach to genetic compatibility in the desmid *Micrasterias thomasiana*. *Brit. Phycol. J.*, **22**, 277–298.

Bold, H.C., and Wynne, M.J. 1985. *Introduction to the Algae*, 2nd ed., Prentice-Hall, Englewood Cliffs, NJ.

Bonome, T.A., Leebens-Mack, J.H., and Hoham, R.W. 2000. The phylogenetic affinities of cold tolerant species of the green algae, *Chloromonas* and *Chlamydomonas* (Chlorophyceae, Volvocales): an analysis using 18S rDNA and the ITS1 and ITS2 regions. *Proc. 57th Annual Eastern Snow Conf.*, (Abstract)

Boutron, C.F., Ducroz, F.M., Görlach, U., Jaffrezo, J.-L., Davidson, C.I., and Bolshov, M.A. 1993. Variations in heavy metal concentrations in fresh Greenland snow from January to August 1989. *Atmosph. Environ.*, **27A**, 2773–2779.

Boutron, C.F., Candelone, J.P., and Hong, S. 1994. The changing occurrence of natural and man-derived heavy metals in Antarctic and Greenland ancient ice and recent snow. *Int. J. Environ. Anal. Chem.*, **55**, 203–209.

Broady, P.A. 1989. Survey of algae and other terrestrial biota at Edward VII Peninsula, Marie Byrd Land. *Antarct. Sci.*, **1**, 215–224.

Broady, P.A. 1996. Diversity, distribution and dispersal of Antarctic terrestrial algae. *Biodivers. Conserv.*, **5**, 1307–1335.

Brock, T.D. 1967. Life at high temperatures. *Science*, **158**, 1012–1019.

Brooks, P.D., Schmidt, S.K., Sommerfeld, R., and Musselman, R. 1993. Distribution and abundance of microbial biomass in Rocky Mountain spring snowpacks. In *Proc. 50th Annual Eastern Snow Conf.*, pp. 301–306.

Buchheim, M.A., Turmel, M., Zimmer, E.A., and Chapman, R.L. 1990. Phylogeny of *Chlamydomonas* (Chlorophyta) based on cladistic analysis of nuclear 18S rRNA sequence data. *J. Phycol.*, **26**, 689–699.

Buchheim, M.A., Buchheim, J.A., and Chapman, R.L. 1997. Phylogeny of *Chloromonas* (Chlorophyceae): a study of 18S ribosomal RNA gene sequences. *J. Phycol.*, **33**, 286–293.

Catranis, C., and Starmar, W.T. 1991. Microorganisms entrapped in glacial ice. *Antarc. J. U.S.*, **26**, 234–236.

216

Chapman, B.E., Roser, D.J., and Seppelt, R.D. 1994. ^{13}C NMR analysis of Antarctic cryptogam extracts. *Antarct. Sci.*, **6**, 295–305.

Clarke, A., Leeson, E.A., and Morris, G.J. 1986. The effects of temperature on ^{86}Rb uptake by two species of *Chlamydomonas* (Chlorophyta, Chlorophyceae). *J. Exp. Bot.*, **37**, 1285–1293.

Clow, G.D. 1987. Generation of liquid water on Mars through the melting of a dusty snowpack. *Icarus*, **72**, 95–127.

Curl, H. Jr., Hardy, J.T., and Ellermeier, R. 1972. Spectral absorption of solar radiation in alpine snow fields. *Ecology*, **53**, 1189–1194.

Czygan, F.-C. 1968. Sekundär-Carotinoide in Grünalgen. I. Chemie, Verkommen und Faktoren, welche die Bildung dieser Polyene beeinflussen. *Arch. Mikrobiol.*, **61**, 69–76.

Czygan, F.-C. 1970. Blutregen und Blutschnee: Stickstoffmangelzellen von *Haematococcus pluvialis* und *Chlamydomonas nivalis*. *Arch. Mikrobiol.*, **74**, 69–76.

Davies, T.D., Delmas, R., Jones, H.G., and Tranter, M. 1989. As pure as the driven snow. *New Sci.*, **122**, 45–49.

Devos, N., Ingouff, M., Loppes, R., and Matange, R. 1998. Rubisco adaptation to low temperatures: a comparative study in psychrophilic and mesophilic unicellular algae. *J. Phycol.*, **34**, 655–660.

Dozier, J. 1987. Recent research in snow hydrology. *Rev. Geophys.*, **25**, 153–161.

Duval, B. 1993. Snow algae in northern New England. *Rhodora*, **95**, 21–24.

Duval, B., and Hoham, R.W. 2000. Snow algae in the northeastern U.S.: photomicrographs, observations and distribution of *Chloromonas* spp. (Chlorophyta). *Rhodora*, **102**, in press.

Duval, B., Duval, E., and Hoham, R.W. 1999. Snow algae of the Sierra Nevada, Spain, and High Atlas Mountains of Morocco. *Int. Microbiol.*, **2**, 39–42.

Duval, B., Shetty, K., and Thomas, W.H. 1999. Phenolic compounds and antioxidant properties in the snow alga *Chlamydomonas nivalis* after exposure to UV light. *J. Appl. Phycol.*, **11**, 559–566.

Dybas, C.L. 1998. Alga rhythms. *Adirond. Life*, **29**(2), 20–25.

Ettl, H. 1970. Die Gattung *Chloromonas* Gobi *emend.* Wille (*Chlamydomonas* und die nächstverwandten Gattungen I). *Nova Hedwigia*, No. 34, 1–283.

Ettl, H. 1976. Die Gattung *Chlamydomonas* Ehrenberg (*Chlamydomonas* und die nchstverwandten Gattungen II). *Nova Hedwigia*, No. 49, 1–1122.

Ettl, H., and Schlösser, U.G. 1992. Towards a revision of the systematics of the genus *Chlamydomonas* (Chlorophyta). 1. *Chlamydomonas applanata* Pringsheim. *Bot. Acta*, **105**, 323–330.

Felip, M., Sattler, B., Psenner, R., and Catalan, J. 1995. Highly active microbial communities in the ice and snow cover of high mountain lakes. *Appl. Environ. Microbiol.*, **61**, 2394–2401.

Feller, G., Thiry, M., and Gerday, C. 1991. Nucleotide sequence of the lipase gene lip2 from the Antarctic psychrotroph *Moraxella* TA144 and site-specific mutagenesis of the conserved serine and histidine residues. *DNA Cell Biol.*, **10**, 381–388.

Feller, G., Narinx, E., Arpigny, J.L., Zekhnini, Z., Swings, J., and Gerday, C. 1994.

217

Temperature dependence of growth, enzyme secretion and activity of psychrophilic Antarctic bacteria. *Appl. Microbiol. Biotechnol.*, **41**, 477–479.

Feller, G., Narinx, E., Arpigny, J.L., Aittaleb, M., Baise, E., Genicot, S., and Gerday, C. 1996. Enzymes from psychrophilic organisms. *Fems Microbiol. Rev.*, **18**, 189–202.

Fiore, D.C., McKee, D.D., and Janiga, M.A. 1997. Red snow – Is it safe to eat? – A pilot study. *Wildern. Environ. Med.*, **8**, 94–95.

Fjerdingstad, Einer, Vanggaard, L., Kemp, K., and Fjerdingstad, Erik. 1978. Trace elements of red snow from Spitsbergen with a comparison with red snow from East-Greenland (Hudson Land). *Arch. Hydrobiol.*, **84**, 120–134.

Fjerdingstad, Erik. 1973. Accumulated concentrations of heavy metals in red snow algae in Greenland. *Hydrologie*, **35**, 247–251.

Fjerdingstad, Erik, Kemp, K., Fjerdingstad, Einer, and Vanggaard, L. 1974. Chemical analyses of red 'snow' from East-Greenland with remarks on *Chlamydomonas nivalis* (Bau.) Wille. *Arch. Hydrobiol.*, **73**, 70–83.

Fogg, G.E. 1967. Observations on the snow algae of the South Orkney Islands. *Philos. Trans. R. Soc. London Ser. B*, **40**, 293–338.

Franzén, L.G., Hjelmroos, M., Kållberg, P., Brorström-Lundén, E., Juntto, S., and Savolainen, A.-L. 1994a. The 'yellow snow' episode of northern Fennoscandia, March 1991-A case study of long-distance transport of soil, pollen and stable organic compounds. *Atmosph. Environ.*, **28**, 3587–3604.

Franzén, L.G., Mattsson, J.O., Mårtensson, U., Nihlén, T., and Rapp, A. 1994b. Yellow snow over the Alps and Subarctic from dust storm in Africa, March 1991. *Ambio*, **23**, 233–235.

Fritsch, F.E. 1912. Freshwater algae collected in the South Orkneys by Mr. R.N. Rudmose Brown, B.Sc., of the Scottish National Antarctic Expedition, 1902–1904. *J. Linn. Soc. Bot.*, **40**, 293–338.

Fukushima, H. 1953. Studies on the cryoalgae of Japan. 1. Notes on *Oocystis lacustris* forma *nivalis*. *Nagaoa*, **3**, 36–40.

Fukushima, H. 1963. Studies on cryophytes in Japan. *J. Yokohama Munic. Univ. Ser. C, Nat. Sci.*, **43**, 1–146.

Gallet, C., and Pellissier, F. 1997. Phenolic compounds in natural solutions of a coniferous forest. *J. Chem. Ecol.*, **23**, 2401–2412.

Gamache, S. 1991. *Influence des Algues sur la Physico-Chimie de la Neige lors de la Fonte Printanière*. M.Sc. thesis, Institut National de la Recherche Scientifique, Ste. Foy, Québec, Canada.

Garric, R.K. 1965. The cryoflora of the Pacific Northwest. *Am. J. Bot.*, **52**, 1–8.

Gerdel, R.W., and Drouet, F. 1960. The cryoconite of the Thule area, Greenland. *Trans. Am. Microscop. Soc.*, **79**, 256–272.

Germain, L. 1991. *Dynamique des Composés Azotés et Activité Microbiologique de la Neige au cours de la Fonte Printanière*. M.Sc. thesis, Institut National de la Recherche Scientifique, Ste. Foy, Québec, Canada.

218

Gerrath, J.F., and Nicholls, K.H. 1974. A red snow in Ontario caused by the dinoflagellate, *Gymnodinium pascheri*. *Can. J. Bot.*, **52**, 683–685.

Giese, A.C. 1973. *Cell Physiology*, 4th ed., W. C. Saunders, Philadelphia.

Gilichinsky, D.A. 1993. Viable microorganisms in permafrost: The spectrum of possible applications to new investigations. In *Joint Russian-American Seminar on Cryopedology and Global Change (1992), Proc. 1st Int. Conf. Cryoped.* Russian Acad. Sci. Publ., Pushchino, Russia, pp. 268–270.

Goodison, B.E., Louie, P.Y.T., and Metcalfe, J.R. 1986. Investigations of snowmelt acidic shock potential in south central Ontario, Canada. In *Modelling Snowmelt-Induced Processes*, E.M. Morris (Ed.). IAHS Publication No. 155, IAHS Press, Wallingford, UK, pp. 297–309.

Gradinger, R., and Nürnberg, D. 1996. Snow algal communities on Arctic pack ice floes dominated by *Chlamydomonas nivalis* (Bauer) Wille. *Proc. NIPR Sympos. Polar Biol.*, **9**, 35–43.

Grinde, B. 1983. Vertical distribution of the snow alga *Chlamydomonas nivalis* (Chlorophyta, Volvocales). *Polar Biol.*, **2**, 159–162.

Grout, B.W.W., Morris, G.J., and Clarke, A. 1980. Changes in the plasma membrane of *Chlamydomonas* following freezing and thawing. *Cryo-Letters*, **1**, 251–256.

Haberle, R.M. 1995. Buried dry ice on Mars. *Nature*, **374**, 595–596.

Häder, D.-P., and Häder, M.A. 1989. Effects of solar U.V.-B irradiation on photomovement and motility in photosynthetic and colorless flagellates. *Environ. Exp. Bot.*, **29**, 273–282.

Handfield, M., Jones, H.G., Letarte, R., and Simard, P. 1992. Seasonal fluctuation pattern of the microflora on Agassiz ice sheet, Ellesmere Island, Canadian Arctic. *Musk-ox*, **39**, 119–123.

Hardy, J.T., and Curl, H. Jr. 1972. The candy-colored, snow-flaked alpine biome. *Nat. Hist.*, **81**(9), 74–78.

Hindák, F., and Komárek, J. 1968. Cultivation of the cryosestonic alga *Koliella tatrae* (Kol) Hind. *Bio. Plant.*, **10**, 95–97.

Hirano, M. 1965. Freshwater algae in the Antarctic regions. In *Biology and Ecology in Antarctica*, J. van Mieghem and P. van Oye (Eds.). W. Junk, The Hague, Monographiae Biologicae, **15**, 127–193.

Hoham, R.W. 1971. *Laboratory and Field Studies on Snow Algae of the Pacific Northwest*. Ph.D. thesis, University of Washington, Seattle. (Libr. Congr. Card No. Mic., 71-28, 423), 207 pp. (Univ. Microfilms, Ann Arbor, Mich. [Diss. Abstr., 32B, 2557B]).

Hoham, R.W. 1973. Pleiomorphism in the snow alga, *Raphidonema nivale* Lagerh. (Chlorophyta), and a revision of the genus *Raphidonema* Lagerh. *Syesis*, **6**, 243–253.

Hoham, R.W. 1974a. New findings in the life history of the snow alga, *Chlainomonas rubra* (Stein & Brooke) comb. nov. (Chlorophyta, Volvocales). *Syesis*, **7**, 239–247.

Hoham, R.W. 1974b. *Chlainomonas kolii* (Hardy et Curl) comb. nov. (Chlorophyta, Volvocales), a revision of the snow alga, *Trachelomonas kolii* Hardy et Curl (Euglenophyta, Euglenales). *J. Phycol.*, **10**, 392–396.

Hoham, R.W. 1975a. Optimum temperatures and temperature ranges for growth of snow algae. *Arctic Alpine Res.*, **7**, 13–24.

Hoham, R.W. 1975b. The life history and ecology of the snow alga *Chloromonas pichinchae* (Chlorophyta, Volvocales). *Phycologia*, **14**, 213–226.

Hoham, R.W. 1976. The effect of coniferous litter and different snow meltwaters upon the growth of two species of snow algae in axenic culture. *Arctic Alpine Res.*, **8**, 377–386.

Hoham, R.W. 1980. Unicellular chlorophytes – snow algae. In *Phytoflagellates*, E.R. Cox (Ed.). Elsevier–North Holland, New York, pp. 61–84.

Hoham, R.W. 1987. Snow algae from high-elevation, temperate latitudes and semi-permanent snow: Their interaction with the environment. In *Proc. 44th Annual Eastern Snow Conf.*, pp. 73–80.

Hoham, R.W. 1989a. Snow microorganisms and their interaction with the environment. In *Proc. 57th Annual Western Snow Conf.*, pp. 31–35.

Hoham, R.W. 1989b. Snow as a habitat for microorganisms. In *Exobiology and Future Mars Missions*, C.P. McKay and W.L. Davis (Eds.). NASA Conf. Publ. 10027. Ames Res. Center, Moffett Field, CA, pp. 32–33.

Hoham, R.W. 1992. Environmental influences on snow algal microbes. In *Proc. 60th Annual Western Snow Conf.*, pp. 78–83.

Hoham, R.W., and Mullet, J.E. 1977. The life history and ecology of the snow alga *Chloromonas cryophila* sp. nov. (Chlorophyta, Volvocales). *Phycologia*, **16**, 53–68.

Hoham, R.W., Kemp, K., Fjerdingstad, Erik, and Fjerdingstad, Einer. 1977. Comparative study of trace elements and nutrients in green snow (Washington, U.S.A.) and red snow (Greenland and Spitzbergen). *J. Phycol.* **13**(Suppl.), 30 (Abstr. 167).

Hoham, R.W., and Mullet, J.E. 1978. *Chloromonas nivalis* (Chod.) Hoh. & Mull. comb. nov., and additional comments on the snow alga, *Scotiella*. *Phycologia*, **17**, 106–107.

Hoham, R.W., and Blinn, D.W. 1979. Distribution of cryophilic algae in an arid region, the American Southwest. *Phycologia*, **14**, 133–145.

Hoham, R.W., Roemer, S.C., and Mullet, J.E. 1979. The life history and ecology of the snow alga *Chloromonas brevispina* comb. nov. (Chlorophyta, Volvocales). *Phycologia*, **18**, 55–70.

Hoham, R.W., Mullet, J.E., and Roemer, S.C. 1983. The life history and ecology of the snow alga *Chloromonas polyptera* comb. nov. (Chlorophyta, Volvocales). *Can. J. Bot.*, **61**, 2416–2429.

Hoham, R.W., and Mohn, W.W. 1985. The optimum pH of four strains of acidophilic snow algae in the genus *Chloromonas* (Chlorophyta) and possible effects of acid precipitation. *J. Phycol.*, **21**, 603–609.

Hoham, R.W., Yatsko, C.P., Germain, L., and Jones, H.G. 1989. Recent discoveries of snow algae in Upstate New York and Québec Province and preliminary reports on related snow chemistry. In *Proc. 46th Annual Eastern Snow Conf.*, pp. 196–200.

Hoham, R.W., Laursen, A.E., Clive, S.O., and Duval, B. 1993. Snow algae and other microbes in several alpine areas in New England. In *Proc. 50th Annual Eastern Snow Conf.*, pp. 165–173.

Hoham, R.W., Kang, J.Y., Hasselwander, A.J., Behrstock, A.F., Blackburn, I.R., Johnson, R.C., and Schlag, E.M. 1997. The effects of light intensity and blue, green and red wavelengths on mating strategies in the snow alga, *Chloromonas* sp.-D, from the Tughill Plateau, New York State. In *Proc. 65th Annual Western Snow Conf.*, pp. 80–91.

Hoham, R.W., Schlag, E.M., Kang, J.Y., Hasselwander, A.J., Behrstock, A.F., Blackburn, I.R., Johnson, R.C., and Roemer, S.C. 1998. The effects of irradiance levels and spectral composition on mating strategies in the snow alga, *Chloromonas* sp.-D, from the Tughill Plateau, New York State. *Hydrol. Processes*, **12**, 1627–1639.

Hoham, R.W., and Ling, H.U. 2000. Snow algae: the effects of chemical and physical factors on their life cycles and populations. In *Journey to Diverse Microbial Worlds: Adaptation to Exotic Environments* (In the book series, Cellular Origin and Life in Extreme Environments), J. Seckbach (Ed.). Kluwer Dordrecht, The Netherlands, pp. 131–145.

Hornbeck, J.W. 1986. Modelling the accumulation and effects of chemicals in snow. In *Modelling Snowmelt-Induced Processes*, E.M. Morris (Ed.). IAHS Publication No. 155, IAHS Press, Wallingford, UK, pp. 325–333.

Horner, R.A. 1985. *Sea Ice Biota*, CRC Press, Boca Raton, FL.

Horner, R.A. 1996. Ice algal investigations: Historical perspective. *Proc. NIPR Symp. Polar Biol.*, **9**, 1–12.

Hoshaw, R.W. 1961. Sexual cycles of three green algae for laboratory study. *Am. Biol. Teach.*, **23**, 489–499.

Huber-Pestalozzi, G. 1961. Das Phytoplankton des Süsswassers Systematik und Biologie. In *Die Binnengewässer*, Vol. 16 (Part 5), A. Thienemann (Ed.). E. Schweizerbart'sche Verlagsbuchhandlung, Stuttgart.

Ishikawa, S., Matsuda, O., and Kawaguchi, K. 1986. Snow algal blooms and their habitat conditions observed at Syowa Station, Antarctica. *Mem. Natl. Inst. Polar Res. Spec. Issue*, **44**, 191–197.

Javornický, P., and Hindák, F. 1970. *Cryptomonas frigoris* spec. nova (Cryptophyceae), the new cyst-forming flagellate from the snow of the High Tatras. *Biologia (Bratislava)*, **25**, 241–250.

Jones, H.G. 1987. Chemical dynamics of snow cover and snowmelt in a boreal forest. In *NATO ASI Series C: Mathemat. Physic. Sci.*, *Vol. 211, Seasonal Snowcovers: Physics, Chemistry, Hydrology*, H.G. Jones and W.J. Orville-Thomas (Eds.). Reidel, Dordrecht, The Netherlands, pp. 531–574.

Jones, H.G. 1991. Snow chemistry and biological activity: a particular perspective on nutrient cycling. In *NATO ASI Series G: Ecol. Sci.*, *Vol. 28, Seasonal Snowpacks, Processes of Compositional Change*, T.D. Davies, M. Tranter, and H.G. Jones (Eds.). Springer-Verlag, Berlin, pp. 173–228.

Judge, B., Scholin, C.A., and Anderson, D.M. 1993. RFLP analysis of a fragment of the large-subunit ribosomal RNA gene of globally distributed populations of the toxic dinoflagellate *Alexandrium*. *Biol. Bull.*, **185**, 329–330.

Kaplan, A., and Reinhold, L. 1999. CO_2-concentrating mechanisms in photosynthetic microorganisms. *Annu. Rev. Plant Physiol.*, **50**, 539–570.

Kappen, L. 1993. Lichens in the Antarctic region. In *Antarctic Microbiology*, E.I. Friedmann (Ed.). Wiley-Liss, New York, pp. 433–490.

Kawecka, B. 1978. Biology and ecology of snow alga *Chlamydomonas nivalis* /Bauer/ Wille/ Chlorophyta, Volvocales/. *Proc. Crypt. Symp.* SAS, 47–52.

Kawecka, B. 1981. Biology and ecology of snow algae. 2. Formation of aplanospores in *Chlamydomonas nivalis* (Bauer) Wille (Chlorophyta, Volvocales). *Acta Hydrobiol.*, **23**, 211–215.

Kawecka, B. 1983/1984. Biology and ecology of snow algae. 3. Sexual reproduction in *Chloromonas rostafiński* (Starmach et Kawecka) Gerloff et Ettl (Chlorophyta, Volvocales). *Acta Hydrobiol.* **25/26**, 281–285.

Kawecka, B. 1986. Ecology of snow algae. *Polish Polar Res.*, **7**, 407–415.

Kawecka, B., and Drake, B.G. 1978. Biology and ecology of snow algae. 1. The sexual reproduction of *Chlamydomonas nivalis* (Bauer) Wille (Chlorophyta, Volvocales). *Acta Hydrobiol.*, **20**, 111–116.

Kepner, R.L. Jr., Wharton, R.A. Jr., and Galchenko, V. 1997. The abundance of planktonic virus-like particles in Antarctic lakes. In *Proc. 1st International Workshop on Polar Desert Ecosystems*, W.B. Lyons, C. Howard-Williams, and I. Hawes (Eds.). Balkema Press, Rotterdam, The Netherlands, pp. 241–250.

Kessler, J.O., Hill, N.A., and Häder, D.-P. 1992. Orientation of swimming flagellates by simultaneously acting external factors. *J. Phycol.*, **28**, 816–822.

Kiener, W. 1944. Green snow in Nebraska. *Proc. Nebraska Acad. Sci.*, **54**, 12.

Kirst, G.O., and Wiencke, C. 1995. Ecophysiology of polar algae. *J. Phycol.*, **31**, 181–199.

Kobayashi, Y. 1967. Coloured snow with *Chlamydomonas nivalis* in the Alaskan Arctic and Spitsbergen. *Bull. Nat. Sci. Mus.*, **10**, 207–210.

Kobayashi, Y., and Fukushima, H. 1952. On the red and green snow newly found in Japan II. *Bot. Mag. Tokyo*, **65**, 128–136.

Kohshima, S. 1984a. A novel cold-tolerant insect found in a Himalayan glacier. *Nature*, **310**, 225–227.

Kohshima, S. 1984b. Living micro-plants in the dirt layer dust of Yala glacier. In *Glacial studies in Langtang Valley*, K. Higuchi (Ed.). Nagoya Data Center Glac. Res., Jap. Soc. Snow Ice Off., Nagoya, Japan, pp. 91–97.

Kohshima, S. 1987a. Glacial biology and biotic communities. In *Evolution and Coadaptation in Biotic Communities*, S. Kawano, J.H. Connell, and T. Hidaka (Eds.). Faculty of Science, Kyoto University, Japan, pp. 77–92.

Kohshima, S. 1987b. Formation of dirt layers and surface dust by micro-plant growth in Yala (Dakpatsen) Glacier, Nepal Himalayas. *Bull. Glac. Res.*, **5**, 63–68.

Kohshima, S. 1994. Ecological characteristics of the glacier ecosystem. *Japanese Journal of Ecology*, **44**, 93–98 (in Japanese).

Kohshima, S., Seko, K., and Yoshimura, Y. 1993. Biotic acceleration of glacier melting in Yala Glacier, Langtang region, Nepal Himalaya. In *Snow and Glacier Hydrology*,

G.J. Young (Ed.). IAHS Publication 218, IAHS Press, Wallingford, UK, pp. 309–316.

Kol, E. 1942. The snow and ice algae of Alaska. *Smithsonian Misc. Collect.*, **101**, 1–36.

Kol, E. 1963. On the red snow of Finse (Norway). *Ann. Hist.-Natur. Mus. Nat. Hung.*, **55**, 155–160.

Kol, E. 1968a. Kryobiologie. Biologie und Limnologie des Schnees und Eises I. Kryovegetation. In *Die Binnengewässer*, Vol. 24, H.-J. Elster and W. Ohle (Eds.). E. Schweizerbart'sche Verlagsbuchhandlung, Stuttgart.

Kol, E. 1968b. A note on red snow from New Zealand. *New Zealand J. Bot.*, **6**, 243–244.

Kol, E. 1971. Green snow and ice from the Antarctica. *Ann. Hist.-Natur. Mus. Nation. Hung. Bot.*, **63**, 51–56.

Kol, E., and Eurola, S. 1973. Red snow in the Kilpisjarvi region, North Finland. *Astarte*, **7**, 61–66.

Kol, E., and Peterson, J.A. 1976. Cryobiology. In *The Equatorial Glaciers of New Guinea, Results of the 1971–1973 Australian Universities' Expeditions to Irian Jaya: Survey, Glaciology, Meteorology, Biology and Palaeoenvironments*, G.S. Hope, J.A. Peterson, U. Radok, and I. Allison (Eds.). A. A. Balkema, Rotterdam, pp. 81–91.

Komárek, J., Hindák, F., and Javornický, P. 1973. Ecology of the green kryophilic algae from Belanské Tatry Mountains (Czechoslovakia). *Arch. Hydrobiol., Suppl. 41, Algol. Stud.*, **9**, 427–449.

Ladygin, V.G., and Shirshikova, G.N. 1993. Effect of carotene composition on resistance of algal cells to UV-C radiation. *Russian Plant Physiol. [Fiziologiya Rastenii]*, **40**, 567–572.

Lagerheim, G. 1892. Die Schneeflora des Pichincha. *Ber. Deut. Bot. Ges.*, **10**, 517–534.

Lee, Y.-K., and Soh, C.-W. 1991. Accumulation of astaxanthin in *Haematococcus lacustris* (Chlorophyta). *J. Phycol.*, **27**, 575–577.

Ling, H.U. 1996. Snow algae of the Windmill Islands region, Antarctica. *Hydrobiologia*, **336**, 99–106.

Ling, H.U., and Seppelt, R.D. 1990. Snow algae of the Windmill Islands, continental Antarctica. *Mesotaenium berggrenii* (Zygnematales, Chlorophyta) the alga of grey snow. *Antarct. Sci.*, **2**, 143–148.

Ling, H.U., and Seppelt, R.D. 1993. Snow algae of the Windmill Islands, continental Antarctica. 2. *Chloromonas rubroleosa* sp. nov. (Volvocales, Chlorophyta), an alga of red snow. *Eur. J. Phycol.*, **28**, 77–84.

Ling, H.U., and Seppelt, R.D. 1998. Snow algae of the Windmill Islands, continental Antarctica – 3 – *Chloromonas polyptera* (Volvocales, Chlorophyta). *Polar Biol.*, **20**, 320–324.

Llano, G. 1962. The terrestrial life of the Antarctic. *Sci. Am.*, **207**, 213–230.

Loppes, R., Devos, N., Willem, S., Barthélemy, P., and Matagne, R.F. 1996. Effect of temperature on two enzymes from a psychrophilic *Chloromonas* (Chlorophyta). *J. Phycol.*, **32**, 276–278.

Lukavský, J. 1993. First record of cryoseston in the Bohemian Forest Mts. (Šumava). *Arch. Hydrobiol., Algol. Stud. Suppl.*, **69**, 83–89.

223

Marcarelli, A.M., Rogers, H.S., Ragan, M.D., Barnes, J.M., Ungerer, M.D., Petre, B.M., and Hoham, R.W. 2000. The transformation of cells from oblong to spherical in the life cycle of the green alga, *Chloromonas* sp.-D (Chlorophyceae, Volvocales). *Proc. 57th Annual Eastern Snow Conf.*, (Abstract)

Marchand, P.J. 1991. *Life in the Cold*, 2nd ed., University Press of New England, Hanover, New Hampshire.

Marchant, H.J. 1982. Snow algae from the Australian Snowy Mountains. *Phycologia*, **21**, 178–184.

Marchant, H.J. 1998. Life in the snow: algae and other microorganisms. In *Snow (A Natural History; an Uncertain Future)*, K. Green (Ed.). Australian Alps Liaison Committee, Canberra, pp. 83–97.

Margesin, R., and Schinner., F. 1994. Properties of cold-adapted microorganisms and their potential role in biotechnology. *J. Biotech.*, **33**, 1–14.

Marshall, W.A., and Chalmers, M.O. 1997. Airborne dispersal of Antarctic terrestrial algae and cyanobacteria. *Ecography*, **20**, 585–594.

McIntyre, N.F. 1984. Cryoconite hole thermodynamics. *Can. J. Earth Sci.*, **21**, 152–156.

McKay, C.P., and Stoker, C.R. 1989. The early environment and its evolution on Mars: Implications for life. *Rev. Geophys.*, **27**, 189–214.

McKay, D.S., Gibson, E.K. Jr., Thomas-Keprta, K.L., Vali, H., Romanek, C.S., Clemett, S.J., Chillier, X.D.F., Maechling, C.R., and Zare, R.N. 1996. Search for past life on Mars: Possible relic biogenic activity in Martian meteorite ALH84011. *Science*, **273**, 924–930.

Melnikov, I.A. 1989. *Ecosystem of the Arctic Sea Ice*, Special Publ., P. P. Shirshov Institute of Oceanology, Moscow, Russia (in Russian).

Mataloni, G., and Tesolin, G. 1997. A preliminary survey of cryobiontic algal communities from Cierva Point (Antarctic Peninsula). *Antarct. Sci.*, **9**, 250–257.

Microbio Resources, Inc. 1989. *Algal Culture from the Test Tube to Commercialized Products*. Microbio Resources, Inc., San Diego, CA.

Miyachi, S., Nakayama, O., Yokohama, Y., Hara, Y., Ohmori, M., Komagata, K., Sugawara, H., and Ugaw, Y. 1989. *World Catalogue of Algae*, 2nd ed., Japan Scientific Societies Press, Tokyo.

Moiroud, A., and Gounot, A.-M. 1969. Sur une bacterie psychrophilile obligatoire isolee de limons glaciaries. *Comp. Rend. Acad. Sci. Paris D*, **269**, 2150–2152.

Moloney, K.A., Stratton, L.J., and Klein, R.M. 1983. Effects of simulated acidic, metal-containing precipitation on coniferous litter decomposition. *Can. J. Bot.*, **61**, 3337–3342.

Moore, J.P. 1899. A snow-inhabiting enchytraeid *Mesenchytraeus solifugus* Emery. *Proc. Acad. Natur. Sci., Philadelphia*, **51**, 125–144.

Morita, R.Y. 1975. Psychrophilic bacteria. *Bact. Rev.*, **39**, 144–167.

Morita, E., Abe, T., Tsuzuki, M., Fujiwara, S., Sato, N., Hirata, A., Sonoike, K., and Nozaki, H. 1998. Presence of the CO_2-concentrating mechanism in some species of the pyrenoid-less free-living algal genus *Chloromonas* (Volvocales, Chlorophyta). *Planta*, **204**, 269–276.

224

Morita, E., Abe, T., Tsuzuki, M., Fujiwara, S., Sato, N., Hirata, A., Sonoike, K., and Nozaki, H. 1999. Role of pyrenoids in the CO_2-concentrating mechanism: comparative morphology, physiology and molecular phylogenetic analysis of closely related strains of *Chlamydomonas* and *Chloromonas* (Volvocales). *Planta*, **208**, 365–372.

Morris, G.J., Coulson, G., and Clarke, A. 1979. The cryopreservation of *Chlamydomonas*. *Cryobiology*, **16**, 401–410.

Morris, G.J., Coulson, G.E., Clarke, K.J., Grout, B.W.W., and Clarke, A. 1981. Freezing injury in *Chlamydomonas*: a synoptic approach. In *Effects of Low Temperatures on Biological Membranes*, G.J. Morris and A. Clarke (Eds.). Academic Press, London, pp. 285–306.

Morris, G.J., Smith, D., and Coulson, G.E. 1988. A comparative study of the changes in the morphology of hyphae during freezing and viability upon thawing for twenty species of fungi. *J. Gen. Microbiol.*, **134**, 2897–2906.

Mosser, J.L., Mosser, A.G., and Brock, T.D. 1977. Photosynthesis in the snow: The alga *Chlamydomonas nivalis* (Chlorophyceae). *J. Phycol.*, **13**, 22–27.

Müller, T., Bleiss, W., Martin, C.-D., Rogaschewski, S., and Fuhr, G. 1998a. Snow algae from northwest Svalbard: their identification, distribution, pigment and nutrient content. *Polar Biol.*, **20**, 14–32.

Müller, T., Schnelle, T., and Fuhr, G. 1998b. Dielectric single cell spectra in snow algae. *Polar Biol.*, **20**, 303–310.

Nagashima, H., Matsumoto, G.I., Ohtani, S., and Momose, H. 1995. Temperature acclimation and the fatty acid composition of an Antarctic green alga *Chlorella*. *Proc. NIPR Symp. Polar Biol.*, **8**, 194–199.

Newton, A.P.W. 1982. Red-coloured snow algae in Svalbard – Some environmental factors determining the distribution of *Chlamydomonas nivalis* (Chlorophyta Volvocales). *Polar Biol.*, **1**, 167–172.

Nozaki, H. 1989. Morphological variation and reproduction in *Gonium viridistellatum* (Volvocales, Chlorophyta). *Phycologia*, **28**, 77–88.

Omelchenko, M.V., Vasilyeva, L.V., and Zavarzin, G.A. 1993. Psychrophilic methanotroph from tundra soil. *Cur. Microbiol.*, **27**, 255–259.

Paerl, H.W. 1980. Attachment of microorganisms to living and detrital surfaces in freshwater systems. In *Adsorption of Microorganisms to Surfaces*, G. Bitton and K.C. Marshall (Eds.). Wiley, New York, pp. 375–402.

Palmisano, A.C., and Garrison, D.L. 1993. Microorganisms in Antarctic sea ice. In *Antarctic Microbiology*, E.I. Friedmann (Ed.). Wiley–Liss, New York, pp. 167–218.

Parker, B.C., Simmons, G.M. Jr., Lowe, F.G., Wharton, R.A. Jr., and Seaburg, K.G. 1982a. Removal of organic and inorganic matter from Antarctic lakes by aerial escape of blue-green algal mats. *J. Phycol.*, **18**, 72–78.

Parker, B.C., Simmons, G.M. Jr., Seaburg, K.G., Cathey, D.D., and Allnut, F.C.T. 1982b. Comparative ecology of planktonic communities in seven Antarctic oasis lakes. *J. Plankton Res.*, **4**, 271–286.

Pollock, R. 1970. What colors the mountain snow? *Sierra Club Bull.*, **55**, 18–20.

Prescott, G.W. 1978. *How to Know the Freshwater Algae*, 3rd ed., W.C. Brown, Dubuque, IA.

Provasoli, L. 1966. Media and prospects for the cultivation of marine algae. In *Culture and Collections of Algae*, A. Watanabe and A. Hattori (Eds.). Hakone, Japan, pp. 63–76.

Rascher, C.M., Driscoll, C.T., and Peters, N.E. 1987. Concentration and flux of solutes from snow and forest floor during snowmelt in the west-central Adirondack region of New York. *Biogeochemistry*, **3**, 209–224.

Richardson, S.G., and Salisbury, F.B. 1977. Plant responses to the light penetrating snow. *Ecology*, **58**, 1152–1158.

Roessler, P.G. 1990. Environmental control of glycerolipid metabolism in microalgae: commercial implications and future research directions. *J. Phycol.*, **26**, 393–399.

Roser, D.J., Melick, D.R., Ling, H.U., and Seppelt, R.D. 1992. Polyol and sugar content of the terrestrial plants from continental Antarctica. *Antarct. Sci.*, **4**, 413–420.

Rothschild, L.J. 1990. Earth analogs for Martian life. Microbes in evaporites, a new model system for life on Mars. *Icarus*, **88**, 246–260.

Savarino, J., Boutron, C.F., and Jaffrezo, J.-L. 1994. Short-term varions of Pb, Cd, Zn and Cu in recent Greenland snow. *Atmosph. Environ.*, **28**, 1731–1737.

Scherer, R.P. 1991. Quaternary and tertiary microfossils from beneath Ice Stream B: Evidence for a dynamic west Antarctic ice sheet history. *Palaeogeogr. Palaeoclimat. Palaeoecol.*, **90**, 395–412.

Schinner, F., Margesin, R., and Pümpel, T. 1992. Extracellular protease-producing psychrotrophic bacteria from high alpine habitats. *Arctic Alpine Res.*, **24**, 88–92.

Schlösser, U.G. 1994. SAG-Sammlung von Algenkulturen at the University of Göttingen, Catalogue of Strains, 1994. *Bot. Acta*, **107**, 113–186.

Schofield, C.L., and Trojnar, J.R. 1980. Aluminum toxicity to brook trout (*Salvelinus fontinalis*) in acidified waters. In *Polluted Rain. Environmental Science Research Servies*, Vol. 17, T.Y. Toribara, M.W. Miller, and P.E. Morrow (Eds.). Plenum Press, New York, pp. 341–366.

Simmons, G.M., Vestal, J.R., and Wharton, R.A. Jr. 1993. Environmental regulators of microbial activity in continental Antarctic lakes. In *Antarctic Microbiology*, E.I. Friedmann (Ed.). Wiley–Liss, New York, pp. 491–541.

Sinclair, N.A., and Stokes, J.L. 1965. Obligately psychrophilic yeasts from the polar region. *Can. J. Microbiol.*, **11**, 259–270.

Starr, R.C., and Zeikus, J.A. 1993. UTEX – The culture collection of algae at the University of Texas at Austin. *J. Phycol. Suppl.* to **29**(2), 1–106.

Stein, J.R. 1963. A *Chromulina* (Chrysophyceae) from snow. *Can. J. Bot.*, **41**, 1367–1370.

Stein, J.R., and Amundsen, C.C. 1967. Studies on snow algae and fungi from the Front Range of Colorado. *Can. J. Bot.*, **45**, 2033–2045.

Stein, J.R., and Bisalputra, T. 1969. Crystalline bodies in an algal chloroplast. *Can. J. Bot.*, **47**, 233–236.

Stein, J.R., and Brooke, R.C. 1964. Red snow from Mt. Seymour, British Columbia. *Can. J. Bot.*, **42**, 1183–1188.

Sugawara, H., Ma, J., Miyazaki, S., Satoru, J., and Takishima, Y. 1993. *World Directory of Collections of Cultures of Microorganisms: Bacteria, Fungi and Yeasts*, 4th ed., WFCC World Data Center on Microorganisms, Wako, Japan.

Sutton, E.A. 1972. *The Physiology and Life Histories of Selected Cryophytes of the Pacific Northwest*. Ph.D. thesis, Oregon State University, Corvallis. (Libr. Congr. Card No. Mic., 73-3995), 98 pp. (Univ. Microfilms, Ann Arbor, Mich. [Diss. Abstr., 33B, 3828B]).

Sze, P. 1993. *A Biology of the Algae*, 2nd ed., W.C. Brown, Dubuque, IA.

Tazaki, K., Fyfe, W.S., Iizumi, S., Sampei, Y., Watanabe, H., Goto, M., Miyake, Y., and Noda, S. 1994a. Aerosol nutrients for Arctic ice algae. *Nat. Geogr. Res. Explorat.*, **10**, 116–117.

Tazaki, K., Fyfe, W.S., Iizumi, S., Sampei, Y., Watanabe, H., Goto, M., Miyake, Y., and Noda, S. 1994b. Clay aerosols and Arctic ice algae. *Clays and Clay Mineral.*, **42**, 402–408.

Tearle, P.V. 1987. Cryptogamic carbohydrate release and microbial response during spring freeze-thaw cycles in Antarctic fellfield fines. *Soil Biol. Biochem.*, **19**, 381–390.

Thomas, W.H. 1972. Observations on snow algae in California. *J. Phycol.*, **8**, 1–9.

Thomas, W.H. 1994. Tioga Pass revisited: Interrelationships between snow algae and bacteria. In *Proc. 62nd Annual Western Snow Conf.*, pp. 56–62.

Thomas, W.H., and Broady, P.A. 1997. Distribution of coloured snow and associated algal genera in New Zealand. *N. Zealand J. Bot.*, **35**, 113–117.

Thomas, W.H., and Duval, B. 1995. Sierra Nevada, California, U.S.A., snow algae: Snow albedo changes, algal-bacterial interrelationships, and ultraviolet radiation effects. *Arctic Alpine Res.*, **27**, 389–399.

Tranter, M., Davies, T.D., Abrahams, P.W., Blackwood, I., Brimblecombe, P., and Vincent, C.E. 1987. Spacial variability in the chemical composition of snowcover in a small, remote Scottish catchment. *Atmosph. Environ.*, **21**, 853–862.

VanWinkle-Swift, K.P., Aliaga, G.R., and Pommerville, J.C. 1987. Haploid spore formation following arrested cell fusion in *Chlamydomonas* (Chlorophyta). *J. Phycol.*, **23**, 414–427.

Viala, G. 1966. L'astaxanthine chez le *Chlamydomonas nivalis* Wille. *Compt. Rend. Hebd. Seances Acad. Sci.*, **263**, 1383–1386.

Vincent, W.F., Howard-Williams, C., and Broady, P. 1993. Microbial communities and processes in Antarctic flowing waters. In *Antarctic Microbiology*, E.I. Friedmann (Ed.). Wiley–Liss, New York, pp. 543–569.

Vuorinen, M., and Kurkela, T. 1993. Concentration of CO_2 under snow cover and the winter activity of the snow blight fungus *Phacidium infestans*. *Eur. J. For. Path.*, **23**, 441–447.

Wailes, G.H. 1935. Notes on the flora and fauna of snow and ice in North West America. *Museum Art Notes, Vancouver City Museum*, **8**(Suppl. 1), 1–4.

Walker, D.A., Krantz, W.B., Lewis, B.E., Price, E.T., and Tabler, R.D. 1993. Hierarchic studies of snow-ecosystem interactions: A 100-year snow-alteration experiment. In *Proc. 50th Annual Eastern Snow Conf.*, pp. 407–414.

Warren, G.J. 1994. A bibliography of biological ice nucleation. *Cryo-letters*, **15**, 323–331.

Warren, S.G. 1982. Optical properties of snow. *Rev. Geophys. Space Phys.*, **20**, 67–89.

Weiss, R.L. 1983. Fine structure of the snow alga (*Chlamydomonas nivalis*) and associated bacteria. *J. Phycol.*, 19, 200–204.

Welch, P.S. 1916. Glacier Oligochaeta from Mt. Rainier. *Science*, 43, 143.

Wharton, R.A. Jr., Vinyard, W.C., Parker, B.C., Simmons, G.M. Jr., and Seaburg, K.G. 1981. Algae in cryoconite holes on Canada Glacier in southern Victorialand, Antarctica. *Phycologia*, 20, 208–211.

Wharton, R.A. Jr., and Vinyard, W.C. 1983. Distribution of snow and ice algae in western North America. *Madroño*, 30, 201–209.

Wharton, R.A. Jr., Parker, B.C., and Simmons, G.M. Jr. 1983. Distribution, species composition and morphology of algal mats in Antarctic dry valley lakes. *Phycologia*, 22, 355–365.

Wharton, R.A. Jr., McKay, C.P., Simmons, G.M. Jr., and Parker, B.C. 1985. Cryoconite holes on glaciers. *Bio. Sci.*, 35, 499–503.

Wharton, R.A. Jr., McKay, C.P., Mancinelli, R.L., Clow, G.D., and Simmons, G.M. Jr. 1989a. The Antarctic dry valley lakes: Relevance to Mars. In *Exobiology and Future Mars Missions*, C.P. McKay and W.L. Davis (Eds.). NASA Conf. Publ. 10027. Ames Res. Center, Moffett Field, CA, p. 62.

Wharton, R.A. Jr., McKay, C.P., Mancinelli, R.L., and Simmons, G.M. Jr. 1989b. Early Martian environments: The Antarctic and other terrestrial analogs. *Adv. Space Res.*, 9, (6)147–(6)153.

Wharton, R.A. Jr., Simmons, G.M. Jr., and McKay, C.P. 1989c. Perennially ice-covered Lake Hoare, Antarctica: Physical environment, biology and sedimentation. *Hydrobiologia*, 172, 305–320.

Wharton, R.A. Jr., McKay, C.P., Clow, G.D., and Andersen, D.T. 1993. Perennial ice covers and their influence on Antarctic lake ecosystems. *Phys. Biogeochem. Proc. Antarct. Lakes, Antarct. Res. Ser.*, 59, 53–70.

Wille, N. 1903. Algologische Notizen XI. *Nyt. Mag. Naturvidensk*, 41, 109–162.

Williams, M.W. 1993. Snowpack storage and release of nitrogen in the Emerald Lake watershed, Sierra Nevada. In *Proc. 50th Annual Eastern Snow Conf.*, pp. 239–245.

Yoshimura, Y., Kohshima, S., and Ohtani, S. 1997. A community of snow algae on a Himalayan glacier: Change of algal biomass and community structure with altitude. *Arctic Alpine Res.*, 29, 126–137.

Zent, A.P. 1996. The evolution of the Martian climate. *Am. Sci.*, 84, 442–451.

5   The Effect of Snow Cover on Small Animals

C.W. AITCHISON

5.1   Introduction

Snow cover affects the lives of all animals in contact with it. The physical, chemical, and microbiological characteristics discussed in the previous chapters may influence to varying degrees the ability of animals to survive in snow–covered environments. Larger animals exposed to conditions above snow must adapt their behaviour, feeding habits, and morphology accordingly, while smaller animals may be below, within, or sometimes on snow cover. The thermal properties of snow cover allow organisms to survive in the relatively benign microenvironment of the subnivean (under the snow) space (Mani, 1962). The distribution of areas with snow cover are discussed with respect to animal habitats, and the physical properties and details of the subnivean and intranivean (within snow) microhabitats are described. Snow cover affects fauna where 15 to 25 cm of snow cover persists for at least 2 to 8 weeks per year. Animals may survive in the subnivean or intranivean spaces (Pruitt, 1978) of the tundra, taiga, deciduous, and mixed woods of the arctic, boreal, and temperate regions, respectively. Most of these areas with snow cover are at higher latitudes and/or altitudes, especially where there is a strong continental climate. Enough precipitation is needed to produce the requisite snow cover to provide insulation for animal survival against extreme temperatures (Mani, 1968; Aitchison, 1978a).

Keränen's (1920) classic synopsis on snow, which discusses in detail the effect of meteorology on the physical characteristics of snow, was followed by Geiger's (1965) extensive meteorological work on microclimates, which describes the subnivean space well. The snow cover is an ecotone between two different environments: (1) the dry, very cold, windy, and changeable atmospheric air and (2) the humid, relatively warm, and stable air of the space underneath (Coulianos and Johnels, 1962; Pruitt, 1970; Aitchison, 1978a).

Pruitt (1970) defined the crucial thickness of snow cover as the "hiemal threshold" (HT), i.e., that snow cover thickness at which the subnivean environment is insulated from diel fluctuations of the ambient air temperature. Snow density and grain

229

characteristics affect its effective thermal conductivity and insulative capacity (Pomeroy and Brun, Chapter 2; Pruitt, 1970). Thicknesses greater than 20 cm are not believed to increase the insulating capacity of the snow cover significantly (Coulianos and Johnels, 1962; Geiger, 1965; Pruitt, 1970). In southern Finland the HT is 15 to 20 cm (Ylimäki, 1962), whereas in colder, northern Finland it is 25 to 30 cm (Keränen, 1920).

The thermal regime of the subnivean microenvironment makes it a favourable habitat for many small animals. Once the HT has been established, the temperature regime, not the overall snow cover thickness, affects animal activity (Novikov, 1940). The subnivean space is a thermally stable place in which the soil surface temperature remains near 0°C, while the ambient air temperature fluctuates (Strübing, 1958; Geiger, 1965; Pruitt, 1970; Aitchison, 1974). Generally the subnivean temperature varies between 0°C and −10°C, depending on snow cover thickness and density (Keränen, 1920; Coulianos and Johnels, 1962; Geiger, 1965; Aitchison, 1974, 1978a, 1984a; Pruitt, 1978). However, in the hard tundra snow conditions, it may drop to less than −20°C (Barrow, Alaska) (MacLean, 1975) or −25°C (Devon Island) (Fuller et al., 1975). Bushy vegetation and/or thick litter under snow cover further insulate the soil surface from fluctuations of ambient air temperature (MacKinney, 1929; Walker, Billings, and de Molenaar, Chapter 6).

The intranivean environment occurs within the snow cover, where temperatures are cooler than those of the subnivean space but generally warmer than the ambient air temperature (Geiger, 1965; Pruitt, 1970; Aitchison, 1974; Andreev and Krechmar, 1976; Korhonen, 1980; Leinaas, 1981a). At an extreme subnivean temperature of −18°C with at least 20 cm of snow, the intranivean temperature was about −26°C, while the ambient air temperature was −48°C in Siberia (Andreev and Krechmar, 1976); however, in the snow cover of a warm, maritime climate, the intranivean temperature may not be so different from either the subnivean or ambient air temperature (Leinaas, 1981a).

Light penetration into the snow cover (Pomeroy and Brun, Chapter 2) can affect the subnivean fauna (via biological clocks and photoperiod or "Zeitgeber") (Evernden and Fuller, 1972; Richardson and Salisbury, 1977). The physical characteristics of snow influence the degree of light penetration. For example, in northern Alberta in January, light did not penetrate 25 cm of fluffy, low density snow cover; however, as the snow cover metamorphosed and became denser in March, light penetrated to 36 cm (Evernden and Fuller, 1972).

Gas concentrations under snow cover can vary appreciably in both space and time (Tranter and Jones, Chapter 3). There is no strong correlation between the subnivean temperature and the concentration of CO_2. Gas concentrations can be quite steady over much of the winter period in relatively shallow cold snow cover (0.03 percent in Alaska; 0.02 percent in Manitoba), with a spring maximum of subnivean CO_2 concentration (over 0.05 percent in Alaska [Kelley, Weaver, and Smith, 1968], or 0.08 percent in Manitoba taiga [Penny and Pruitt, 1984]). By late winter, the CO_2 concentration can rise under ice layers within the snow cover. This is partly due to accumulated biological respiration from animals (Bashenina, 1956; Kelley et al., 1968; Batzli et al., 1980; Korhonen, 1980a; Penny and Pruitt, 1984), and partly due to soil microbial activity at $-8°C$ or less (Tranter and Jones, Chapter 3). Mammals respond to increased CO_2 by avoiding areas where concentrations are high. When small mammal population densities are high, ventilation shafts, or snow chimneys, release the gas from under the snow (Batzli et al., 1980). Bashenina (1956) and Penny and Pruitt (1984) found more CO_2 accumulated in low-lying areas.

The nitrogenous excretory products from invertebrates under, within, and on the snow cover are dissolved in spring meltwater (Tranter and Jones, Chapter 3). Generally the metabolic activity of winter-active invertebrates is poorly known (Zinkler, 1966; Steigen, 1975a, 1975b).

5.2   Invertebrates

5.2.1   Nival and Aeolian Fauna
Nival fauna, the permanent residents of snow regions (Edwards, 1972), includes oligochaetes, snails, centipedes, mites (especially oribatids), spiders (especially linyphiids and erigonids), phalangids, pseudoscorpions, snow scorpionflies, springtails, bristletails, and butterflies (Bäbler, 1910; Kaisila, 1952; Janetschek, 1955; Swan, 1961; Mani, 1962, 1968; Edwards, 1972, 1987; Masutti, 1978; Ashmole et al., 1983; Thaler, 1988, 1989, 1992; Janetschek, 1995; Meyer and Thaler, 1995), all commonly seen up to 5,150 m (Swan, 1961) (Table 5.1). Generally the nival fauna is cold stenothermic (cold loving), small, and melanistic, with predatory or scavenging food habits (Edwards, 1987). The nival foragers include flies, springtails, carabid and staphylinid beetles, phalangids, grylloblattids, and ice worms (Mann, Edwards, and Gara, 1980; Ashmole et al., 1983). Even at 6,667 m in the Himalayas, the jumping spider *Euophrys omnisuperstes* was active (Swan, 1961; Wanless, 1975) (Table 5.1). Many of these invertebrates, other than springtails, are predatory (Bäbler, 1910; Edwards, 1987). The true nival fauna contains high-altitude species adapted to permanent habitats and with no adaptations for long-distance migration; 130 species have been collected in the Cairngorms in Scotland (Ashmole et al., 1983). Animal life

231

Table 5.1. *Endemic nival fauna from various countries and the altitudes at which the specimens were collected.*

Species	Country	Altitude (m)	Reference
Acari			
Pergamasus franzi	Austria	2,600	Janetschek (1993)
Bdella iconica	Austria	2,600	Janetschek (1993)
Araneae			
Euophrys omnisuperstes	Nepal	6,700	Janetschek (1990), Wanless (1975)
Diplocephalus rostratus	Austria	3,100+	Thaler (1988)
Erigone tirolensis	Austria	3,100	Thaler (1988)
Hilaira montigera	Austria	3,100	Thaler (1988)
Lepthyphantes armatus	Austria	3,100	Thaler (1988)
Lepthyphantes baebleri	Austria	3,100	Thaler (1988)
Xysticus bonneti	Austria	3,100	Thaler (1988)
Collembola			
Hypogastrura himalayana*	Nepal	5,500	Janetschek (1990)
Tomocerus nepalicus*	Nepal	5,600	Janetschek (1990), Mani (1968)
Isotoma mazda	Nepal	6,000	Janetschek (1990), Mani (1968)
Isotoma saltans*	Switzerland	4,600	Bäbler (1910)
Isotoma saltans	Austria	2,600	Kopeszki (1988), Janetschek (1993)
Isotomurus palliceps*	Austria	2,600	Janetschek (1993)
Coleoptera			
Bembidion glaciale	Austria	2,600	Janetschek (1993)
Nebria castanea	Austria	2,600	Janetschek (1993)
Atheta sp.	Nepal	4,875	Mani (1962)
Lepidoptera			
Boloria pales	Austria	2,600	Janetschek (1993)

*On ice and glaciers.

concentrates near the soil surface under stones and in silken chambers (Wanless, 1975; Meyer and Thaler, 1995). In the Italian Alps, winter activity of nival invertebrates was monitored, with typical winter species of springtails, stoneflies, flies, and snow scorpionflies on snow; winter activity was also seen in wasps, some butterflies, and beetles (Masutti, 1978).

The aeolian fauna, passively deposited on alpine snow fields by updrafts of wind from lower elevations, provides a food source for the resident nival arthropods, birds, and mammals (Mann et al., 1980). Numbers are highest in July, and the specimens mainly consist of dipterans (flies), homopterans (true bugs), coleopterans (beetles), and hymneopterans (wasps), all of which are characterised by dispersal flights (Kaisila, 1952; Edwards, 1972, 1987; Ashmole et al., 1983; Heiniger, 1989; Edwards and Sugg, 1993). The diversity of this fauna decreases with increasing latitude, and when there is little dust and/or vegetation on the snow, there are also few arthropods (Meyer and Thaler, 1995).

Activity by invertebrates at low temperatures may be maintained by following a thermal gradient that is within their preferred temperature range. It has been suggested that they have little competition for food (Viramo, 1983), that the snow surface simplifies mate location (Leinaas, 1981a), and that they have fewer predators (Hågvar and Østbye, 1973; Jonsson and Sandlund, 1975; Leinaas, 1981a; Aitchison, 1984a).

5.2.2   Subnivean, Intranivean, and Supranivean Fauna

During winter many invertebrates inhabit the stable microenvironment of the subnivean space. The subnivean fauna include oligochaetes, molluscs, centipedes, pseudoscorpions, phalangids, spiders, mites, springtails, beetles, flies, wasps, and other insects (Holmquist, 1926; Palmén, 1948; Mezhzherin, 1958; Mani, 1962; Polenec, 1962; Näsmark, 1964; Huhta, 1965; Kawakami, 1966; Kühnelt, 1969; Oswald and Minty, 1970; Berman et al., 1973; Willard, 1973; Thaler and Steiner, 1975; Granström, 1977; Olynyk and Freitag, 1977; Aitchison, 1978a, 1978b, 1979a, 1979b, 1979c, 1979d, 1979e, 1979f, 1983, 1984a, 1984b, 1984c, 1987, 1989; Flatz, 1979; Puntscher, 1979; Flatz and Thaler, 1980; Leinaas, 1981a; Green, 1982; Merriam, Wegner, and Caldwell, 1983; Viramo, 1983; Schmidt and Lockwood, 1992).

Small arthropods move about in their protected microenvironment in the temperature range of $0°C$ to $-10°C$ (Näsmark, 1964; Aitchison, 1974, 1978a; Schmidt and Lockwood, 1992). Most of these invertebrates are not active below $-5°C$ (Huhta, 1965; Aitchison, 1979b, 1979c, 1979d, 1979e, 1979f; Leinaas, 1981a; Merriam et al., 1983), with the exception of mites, spiders, springtails, and some wasps (Aitchison, 1978a, 1979a, 1979c, 1979d).

The major winter-active groups to be discussed are the mites, spiders, springtails, beetles, flies, and wasps. Probably long before the subnivean fauna was discovered, insects and spiders were noticed on snow. At the Arctic Circle in the former Soviet Union (FSU), Novikov (1940) observed active insects and spiders on snow at $-1°C$ and concluded that their activity on snow depended on the temperature regime of

233

the subnivean space and not on the thickness of the snow cover. Many have noticed this supranivean invertebrate activity (Chapman, 1954; Wolska, 1957; Buchar, 1968; Østbye, 1966; Hågvar, 1971, 1973; Huhta and Viramo, 1979; Koponen, 1983).

Mites (order Acari) are ubiquitous small arachnids, which live predominantly on the soil surface and within the soil. They have even been collected at 5,000 m in the Himalayas (Mani, 1962), on the sub-Antarctic islands (Dalenius and Wilson, 1958), and in the Antarctic (Sømme, 1993), often at subzero temperatures (Dalenius and Wilson, 1958; Sømme, 1993). They move into the humus as temperatures cool in autumn. A number have low thermal preferences (Wallwork, 1970). On the Canadian prairies, Willard (1973) and Aitchison (1979c) found increases in subnivean numbers in February and March. This is a time of increased snow density and metamorphosis, which allows greater light penetration (Geiger, 1965; Evernden and Fuller, 1972), and possibly a response to increased intensity of incident solar radiation from 19 to 55 W/m^2 (Budyko, 1963). Kevan (1962) also noted that mites were active under snow. Likewise, in Australia and Finland subnivean pitfall traps collected many mites (Green, 1982; Viramo, 1983). Although the ecology of the family Rhagidiidae is virtually unknown (Zacharda, 1980), this family is one of the most abundant subnivean, winter-active acarine groups in North America. Up to 4,800 m in the Himalayas rhagidiids have been collected from the upper tree line and at the foot of a glacial moraine (Zacharda and Daniel, 1987). The families Eupodidae and Parasitidae are also common winter-active groups (Aitchison, 1979c; Schmidt and Lockwood, 1992). Table 5.2 shows examples of nival fauna from different parts of the world.

Spiders (order Araneae) are quite active in the subnivean space, down to −8°C (Aitchison, 1978a). Their activity at subzero temperatures in the subnivean micro-climate has been documented in Europe (Mezhzherin, 1958; Näsmark, 1964; Huhta, 1965; Granström, 1977; Schaefer, 1977a; Flatz, 1979; Puntscher, 1979; Flatz and Thaler, 1980) and in North America (Holmquist, 1926; Olynyk and Freitag, 1977; Aitchison, 1978a, 1984b; Merriam et al., 1983; Schmidt and Lockwood, 1992). In south central Canada five families (Linyphiidae, Erigonidae, Lycosidae, Clubionidae, and Thomisidae) are winter active, especially the small erigonid spiders, such as *Ceraticelus* spp., comprising 34.6 percent of the catch and being active down to −8°C (Aitchison, 1978a, 1984b). Immature *Agroeca* and *Pardosa* and adult linyphiids maintained activity during winter months (Holmquist, 1926; Wolska, 1957; Almquist, 1969; Huhta, 1971; Hågvar, 1973; Thaler and Steiner, 1975; Aitchison, 1978a).

Spiders are often encountered on snow surfaces (Levander, 1913; Chapman, 1954; Mani, 1962; Thaler and Steiner, 1975; Huhta and Viramo, 1979; Koponen, 1983, 1989; Fox and Stroud, 1986; Janetschek, 1993). Figure 5.1 shows *Bolyphantes index* in its

234

Table 5.2. *Arachnids associated with snow cover in different countries, location in the snow cover and temperature range of activity.*

Species	Country	Location*	Temperature (°C)	Reference
Acari				
Rhagidia sp.	Nepal	SP?	>−10	Mani (1962)
Rhagidia sp	Canada	SB	>−10	Aitchison (1979c)
Bdella sp.	Nepal	SP?	>−10	Mani (1962)
Bdella sp	Canada	SB	>−10	Aitchison (1979c)
Parasitus sp.	Nepal	SP?	>−10	Mani (1962)
Parasitus sp.	Canada	SB	>−10	Aitchison (1979c)
Parasitus sp.	USA	SB	>−5	Schmidt and Lockwood (1992)
Eupodes sp.	USA	SB	>−5	Schmidt and Lockwood (1992)
Evadorhagidia sp.	USA	SB	>−5	Schmidt and Lockwood (1992)
Araneae				
Ceraticelus spp.	Canada	SB	>−8	Aitchison (1978a, 1984b)
Agroeca spp.	Canada	SB	>−6.8	Aitchison (1978a)
Pardosa spp.	Canada	SB	>−6.2	Aitchison (1978a)
Lepthyphantes cristatus[†]	Czechoslovakia	SP	0	Buchar (1968)
Centromerus incilium[†]	Norway	SP	0	Hågvar (1973)
Bolyphantes index[†]	Norway	SP	>−5	Østbye (1966)
Bolyphantes index	Finland	SP	0	Huhta and Viramo (1979), Koponen (1983, 1989)
Macrargus rufus	Finland	SP	+2 to −2	Huhta and Viramo (1979), Koponen (1983, 1989)
Tmeticus affinis	Finland	SP	+2 to −2	Huhta and Viramo (1979), Koponen (1983, 1989)
Helophora insignis	USA	SB	−4 to −5	Schmidt and Lockwood (1992)
Gnaphosa intermedia	Finland	SP	0	Koponen (1983, 1989)
Pseudoscorpionida				
Neobisium spp.	Spain, Germany	SP	−1 to −3	Schwaller (1980)
Microbisium spp.	Canada	SB	>−4	Aitchison (1979f)
Phalangida				
Undetermined species	Finland	SP	>−1.5	Levander (1913), Viramo (1983)

*SB – subnivean; SP – supranivean.

[†] Webs over snow depressions.

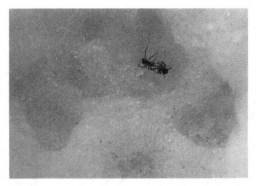

Figure 5.1. A pair of *Bolyphantes index* in a web
constructed over a fox footprint in snow
(Hågvar, 1973).

web over a fox footprint in the snow cover. The families Linyphiidae, Erigonidae, Thomisidae, Gnaphosidae, Lycosidae, and Salticidae frequent the snow surface as well as the subnivean space (Buchar, 1968; Mani, 1968; Hågvar, 1973; Thaler and Steiner, 1975; Aitchison, 1978a, 1984c; Huhta and Viramo, 1979). Subnivean trapping produced a richer fauna, as some species never migrate to the snow surface (Koponen, 1989). Table 5.1 provides some details about winter-active spiders.

Other arachnids on the snow surface include pseudoscorpions (Schwaller, 1980) and phalangids (harvestmen) (Levander, 1913; Viramo, 1983). The former also are active in the subnivean space (Aitchison, 1979f) (Table 5.1).

Springtails (order Collembola) are a major group of winter-active insects that are active in the subnivean space at temperatures down to $-7.8°$C (Kevan, 1962; Aitchison, 1979a) (see also Table 5.3). Activity peaks were seen in late winter (February and March), as with the mites (Wallwork, 1970; Willard, 1973), perhaps because of greater light penetration through the denser snow cover (Geiger, 1965; Evernden and Fuller, 1972). The family Isotomidae is dominant in the Arctic, with cold-hardy *Isotoma viridis* active on snow (Agrell, 1941; Chapman, 1954). This is also a common winter-active, subnivean species in continental North America (Aitchison, 1979a; Schmidt and Lockwood, 1992), along with other species of *Isotoma*. Isotomids were also collected in subnivean pitfall traps in the Australian Alps (Green, 1982).

Collembolans are commonly seen on snow at about $0°$C (Table 5.2) and are abundant at high altitude, with some species even living on permanent snow and ice at 6,000 m in the Himalayas (Mani, 1962, 1968). In the European mountains *Isotoma saltans* (see Figure 5.2) is associated with glacier ice and is often called "Gletscherfloh," or glacier flea (Strübing, 1958). Especially abundant on melting snow (Chapman, 1954),

Table 5.3. *Collembolans associated with snow cover in different countries, location in the snow cover, and temperature range of activity.*

Species	Country	Location*	Temperature (°C)	Reference
Dicyrtomina rufescens	Japan	SP	eats at −1	Uchida and Fujita (1968)
Hypogastrura socialis	USA	SP	eats at 0	MacNamara (1924)
Hypogastrura socialis	Finland	SP	0	BK & BK (1980)[†]
Hypogastrura socialis	Norway	SP	migrates at 0	Hågvar (1995)
Hypogastrura spp.	Nepal	SP	0	Mani (1962)
Hypogastrura spp.	Norway	SP	0	Østbye (1966)
Hypogastrura spp.	Finland	SP	0	Levander (1913), Koponen (1983)
Hypogastrura spp.	Germany	SP	0	Strübing (1958)
Entomobrya nivalis	Finland	SP	0	BK & BK (1980)[†]
Entomobrya spp.	USA	SB	0	Holmquist (1926)
Tomocerus flavescens	USA	SB	0	Holmquist (1926)
Tomocerus flavescens	Finland	SP	0	BK & BK (1980)[†]
Tomocerus flavescens	Canada	SB	eats at −2	Aitchison (1983)
Tomocerus spp.	USA	SP	eats at 0	Knight (1976)
Lepidocyrtus cyaneus	Canada	SB	eats at −2	Aitchison (1983)
Lepidocyrtus lignorum	Finland	SP	0	BK & BK (1980)[†]
Orchesella bifasciata	Finland	SP	0	BK & BK (1980)
Orchesella ainslei	Canada	SB	eats at −2	Aitchison (1983)
Isotoma alpa	USA	SB	>−4	Schmidt and Lockwood (1992)
Isotoma gelida	USA	SB	>−4	Schmidt and Lockwood (1992)
Isotoma hiemalis	Finland	SP	0	BK & BK (1980)[†]
Isotoma hiemalis	Switzerland	SP	>−3	Zettel (1984)
Isotoma olivacea	Poland	SP	+5 to −5	Wolska (1957)
Isotoma saltans	Poland	SP	0 to −4	Wolska (1957), Wojtusiak (1951)
Isotoma saltans	Germany	SP	+5 to −5	Strübing (1958)
Isotoma viridis	Sweden	SP	>−8	Agrell (1941)
Isotoma viridis	Canada	SB	eats at −2	Aitchison (1983)
Isotoma viridis	USA	SB	>−4	Schmidt and Lockwood (1992)

*SB – subnivean; SP – supranivean.
[†]BK & BK, Brummer-Korvenkontio and Brummer-Korvenkontio.

237

Figure 5.2. Collembolans *Isotoma saltans* on snow
(Schaller, Vienna).

collembolan aggregations produce sooty, red, or gold snow, depending upon the colour
of the aggregating species (Strübing, 1958; Mani, 1962). In Norway, *Hypogastrura
socialis* masses on snow in mild conditions, possibly to migrate to snow-free patches
for early reproduction. This species can travel 200 to 300 m/day; or 2 to 3 km in
10 days of fair weather, by steadily jumping in a certain direction (Hågvar, 1995).
In mild conditions entomobryids and isotomids are on snow cover, moving into it to
avoid freezing soil or flooding (Brummer-Korvenkontio and Brummer-Korvenkontio,
1980; Leinaas, 1983).

Solar radiation and temperature conditions affect collembolan activity on the snow
cover; the former can raise the body temperature of *Isotoma violacea* several degrees
Celsius (Fox and Stroud, 1986). Nival collembolans have many pigment granules in
their epidermal cells as protection against strong solar radiation (Eisenbeis and Meyer,
1986; Kopeszki, 1988). The activity of some of these animals has been documented
down to −6°C (Levander, 1913; Tahvonen, 1942; Chapman, 1954) and to −9.8°C
(Tahvonen, 1942). The snow surface activity by *Isotoma hiemalis* is limited by temperature (down to −3°C) and changing barometric pressure, causing the species to
aggregate in depressions with warmer microclimates (Zettel, 1984). In the maritime
climate of southern Norway springtails can be found throughout the snow cover, which
has a minor temperature gradient of only a degree or so (Leinaas, 1981a, 1983). Ice
layers within the snow can prohibit migration to the surface by springtails, which
constantly move up and down in the snow cover (Zettel, 1984; Kopeszki, 1988).

238

Some beetles (order Coleoptera) are active at temperatures near 0°C (Holmquist, 1926; Chapman, 1954; Renken, 1956; Wolska, 1957; Näsmark, 1964; Aitchison, 1979b). The winter-active families include Staphylinidae, Carabidae, and Cantharidae, especially the small species of *Atheta*, *Bembidion*, and *Philonthus* (Chapman, 1954; Heydemann, 1956; Renken, 1956; Näsmark, 1964; Wallwork, 1970; Aitchison, 1979b; Viramo, 1983; Schmidt and Lockwood, 1992). Active cantharid larvae, often associated with winter, have been called "Schneewürme" (Renken, 1956) and prefer cool temperatures (Wolska, 1957). Heydemann (1956) found that below −4°C beetles ceased moving. In Finland, Palmén (1948) noted survival of staphylinids and carabids below snow from 0° to −5.5°C. In the subnivean microclimate carabids stay active, with some inactivity of adults in midwinter (Holmquist, 1926; Flatz and Thaler, 1980; Green, 1982).

The beetles may also be active on the snow surface, especially the carabids and staphylinids, which often prefer low temperatures (Tahvonen, 1942; Chapman, 1954; Wolska, 1957; Viramo, 1983). Likewise, the nival coleopterans, consisting of 45 percent carabids and 17 percent staphylinids, are mostly small and peculiar to the nival zone above 4,000 m but not to the taiga zone (Mani, 1962). In the Himalayas, *Bembidion* was the most important carabid genus (over 25 species from the nival zone), and *Atheta* was the dominant staphylinid genus (Mani, 1962). The same genera are active in the subnivean space in Canada (Aitchison, 1979b) (Table 5.1).

A number of flies (order Diptera), especially the tipulid genus *Chionea*, are associated with snow in the arctic and alpine regions – e.g., the families Anthomyidae, Chironomidae, Simuliidae, Stratiomyidae, Tipulidae, Trichoceridae, Mycetophilidae, Phoridae, Sciaridae, and Sphaeroceridae (Holmquist, 1926; Tahvonen, 1942; Wojtusiak, 1951; Chapman, 1954; Renken, 1956; Mani, 1962; Dahl, 1969; Hågvar, 1971; Hågvar and Østbye, 1973; Jonsson and Sandlund, 1975; Aitchison, 1979d). Unidentified, red cecidomyiid larvae crawled around on the snow surface at about 0°C on windless days (Aitchison unpublished data). Hågvar (1971) recorded a supercooling point of −7.5°C in *Chionea araneiodes*. *Chionea lutescens* has been frequently collected on snow (Novikov, 1940; Tahvonen, 1942; Mani, 1962; Hågvar, 1971, 1976; Mendl, Müller, and Viramo, 1977; Koponen, 1983), even at −10°C (Brummer-Korvenkontio and Brummer-Korvenkontio, 1980), and in subnivean pitfall traps (Broen and Mohrig, 1964; Itämies and Lindgren, 1985) (Table 5.4). Species of the family Chironomidae of the genus *Diamesa* have been encountered walking on the snow surface and occasionally flying about at around −1°C on windless, cloudy days (Hågvar and Østbye, 1973; Jonsson and Sandlund, 1975). Chironomids have a thermal preference between 0° and −2°C (Hågvar and Østbye, 1973).

Table 5.4. *Other insects and oligocheates associated with snow cover in different countries, location in the snow cover, and temperature range of activity.*

Species	Country	Location*	Temperature (°C)	Reference
Diptera				
Chionea araneiodes	Norway	SP	0 to –4	Sømme and Østbye (1969)
Chionea lutescens	Poland	SP	>–5	Wolska (1957)
Chionea lutescens	Germany	SP	>–4	Strübing (1958)
Chionea lutescens	Finland	SP	–2.3	Tahvonen (1942)
Chionea lutescens	Finland	SB	>–10	Itämies and Lindgren (1985)
Chionea spp.	Finland	SP	>–2.3	Tahvonen (1942)
Chionea sp.	Poland	SP	>–5.6	Wojtusiak (1951)
Chionea sp.	USA	SP	>–5.6	Chapman (1954)
Chionea sp.	FSU	SP	>–6	Novikov (1940)
Chionea sp.	Norway	SP	>–6	Hågvar (1971)
Chionea sp.	Nepal	SP	>–8	Mani (1962)
Exechia sp.	Sweden	SB	0	Plassman (1975)
Mycetophila sp.	Sweden	SB	0	Plassman (1975)
Mycetophila sp.	Canada	SB	>–4	Aitchison (1979d)
Suilla longipennis	Canada	SB	>–4	Aitchison (1979d)
Megaselia spp.	Canada	SB	>–4	Aitchison (1979d)
Bradysia spp.	Canada	SB	>–4	Aitchison (1979d)
Leptocera spp.	Canada	SB	>–4	Aitchison (1979d)
Diamesa spp.	Nepal	SP	0 to –7.2	Kohsima (1984)
Diamesa spp.	Nepal	SP	–16	Kohsima (1984)
Musidora lutea	Finland	SP	–6	Tahvonen (1942)
Mecoptera				
Boreus spp.	Finland	SP	–1	Tahvonen (1942), Viramo (1983)
Boreus spp.	Germany	SP	–1	Strübing (1958)
Boreus spp.	Norway	SP	–5.5	Fjellberg and Greve (1968)
Oligochaeta				
Lumbricids and enchytraeids	FSU	SB		Berman et al. (1973)
Lumbricids	Canada	SB	>+2	Aitchison (1979e)

*SB – subnivean; SP – supranivean.

In the subnivean space, flies of the families Anthomyidae, Mycetophilidae, Phoridae, Sciaridae, and Sphaeroceridae (species of *Suilla*, *Mycetophila*, *Megaselia*, and *Bradysia*) are winter-active (Holmquist, 1926; Broen and Mohrig, 1964; Plassman, 1975; Aitchison, 1979d; Merriam et al., 1983; Viramo, 1983). All are cold resistant (Renken, 1956), and some are active to −5°C (Dahl, 1969). In southern Canada this order ceased subnivean movement below −3.5°C (Aitchison, 1979d) (Table 5.4).

The winter-active wasps (order Hymenoptera) include the families Ceraphronidae, Diapriidae, and Scelionidae, which cease movement at −5.5°C (Aitchison, 1979d). Winter-active genera include *Ceraphron*, *Dendrocerus*, *Belyta*, *Scelio*, and *Trimorus* spp. (Aitchison, 1979d). Of other insects, one group that is commonly winter-active, even under the snow cover, is the true bug family Cicadellidae (leafhoppers) (order Homoptera) (Holmquist, 1926; Aitchison, 1978b; Green, 1982; Viramo, 1983), whose nymphs are active at subnivean temperatures of −1.5°C to −4.5°C (Aitchison, 1978b).

Another well-known winter insect is the snow scorpionfly *Boreus* spp. (order Mecoptera), whose cold-resistant adults were observed jumping on snow (Tahvonen, 1942; Strübing, 1958; Fjellberg and Greve, 1968; Viramo, 1983) (Table 5.4). On the snow cover these dark insects absorb solar radiation to increase their body temperature on sunny days as much as 4°C above the ambient air temperature. When the air temperature drops below −3°C, the insects retreat to the warmer subnivean space (Shorthouse, 1979; Courtin, Shorthouse, and West, 1984). The thermal preferendum of this genus is between 0° and −1°C (Wojtusiak, 1951; Fjellberg and Greve, 1968).

Figure 5.3 and its commentary delightfully depict the snow fauna commonly seen in boreal situations in Canada. In this fantasy by Flahey, different species of collembolans act as sled dogs, with *Chionea* and *Boreus* acting as "mushers," while mites cheer them on in their race.

The rock crawlers of the genus *Grylloblatta* (order Grylloblattodea) are flightless carnivores and scavengers from mountains and are associated with snow and low temperatures down to −6°C (Chapman, 1954; Strübing, 1958; Pritchard and Scholefield, 1978; Edwards, 1987). These insects are darkly coloured and readily absorb the sun's rays on the snow surface (Pritchard and Scholefield, 1978).

Within glaciers, *Mesenchytraeus solifugus* or "ice worms" of the Oligochaete family Enchytraeidae survive year round at approximately 0°C in polar and temperate regions, appearing on the snow surface at dusk and burrowing into ice by day (Janetschek, 1955; Goodman, 1971).

241

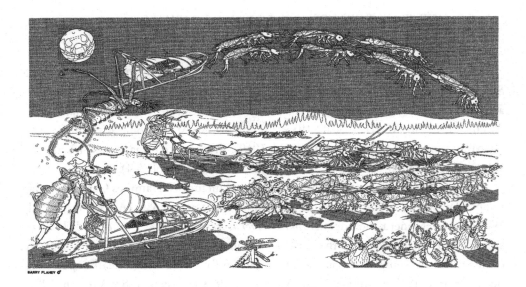

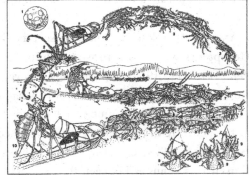

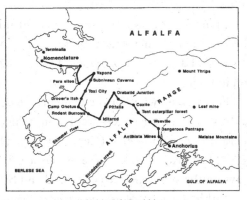

I would like to thank the following, for without their unwitting participation this card would not have been possible. Not all the insects are drawn to scale, but are depicted as accurately as possible with the aid of stereo and compound microscopes. They are listed under Common name, Scientific name, Family name, Order under which they are classified and location where all were collected.

Je tiens à remercier les participants (voluntaires ou non) suivants sans lesquels la réalisation de cette carte aurait été impossible. Les insectes ainsi que l'équipement entomologique ne n'ont pas tous été dessinés à l'échelle mais ont été représentés avec autant de précision que possible à l'aide du microscope et à la loupe binoculaire. Ils ont été décrits d'après leur nom commun, leur nom scientifique, leur nom de famille et le lieu de récolte de la majorité des échantillons.

1 Winter stone fly, Taeniopteryx nivalis, Nemouridas, Plecoptera, Orleans Island, Quebec.
2 Winter snow fly, Chionea alexandriana, Tipulidae, Diptera, Gatineau Park, Quebec.
3 Springtail, Isotoma viridis, Isotomidas, Collembola, Belleville, Ontario.
4 Leiodid beetle, Agyrtes longulus, Agyrtidae, Coleoptera, Salmon Arm, B.C.
5 Winter tick, Dermacentor albipictus, Ixodidae, Acari, Fort Smith, N.W.T.
6 Dog flea, Ctenocephalides canus, Ctenocephalidae, Siphonaptera, Vancouver, B.C.
7 Springtail, Tomocerus flavescens, Tomoceridae, Collembola, Ottawa, Ontario.
8 Springtail, Hypogastrura nivicola, Poduridae, Collembola, Arnprior, Ontario.
9 Predatory mite, Pergamasus crassipes, Parasitidae, Acari, St.John's, Newfoundland.
10 Snow scorpionfly, Boreus nivoriundus, Boreidae, Mecoptera, Ithaca, N.Y.

In the winter of 1925, on the frozen north-west shore of Alaska's Norton sound, the inhabitants of Nome were smitten by a diphtheria epidemic. Because the closest community was 1200 miles away in the gold-rush town of Iditarod, the problem of obtaining life-saving serum was almost insurmountable. The formidable terrain and high Arctic winds allowed only one possibility, that of sending the serum by a relay of dog sled teams. To commemorate those heroic mushers and dogs who saved the people of Nome, the Iditarod dog sled race is held annually, with over 20 checkpoints along the route.

The "Arthropod Iditarod" sled race is also held annually over quite a different route (see map). It commemorates the transportation of the antidote Atropine from the colony of Anchorius (one hundred million inhabitants per cubic metre) to rescue the small arthropod community of Nomenclature. In 1956, an intoxicated entomologist, of no fixed address, inadvertently sprayed the local insect population with an experimental organophosphate, causing outbreaks of involuntary stridulation, systemic agoraphobia, and permanent diapause.

As in the rescue attempt, most sled teams are comprised principally of winter snow fleas and exhibit the multi-directional and unpredictable movements over the snow typical of their breed. Despite the efforts of their mushers, checkpoints are rarely found. Constantly harassed by roving bands of predaceous mites, the race often takes many moults to complete.

A l'hiver de 1925, au large de la côte glacée nord-ouest de Norton Sound en Alaska, les habitants de Nome étaient pris d'une épidemie de diphtérie. La communauté la plus rapprochée, étant à 1200 miles de la ville Iditarod, ville connaissant la ruée vers l'or, posait un obstacle infranchissable: comment se procurer d'un sérum qui leur sauverait la vie? Ce terrain redoutable, joint aux vents élevés arctiques, ne permettaient qu'une possibilité: il faudrait donc expédier le sérum par un relais de traîneaux tirés par des chiens. Afin de commémorer les chauffeurs de traîneaux et ces chiens héroïques qui ont sauvé la vie aux habitants de Nome, la course de traîneaux tirés par chiens de Iditarod est devenue internationale et a lieu annuellement avec plus de vingt points de contrôle placés au long de la route.

La course de traîneaux "Arthropode Iditarod" a lieu annuellement sur une route très différente (Voir carte). Celle-ci commémore la transportation de l'antidote Atropine de la colonie Anchorius, (population de cent millions habitants par mètre carré) afin de sauver la petite communauté d'arthropodes à Nomenclature. En 1956, un entomologiste ivre, sans domicile connu, pulvérisa la population d'insectes locale avec un organophosphate, provoquant chez ces derniers des manifestations de stridulation involuntaires et d'agorophobie systématique.

Contenant principalement des puces d'hiver, la plupart des traîneaux tirés par chiens manifestent des mouvements sur la neige totalement multi-directionnels et irrationnels. En dépit des efforts des chauffeurs, les points de contrôles sont rarement trouvés. Harceler incessamment par des bandes errantes de mites prédatrices, la course requiert souvent plusieurs mues pour compléter.

Arthropod 'Iditarod' International Sled Race
Course internationale de collemboles d'Iditarod

Ant, cipated route of Arthropod sled race

Information regarding cards and posters in this series can be obtained by writing to the artist; Barry Flahey, P.O. Box 298, Manotick, Ontario, Canada, K4M 1A3.

Figure 5.3. Fantasy card of winter-active invertebrates by Barry Flahey, with a description of the race involved.

5.2.3   Physiological and Morphological Mechanisms

There are several physiological mechanisms by which invertebrates can tolerate the low temperatures of winter: cold-hardiness and thermal hysteresis (the difference between the melting and freezing points of the haemolymph). Factors determining cold-hardiness are the ability 1) to resist freezing, 2) to supercool (avoiding freezing below the freezing point of haemolymph by introducing "antifreeze" agents), and 3) to survive long exposures to low temperatures (Asahina, 1966; Leinaas, 1983; Sømme, 1989). Invertebrates fall into two groups: the freezing-resistant group, which tolerates intracellular freezing of body tissues (e.g., insect larvae and pupae), and the freezing-sensitive group, which avoids intracellular freezing but tolerates extracellular freezing by means of lowering the freezing point and supercooling points of the haemolymph and/or the accumulation of low molecular weight, cryoprotective polyhydric alcohols such as glycerol (Salt, 1961; Asahina, 1966; Storey, 1984; Sømme, 1989; Lee, 1991). Inactive invertebrates in diapause tolerate low temperatures by supercooling (accumulating cryoprotective compounds that lower freezing and supercooling points) and by dehydration, both of which increase haemolymph viscosity (MacLean, 1975; Husby and Zachariassen, 1980; Sømme, 1989). The supercooling point is also affected by the amount of food in the gut and the presence of moisture (Sømme, 1989). In *Entomobrya nivalis* cold-hardiness is influenced by temperature and photoperiod (Zettel and Allmen, 1982). Invertebrates are capable of muscular and metabolic activity when supercooled (MacLean, 1975). Winter-active invertebrates do not appear to supercool; it seems that thermal-hysteresis proteins permit them to move (Duman, 1979; Husby and Zachariassen, 1980).

Another physiological adaptation found in this fauna is that of anaerobiosis. There are times when insects become completely encased in ice, resulting in periods of oxygen deficiency. For example, in Norwegian mountains the carabid beetle *Pelophila borealis* can live without oxygen for 6 months; arctic and polar mites, dipterans, and springtails also can undergo anaerobiosis (MacLean, 1975; Sømme, 1993).

Feeding may be deleterious for winter-active invertebrates that ingest food particles containing dust that act as potential ice nucleators (Salt, 1953, 1961). For many insects at the start of winter, cessation of feeding and emptying the gut of ice crystal nucleators enhances their ability to supercool; with resumption of feeding in the spring, the insects lose the capability to supercool (Ohymama and Asahina, 1972; Østbye and Sømme, 1972; Sømme, 1989). The carabid beetle *Pterostichus brevicornis* changed its diet in autumn from insects to dry wood and then ceased feeding in early winter at about $-4°C$ (Kaufmann, 1971).

In south central Canada winter-active, subnivean invertebrates were all frost sensitive, with freezing points of haemolymph between $-7°C$ and $-8°C$ (Aitchison and
Hegdegar, 1982). These animals were prohibited in their movements, as they did not
have cryoprotective compounds in their haemolymph (Husby and Zachariassen, 1980).
It appears that they use thermal hysteresis. Most winter-active collembolan specimens,
however, do have empty guts below $0°C$ (Leinaas, 1981a; Aitchison, 1983).

Most of these cold-active invertebrates exhibit a low temperature preference, often
associated with a low supercooling point. The winter-active spider *Bolyphantes index*
preferred $4.1°C$, maintained normal activity down to $-5°C$, and had a chill-coma
temperature of $-9.3°C$ and a supercooling point of $-15.3°C$ (Hågvar, 1973). The
preferred temperatures are above the chill-coma values. Likewise, the springtail *I.*
hiemalis, which preferred $-2.5°C$ and above, was active down to $-6°C$, experienced
chill-coma at $-8°C$, and had a supercooling point of $-15°C$ (Zettel, 1984). *Chionea*
collected on snow between $0°C$ and $-6°C$ had a supercooling temperature of $-7.5°C$
(Sømme and Østbye, 1969). Mani (1962) noted pronounced cold stenothermy in nival
insects, around $0°C$, often with normal development, growth, and metamorphosis
between $-1.5°C$ and $1.7°C$. The mites and springtails are the arthropods most tolerant
of cold (Mani, 1962; Sømme, 1993).

Some springtails undergo a unique morphological change from summer to
winter, which reverses in the spring, called cyclomorphosis (Fjellberg, 1976, 1978a;
Leinaas, 1981b; Zettel, 1984). In late autumn during a moult (ecdysis) to the winter
morph, the locomotory furcula and tibiotarsae change from the simple summer
form to a more complex winter form, which is enlarged and has more setae (see
Figures 5.4 and 5.5). Zettel (1984) noted that only those specimens of *I. hiemalis* that
undergo cyclomorphosis (morph *I. h. hiemalis*) move to the snow surface, while those

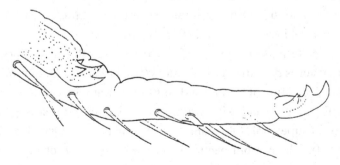

Figure 5.4. Furca (dens and mucro) of a specimen of the
collembolan *Isotoma hiemalis* in ecdysis showing transformation
from the *mucronata* form to the *hiemalis* form (from Fjellberg, 1976).

244

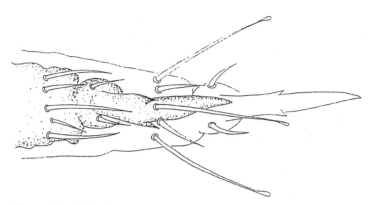

Figure 5.5. Collembolan *Isotoma nivea*: tibiotarsus II of a specimen in ecdysis, showing transformation from the *Vertagopus* form (clavated spur hairs) to the normal form (dotted) (from Fjellberg, 1978a).

that do not change (morph *I. h. mucronata*) stay in the subnivean space. Fjellberg (1978a) observed an even more pronounced change, a taxonomic artefact: an autumn specimen of an unknown species of springtail *Vertagopus* appeared very similar to *Isotoma nivea* with different tibiotarsal spur hairs. In spring, specimens of this "species" were found in ecdysis, changing from *Vertagopus* to *I. nivea* (Fjellberg, 1978a)! In Manitoba, winter specimens of *Isotoma manitoba* and *Isotoma blufusata* underwent cyclomorphosis (Fjellberg, 1978b).

5.2.4   Food Webs

The aeolian fauna, deposited onto snow fields from wind updrafts from lower elevations, contain wind-borne pollen grains, microorganisms, and arthropod fallout, an important nutrient source ("manna") in montane systems for resident nival arthropods, birds, and mammals (Swan, 1961; Mann et al., 1980; Edwards, 1987). On the snow surface food chains are short with few specific interactions. Most other snow insects are scavengers or carnivores that could feed on animals such as springtails. In Chapter 4 Hoham and Duval discuss a microbial food web that may serve as a basis for the higher invertebrate component of a nival food web. Snow communities have many carnivorous and predatory members.

Certainly many of these animals do feed at temperatures near $0°C$ – e.g., springtails (MacNamara, 1924; Mani, 1962; Uchida and Fujita, 1968; Edwards, 1972; Eisenbeis and Meyer, 1986). Cold room experiments with winter-collected specimens fed on fungal hyphae at $2°C$ and $-2°C$ (Aitchison, 1983) corroborate this. These collembolans may well form the basis of a subnivean and/or nival food chain (Mani, 1962; Aitchison,

245

1984b). The herbivorous chironomid fly *Diamesa* feeds on algae in a high Nepalese glacier (Kohshima, 1984). When food is scarce, the nocturnal nival fauna is active for longer periods. Generally the small invertebrates are active by day and the larger ones are active by night, thus avoiding predation (Mann et al., 1980). Fox and Stroud (1986) found a strong correlation between the numbers of spiders and their collembolan prey on the snow surface. There are very few beetles in this habitat (Thaler, 1989; Janetschek, 1990, 1993). Common prey of spiders, springtails (Huhta, 1965) were fed upon by winter-active spiders in laboratory experiments (Aitchison, 1984a). Other possible predators include grylloblattids, which feed at low temperatures (Mann et al., 1980), carabid beetles, and mites (Mani, 1962).

The arachnids are detritivous (mites) or predatory (mites and spiders). Mites in large numbers (e.g., genus *Rhagidia*) wander on the snow surface and usually are associated with springtails (Mani, 1962), on which they probably prey (see Tables 5.1 and 5.2). Winter-active spiders, such as *Lepthyphantes cristatus*, feed at −2°C on surface-active springtails and small flies (Polenec, 1962; Buchar, 1968; Hågvar, 1973; Schaefer, 1976, 1977a; Aitchison, 1984b, 1989) (Table 5.1). Fox and Stroud (1986) found a strong correlation between spider numbers and their collembolan prey, while Aitchison (1989) presented data on winter-active collembolans and spiders, implicating a possible sub-nivean food chain.

Springtails are capable of feeding at temperatures near 0°C, eating blue-green algae from tree trunks (Uchida and Fujita, 1968), red snow algae (Kopeszki, 1988), coniferous pollen on the snow surface (MacNamara, 1924), organic debris (Knight, 1976), kryokonite (fine micaceous particles with algae, protozoans, rotifers, tardigrades, pollen grains, and detritus) (Eisenbeis and Meyer, 1986; Schaller, 1992), and fungal hyphae (Aitchison, 1983) (Table 5.2). Snow algae in the snow cover (Hoham and Duval, Chapter 4) could provide a good food source for intranivean collembolans. At subzero temperatures the lack of feeding avoids the problem of ingesting dust particles into the gut, which may act as ice nucleators causing freezing and subsequent death (Salt, 1953; Joosse and Testerink, 1977). Subnivean onychiurids had no gut contents (Aitchison, 1983), and *Onychiurus subtenuis* under snow consumed little, but once freed of the snow cover fed voraciously (Whittaker, 1981).

Other winter-active invertebrates also feed on a variety of substances. With a carnivorous summer diet, the carabid *Pterostichus brevicornis* fed on rotten wood down to −4°C during winter. This dietary change from insects to dry wood possibly avoids inoculation of the gut with ice nucleators (Kaufmann, 1971). Nepalese *Diamesa* species may feed down to −7.2°C on blue-green algae (Kohshima, 1984) (Table 5.4). Winter-active wasps are parasitic and possibly in search of hosts (usually dipteran). The

246

insects of the genus *Grylloblatta* are voracious at 4°C (Strübing, 1958) and forage on snow fields (Mann et al., 1980), where they feed on species of *Chionea* (in Alberta) (Pritchard and Scholefield, 1978). Ice worms eat cryobiont snow algae and have a low rate of respiration (Goodman, 1971; Mann et al., 1980) (Table 5.4).

5.2.5   Life Cycles and Development

Although they may be active and feed between −2°C and −4°C (Aitchison, 1983, 1984b), most invertebrates do not develop or reproduce at subzero temperatures (Leinaas, 1981a; Aitchison, 1984b, 1984c; Schmidt and Lockwood, 1992). No winter development in collembolans has been noted (Leinaas, 1981a, 1983; Aitchison, 1984c), with the exception of *Isotoma saltans*, which develops between 0°C and 2°C (Strübing, 1958). The population densities of winter-active invertebrates are poorly known (Hågvar, 1971, 1973; Hågvar and Østbye, 1973; Aitchison, 1979a, 1979b, 1979c, 1979d).

At cool winter temperatures, there was no growth in spiders and their life cycles were prolonged (Huhta, 1965; Almquist, 1969; Schaefer, 1976, 1977a, 1977b; Aitchison, 1984c), as they are at high altitudes (Schmoller, 1970). Those that mate on the snow surface during winter frequently do so in webs over depressions (Buchar, 1968; Hågvar, 1973) – e.g., *Bolyphantes index* (see Figure 5.1). Other species, like the linyphiid *Centromerus bicolor*, also reproduce in winter (Schaefer, 1976, 1977a, 1977b; Flatz and Thaler, 1980). Winter-active species, such as *Centromerus sylvaticus*, have high winter mortality compared with species not active during winter (Schaefer, 1977b).

The females of the dipteran *Chionea araneiodes* have fully developed eggs in January and lay eggs at 0°C in winter (Hågvar, 1976). The genus *Diamesa* is active on the snow surface at around 0°C, sometimes *in copula*, and its aquatic larvae develop only at temperatures less than 5°C (Hågvar and Østbye, 1973; Jonsson and Sandlund, 1975).

Densities of supranivean *Boreus* spp. are high in January and February (Fjellberg and Greve, 1968). In late winter on sunny days with ambient air temperatures at about 0°C, copulating adults are often on the snow surface (Wojtusiak, 1951; Chapman, 1954; Shorthouse, 1979) (Table 5.3).

5.3   Vertebrates

5.3.1   Subnivean Vertebrates

Vertebrates are also active under, in, and on the snow cover. Most subnivean animals are small mammals weighing less than 250 g (Pruitt, 1984), such as the

247

microtines and insectivores, together with their predators, mustelid weasels. Before the hiemal threshold is established, shrews and the microtine *Clethrionomys* are active on the snow surface; after its establishment, they go underneath (Soper, 1944; Pruitt, 1972). Under insulative, thick snow cover and with an adequate food supply, the brown lemming, *Lemmus sibericus*, at Barrow, Alaska, can reproduce in winter months (Batzli et al., 1980). On Hokkaido, the northern island of Japan, Kawakami (1966) observed the Alpine Salamander *Hynobius retardatus* as extremely active down to −8°C between December and March under snow. These animals were found under 1 m of snow in burrows 50 cm deep. In early spring they briskly ran up onto the snow cover to ponds and laid eggs between 0°C and −4°C (Kawakami, 1966).

Of the small mammals, the more omnivorous cricetid rodents tend to be hibernators or in torpor (a more southerly phenomena), exhibiting little winter activity (e.g., torpid *Peromyscus* in aggregations with food caches) (Howard, 1951), while the herbivorous microtine rodents are active in the subnivean space in more northerly climes (Bergeron, 1972). Only in the subnivean space can these small mammals survive the thermal rigours of winter, with large, insulated nests in the softer pukak (Zonov, 1982; Pruitt, 1984). Many voles store a food cache up to 3 kg, so that they do not need to forage widely (pikas can store up to 20 kg); the cache size depends on the biotope (Zonov, 1982).

The region of snow cover and its density affect how the animals respond to snow. In steppes or prairies with thin snow cover, the nests are smaller (Zonov, 1982). Voles avoid areas where the snow density exceeds 150 kg m^{-3} (Spencer, 1984). Also snowmobile tracks become barriers to small mammals because of a collapsed subnivean space, and the animals cross over them rather than burrow through the dense snow (up to 400 kg m^{-3}) (Schmid, 1971).

One way voles reduce heat and moisture loss is by huddling in communal nests (West, 1977; Madison, 1984; Wolff, 1984). During winter months there is a reduction of intraspecific aggression, which allows huddling (West, 1977), especially in the genera *Microtus* and *Clethrionomys*. This mechanism is also good for predator avoidance and defense, and foraging is easier (Madison, 1984). Taiga voles consume about 90 percent of their food from winter caches, foraging one at a time for the remainder. Warmer nest temperatures, with about seven individuals per nest, decrease food requirements (Wolff, 1984). Also, those areas with vole aggregations have significantly thicker moss layers and thus more insulation (West, 1977).

Insectivores have high metabolic rates and therefore must feed almost constantly to survive (Seton, 1909; Aitchison, 1987). Shrews also favour varied habitats with litter, deep humus, or snow cover as protection (Yudin, 1964; Ackefors, 1964; Pernetta, 1976). Some behavioural adaptations in the shrews during winter are construction

248

of nests, reduced activity, and hoarding of food (Crowcroft, 1957; Dehnel, 1961; Churchfield, 1982b; Zonov, 1982). Winter foraging trips are short (about 3 min for *Sorex araneus*), with long stays in the nest and subsequent reduction of energy needs (Churchfield, 1982b; Merritt, 1986). For example, the short-tailed shrew *Blarina brevicauda* and other large shrews build elaborate winter nests (Crowcroft, 1957; Dehnel, 1961; Churchfield, 1982a). The nests and relative inactivity further reduce the metabolic rate and food requirements (Churchfield, 1982a; Aitchison, 1987).

The shrews also respond to the presence of snow cover in their microenvironment. Before the HT is established (snow cover less than 20 cm), *Sorex* and *Microsorex* are active on or near the snow surface and around logs, sometimes producing slight ridges near the surface (Seton, 1909). Once the HT is established, the animals remain in the subnivean space (Pruitt, 1970). In early spring, as the thermal character of the snow cover changes, *Sorex cinereus* may go onto the snow, at times dying of starvation or exposure when it cannot find a ventilation shaft by which to return to the subnivean space (Seton, 1909) (see Figure 5.6). In the FSU, the northern limit of shrew distribution is the −30°C mean January isotherm (Yudin, 1964), and those found in the coldest areas are the smallest *Sorex* species – e.g., *Sorex minutissimus* (mean weight about 4 g) (Mezhzherin, 1964).

Another group of vertebrates using the intranivean environment of the snow cover is birds. In extreme ambient air temperatures, small and larger gallinaceous birds will take refuge within the snow (Novikov, 1972; Korhonen, 1980b). Grouse will burrow into the

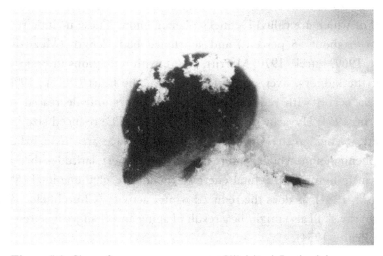

Figure 5.6. Shrew *Sorex araneus* on snow at Kilpisjärvi, Lapland, in northern Finland (A. Kaikusalo).

249

top 10 cm or so of the snow cover, a thermoneutral environment which usually does not drop below −11°C (Korhonen, 1980b); more snow mass means more insulation (Novikov, 1972). Smaller birds, such as sparrows, shelter under snow to survive within holes, subnivean tunnels, and even in subnivean vole nests (Novikov, 1972).

5.3.2 Physiological and Morphological Mechanisms

Small mammals also have some physiological constraints at low temperatures. The minute size of soricine shrews (less than 10 g) approaches the critical mass for maintaining endothermy during winter when much energy is lost as heat, putting the animal at risk and in need of accelerating its metabolic rate (Mezhzherin, 1964; McNab, 1983). In autumn as temperatures fall below the thermoneutral zone, relative food requirements and metabolic rates increase. Some of the specific physiological factors affecting the animal's activity during winter include inactivation of the thyroid (Hyvärinen, 1969), pituitary, adrenals, and parathyroid as well as changes in salivary glands and in brown adipose tissue, which is converted to heat (Rudge, 1968; Pucek, 1970; Hyvärinen, 1984), and all of which reduce metabolism and activity (Aitchison, 1987). In *Sorex araneus* almost all the endocrine system is inactivated during winter (Hyvärinen, 1984). There are great changes in body weight, more in Finnish specimens than in Polish ones (Hyvärinen, 1984). Also the larger shrew, *B. brevicauda* (about 20 g), has an increased thermogenic capacity during winter months at a constant subnivean temperature by means of nonshivering thermogenesis; physiologically this occurs through metabolism of brown adipose tissue (Merritt, 1986).

There are also some morphological adaptations to winter peculiar to the soricine shrews, all of which are called Dehnel's phenomenon. These include reduced body weight (up to about 35 percent) and shortened body length (Mezhzherin, 1964; Hyvärinen, 1969; Pucek, 1970; Merritt, 1986), with reductions in brain volume and weights of the kidneys, liver, and spleen but not the heart (Pucek, 1970). Skeletal changes also occur, with reduced intervertebral discs and decreased skull height (Mezhzherin, 1964; Hyvärinen, 1969; Pucek, 1970). The reduced size increases the hair density, giving greater insulation per unit surface area (Mezhzherin, 1964). Dehnel's phenomenon, which is more marked at northern latitudes than at southern ones, aids in minimising the total energy expenditure of the animal (Mezhzherin, 1964; Hanski, 1984), as does the reduced winter activity (Churchfield, 1982a). This shrinkage in the skull also might be a result of aging and changes in shrew population dynamics (Pruitt, 1954).

Another morphological adaptation occurs in the lemming genus *Dicrostonyx*. This animal has enlarged winter claws (see Figures 5.7 and 5.8) (Formosov, 1946) with which

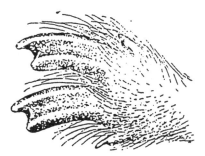

Figure 5.7. Front paw of a collared lemming, *Dicrostonyx*, with winter claws (Yamal Peninsula) (from Formosov, 1946).

it can dig; this genus made tunnels through the softer layers (20 kN m^{-2}) of the hard tundra snow of Alaska (Pruitt, 1984). The large winter claws on the forefeet (middle two claws) are highly modified for a fossorial life in snow and are present during the winter only. Lemmings of the genus *Lemmus* do not have these winter claws and stay in snow, which is much less hard and dense (Sutton and Hamilton, 1932).

5.3.3 Subnivean Food Webs

The Alpine salamander in Japan feeds on earthworms, isopods, snails, and slugs found under the snow cover (Kawakami, 1966).

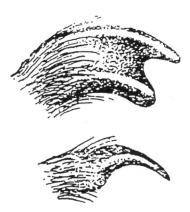

Figure 5.8. Winter and summer claws of the third digit of a *Dicrostonyx* specimen from the Yamal Peninsula (Formosov, 1946).

251

In the Arctic at Devon Island (75°N), winter nests of the collared lemming *Dicrostonyx groenlandica* are found in areas of the deepest snow accumulation. These animals ate *Dryas, Saxifraga,* and *Pedicularis* in autumn and were prey of *Mustela erminea* (Fuller et al., 1975). In taiga, grey and wood voles feed under the snow. Areas with no subnivean caches have extremely active voles; for example, root voles have no caches and forage 60 to 120 m under snow to green grass (Zonov, 1982). In alpine tundra, northern pocket gophers (*Thomomys talpoides*) have 90 percent of their activity under moderately thick alpine snow cover, consuming up to 80 percent of the available vegetation (Walker, Billings, and de Molenaar, Chapter 6).

The insectivores, especially shrews of the genera *Sorex* and *Blarina*, forage continuously in the subnivean space, eating whatever they encounter and can subdue, including voles and other shrews (Crowcroft, 1957; Dehnel, 1961; Yudin, 1962; Churchfield, 1982a). In Sweden a direct correlation between the diversity of winter-active subnivean invertebrates taken in pitfall traps (Näsmark, 1964) and the diversity of stomach contents of shrews was found (Ackefors, 1964). The soricine shrews mainly feed on common winter-active invertebrates (mites, spiders, springtails, and beetles) during winter months (Mezhzherin, 1958, 1964; Ackefors, 1964; Pernetta, 1976; Aitchison, 1984a). They eat most invertebrate prey and, when available, other food such as coniferous seeds (Yudin, 1962; Zonov, 1982). Larger predators take larger prey, and smaller ones take smaller prey (Buckner, 1964). The winter shrews can be very small (mean weight 2.6 g for a winter *Sorex cinereus*) (Aitchison, 1987). The winter-active invertebrates are mainly small (1 to 5 mm) (Aitchison, 1984a) and more energy-rich than larger ones (Hanski, 1984). In a subnivean food chain, shrews are probably the major predators – e.g., *Sorex* in the northern hemisphere (Aitchison, 1984a, 1987) and the small marsupial insectivore *Antechinus* in Australia (Green, 1982). Some species of larger shrews, such as *Blarina brevicauda*, hoard food (e.g., beechnuts, earthworms [up to 40 percent], insects [up to 95.5 percent], snails [up to 1.8 percent], plant material [up to 25.3 percent], and small mammals [up to 60.5 percent]) (Crowcroft, 1957; Dehnel, 1961; Churchfield, 1982a; Aitchison, 1987). Shull (1907) reported *B. brevicauda* hoarding and cacheing snails immobilised by the shrew's toxic salivary venom (Shull, 1907; Merritt, 1986). Table 5.5 compares winter diets of *Sorex minutus* in snow-free Ireland (Grainger and Fairley, 1978) and in snow-covered Siberia where the more common winter-active groups were prey – e.g., spiders, 47.6 percent; beetle adults, 58.5 percent (Yudin, 1962).

The rodents, especially the microtines, and the insectivores are all potential prey for the small weasels. During winter months in Sweden *Mustela nivalis* consumed 48 percent *Microtus* species, 28 percent other rodents, 19 percent rabbits, and 5 percent shrews (Erlinge, 1975). In North America the small-sized mustelids, such as the

Table 5.5. *Percentage frequency of occurrence of winter food items in the guts of S. minutus (after Grainger and Fairley, 1978; Yudin, 1962).*

Food item	Ireland ($n = 87$)	Siberia ($n = 35$)
Lumbricids	0	5.6
Molluscs	0	19.6
Isopods	69	0
Araneae	21	47.6
Acari	42	5.6
Opilionids	30	0
Collembola	0	11.2
Hemiptera	37	17.0
Carabidae	0	42.0
Staphylinidae	0	19.6
Chrysomelidae	0	14.0
Coleopteran adults	90	58.8
Coleopteran larvae	37	0
Hymenoptera	0	16.8
Diptera	49	16.8
Vegetation	53	0

shorttail weasel *Mustela erminea* and the least weasel *Mustela rixosa* are well adapted to areas with prolonged snow cover and abundant prey species. Some female *M. erminea* can easily pass through vole tunnels (Simms, 1979) or dive down through the snow to the subnivean space to prey on voles (Formosov, 1946). As well, ventilation shafts serve as access for these small predators to the subnivean space (Madison, 1984). The small weasel with its long, slender body and large surface-to-volume ratio can rapidly chill and therefore spends most of the winter in the subnivean space in search of voles. Its forays to the snow surface are short (Formosov, 1946). In stressful winter conditions (temperatures below $-40°C$) and with 60 to 90 cm of snow cover in Minnesota, *M. erminea* was found to feed on 22 percent (by volume) shrews (Aldous and Manweiler, 1942).

A possible subnivean food chain in the southern boreal forest could easily begin with soil and snow fungi (Aitchison, 1983) and snow algae (Kopeszki, 1988; Hoham and Duval, Chapter 4) as primary producers. The fungivorous and detritivorous springtails are potential consumers of fungi and algae and in turn may be preyed upon by winter-active mites and spiders (Buchar, 1968; Håvgar, 1973; Aitchison, 1984a, 1984b, 1989). Winter-active subnivean spiders in turn are readily consumed by insectivores, such as the ever-hungry shrews (Yudin, 1962; Ackefors, 1964; Pernetta,

1976). Weasels are known to kill shrews but not necessarily to eat them; however, in some areas shrews are consumed (Aldous and Manweiler, 1942; Hamilton, 1933; Simms, 1969; Aitchison, 1987). Also raptors such as the great grey owl *Strix nebulosa* dive into snow cover to retrieve small mammals, including soricine shrews (Nero, 1969; Collins, 1980), taking a possible secondary predator out of the subnivean space. This is summarised in Figure 5.9.

5.4   Recommendations for Future Research

The lack of economic importance of this fauna and the technical difficulties of analysis associated with subnivean and other snow studies have resulted in a natural history approach to this fauna. There is considerable potential in this field for analytical approaches to energetics and trophic and population dynamics. There are a number of unknowns relating to the subnivean, intranivean, and supranivean micro- and mesofauna: for example, many abiotic components could be better studied, such as the effects of temperature, snow density, snow hardness, light penetration, and CO_2 concentration on the flora and fauna. With regard to CO_2, there is a need to understand its sources (bacteria, fungi, plants, invertebrates, and mammals) (Tranter and Jones, Chapter 3). The snow cover itself could be modified by 1) spraying it with water to produce an ice layer; 2) stirring it to produce a denser and harder surface; and 3) covering it with plastic to create an artificial ice layer. Then faunal reactions, both invertebrate and vertebrate, could be monitored.

Although many experiments have been done on mammals, not that many have been done with invertebrates. There is a need for more research, possibly using the methods previously mentioned, to simulate and monitor natural parameters and to correlate them to faunal activity.

Trophic relations within this microenvironment are only partially known. Radio-tracers or stable isotopes might elucidate these pathways, especially in the invertebrates. The supposition that faunal movements, both within and onto the snow cover, are triggered by thermal changes needs to be confirmed.

Further studies on life histories of winter-active invertebrate species would determine overwintering stages and possible cyclomorphic-type changes. The taxonomy of many of these groups is poorly known (Aitchison, 1979a, 1979b, 1979c, 1979d; Zacharda, 1980). Any morphological, physiological, and behavioural changes of this fauna that enhance their winter activity and survival might help explain the mechanisms of low temperature tolerance.

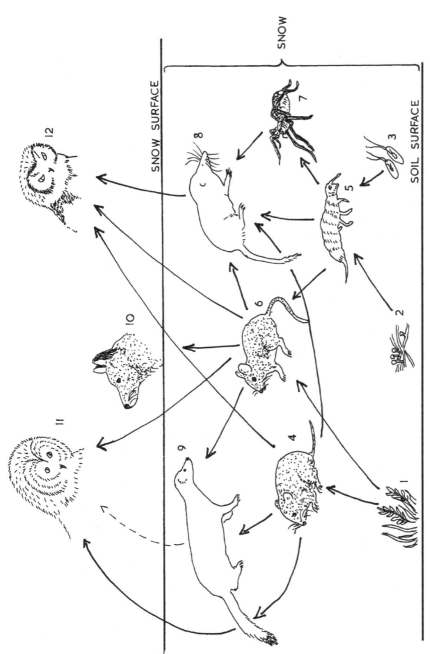

Figure 5.9. Hypothetical winter food web in a boreal region: 1) vegetation, 2) *Cladosporium* fungi, 3) snow algae (*Chloromonas* sp.), 4) red-backed vole *Clethrionomys gapperi*, 5) collembolan *Isotoma* sp, 6) deermouse *Peromyscus maniculatus*, 7) wolf spider *Pardosa* sp, 8) masked shrew *Sorex cinereus*, 9) shorttail weasel *Mustela erminea*, 10) red fox *Vulpes fulva*, 11) great grey owl *Strix nebulosa*, and 12) boreal owl *Aegolius funereus*.

5.5   Acknowledgments

I thank G. D. Farr, T. Seastedt, J. S. Edwards, K. Thaler, and D. A. Walker for constructive comments and suggestions about the manuscript.

5.6   References

Ackefors, H. 1964. Vinteraktiva näbbmöss under snöu. *Zool. Revy.*, **26**, 16–22.

Agrell, I. 1941. Zur Ökologie der Collembolen. Untersuchungen in Swedisch Lappland. *Opuscula ent. (Lund)*, **Suppl. 3**, 1–236.

Aitchison, C.W. 1974. A sampling technique for active subnivean invertebrates in southern Manitoba. *Manitoba Entomol.*, **8**, 32–36.

Aitchison, C.W. 1978a. Spiders active under snow in southern Canada. *Symp. Zool. Soc. Lond.*, **42**, 139–148.

Aitchison, C.W. 1978b. Notes on low temperature and winter activity of Homoptera in Manitoba. *Manitoba Entomol.*, **12**, 58–60.

Aitchison, C.W. 1979a. Winter-active subnivean invertebrates in southern Canada. I. Collembola. *Pedobiologia*, **19**, 113–120.

Aitchison, C.W. 1979b. Winter-active subnivean invertebrates in southern Canada. II. Coleoptera. *Pedobiologia*, **19**, 121–128.

Aitchison, C.W. 1979c. Winter-active subnivean invertebrates in southern Canada. III. Acari. *Pedobiologia*, **19**, 153–160.

Aitchison, C.W. 1979d. Winter-active subnivean invertebrates in southern Canada. IV. Diptera and Hymenoptera. *Pedobiologia*, **19**, 176–182.

Aitchison, C.W. 1979e. Notes on low temperature activity of oligochaetes, gastropods and centipedes in southern Canada. *Am. Midl. Nat.*, **102**(2), 399–400.

Aitchison, C.W. 1979f. Low temperature activity of pseudoscorpions and phalangids in southern Manitoba. *J. Arachnol.*, **7**, 85–86.

Aitchison, C.W. 1983. Low temperature and preferred feeding by winter-active Collembola (Insecta, Apterygota). *Pedobiologia*, **25**, 27–36.

Aitchison, C.W. 1984a. A possible subnivean food chain. In *Winter Ecology of Small Mammals*, J.F. Merritt (Ed.). Special Publication No. 10, Carnegie Museum of Natural History, Pittsburgh, pp. 363–372.

Aitchison, C.W. 1984b. Low temperature feeding by winter-active spiders. *J. Arachnol.*, **12**, 297–305.

Aitchison, C.W. 1984c. The phenology of winter-active spiders. *J. Arachnol.*, **12**, 249–271.

Aitchison, C.W. 1987. Review of winter trophic relations of soricine shrews. *Mammal Rev.*, **17**(1), 1–24.

Aitchison, C.W. 1989. The ecology of winter-active collembolans and spiders. *Aquilo Ser. Zool.*, **24**, 83–89.

Aitchison, C.W., and Hegdegar, B. 1982. Cryoprotectants in winter-active collembolans and spiders. *Symposium on Invertebrate Cold Hardiness*, July 1982, Oslo, abstr.

Aldous, S.E., and Manweiler, J. 1942. The winter food habits of the short-tailed weasel in northern Minnesota. *J. Mamm.*, **23**, 250–255.

Almquist, S. 1969. Seasonal growth of some dune-living spiders. *Oikos*, **20**, 392–408.

Andreev, A.V., and Krechmar, A.V. 1976. Radiotelemetric study of the microclimate in snow burrows of the hazel hen, *Tetrastes bonasia sibiricus*. [In Russian] *Zool. Zh.*, **55**, 1113–1114.

Asahina, E. 1966. Freezing and frost resistance in insects. In *Cryobiology*, H.T. Meryman (Ed.). Academic Press, London, pp. 451–486.

Ashmole, N.P., Nelson, J.M., Shaw, M.R., and Garside, A. 1983. Insects and spiders on snow fields in the Cairngorms, Scotland. *J. Natl. Hist.*, **17**, 599–613.

Bäbler, E. 1910. Die wirbellose, terrestrische Fauna der nivalen Region. *Rev. Suisse Zool.*, **18**, 761–915.

Bashenina, N.V. 1956. The influence of features of the subnivean atmosphere on the distribution of winter nests of *Microtus*. [In Russian] *Zool. Zh.*, **35**(6), 940–942.

Batzli, G.O., White, R.G., MacLean, S.F. Jr., Pitelka, F.A., and Collier, B.D. 1980. The herbivore-based trophic system. In *An Arctic Ecosystem: The Coastal Tundra at Barrow, Alaska*, US/IBP Synthesis Series 12, J. Brown, P.C. Miller, L.L. Tieszen, and F.L. Bunnell (Eds.). Dowden, Hutchinson & Ross, Stroudsburg, PA, Chap. 10, pp. 335–410.

Bergeron, J.-M. 1972. *The Role of Small Mammals in the Population Dynamics of the Semiothisa Complex, Lepidoptera: Population Dynamics of the Semiothisa Complex, Lepidoptera: Geometridae: Ennominae*. Ph.D. thesis, University of Manitoba, Canada.

Berman, D.I., Kononenko, A.P., Sarviro, V.S., and Trofimov, S.S. 1973. Winter activity of soil invertebrates in the taiga of the Gornaya Shoriya. *Soviet J. Ecol.*, **4**(3), 263–265.

Broen, B.v., and Mohrig, W. 1964. Zur Frage der Winteraktivität von Dipteren in der Bodenstreu. *Deut. Ent. Z.*, **12**, 303–310.

Brummer-Korvenkontio, M., and Brummer-Korvenkontio, L. 1980. Springtails (Collembola) on and in snow. *Mem. Soc. Fauna Flora Fennica*, **56**, 91–94.

Buchar, J. 1968. A winter trip among spiders [In Czech]. *Ziva*, **16**, 24–25.

Buckner, C.H. 1964. Populations and ecological relationships of shrews in tamarack bogs of southeastern Manitoba. *J. Mamm.*, **47**, 181–194.

Budyko, M.I. 1963. Atlas of the heat balance of the surface of the earth. Leningrad. *Weather and Climate*, H. Flohn (Ed.). McGraw-Hill, New York, Table 6.

Chapman, J. 1954. Observations of snow insects in western Montana. *Can. Entomol.*, **86**, 357–363.

Churchfield, S. 1982a. Food availability and the diet of the common shrew, *Sorex araneus*, in Britain. *J. Anim. Ecol.*, **51**, 15–28.

Churchfield, S. 1982b. The influence of temperature on the activity and food consumption of the common shrew. *Acta Theriol.*, **27**, 295–304.

Collins, K.M. 1980. *Aspects of the Biology of the Great Gray Owl Strix nebulosa Forster*. M.Sc. thesis, University of Manitoba, Canada.

Coulianos, C.-C., and Johnels, A.G. 1962. Note on the subnivean environment of small mammals. *Arkiv. Zoologi.*, **15**(4), 363–370.

Courtin, G.M., Shorthouse, J.D., and West, R.J. 1984. Energy relations of the snow scorpionfly *Boreus brumalis* (Mecoptera) on the surface of the snow. *Oikos*, **43**, 241–245.

Crowcroft, P. 1957. *The Life of the Shrew*. Max Reinhardt, London.

Dahl, C. 1969. The influence of light, humidity and temperature on Trichoceridae (Diptera). *Oikos*, **20**, 409–430.

Dalenius, P., and Wilson, O. 1958. On the soil fauna of the Antarctic and of the Sub-Antarctic Islands. The Oribatidae (Acari). *Arkiv. Zool.*, **11**, 393–425.

Dehnel, A. 1961. Aufspeichung von Nahrungsvorräten durch *Sorex araneus* Linnaeus 1758. *Acta Theriol.*, **4**, 265–268.

Duman, J.G. 1979. Subzero temperature tolerance in spiders: the role of thermal-hysteresis-factors. *J. Comp. Physiol.*, **131**, 347–352.

Edwards, J.S. 1972. Arthropod fallout on Alaskan soil. *Arctic Alpine Res.*, **4**(2), 167–176.

Edwards, J.S. 1987. Arthropods of alpine aeolian ecosystems. *Annu. Rev. Entomol.*, **32**, 163–179.

Edwards, J.S., and Sugg, P. 1993. Arthropod fallout as a resource in the recolonization of Mount St. Helens. *Ecology*, **74**(3), 954–958.

Eisenbeis, G., and Meyer, E. 1986. Some ultrastructural features of glacier Collembola, *Isotoma* 'sp. G' and *Isotomurus palliceps* (Uzel, 1891) from the Tyrolean Central Alps. *Proc. 2nd Int. Seminar on Apterygota*, University of Siena, Siena Italy; R. Dallai (Ed.). pp. 257–272.

Erlinge, S. 1975. Feeding habits of the weasel *Mustela nivalis* in relation to prey abundance. *Oikos*, **26**, 378–384.

Evernden, L.N., and Fuller, W.A. 1972. Light alteration caused by snow and its importance to subnivean rodents. *Can. J. Zool.*, **50**, 1023–1032.

Fjellberg, A. 1976. Cyclomorphosis in *Isotoma hiemalis* Schött, 1893 (*mucronata* Axelson, 1900) syn. nov. (Collembola, Isotomidae). *Rev. Ecol. Biol. Sol.*, **13**(2), 381–384.

Fjellberg, A. 1978a. Generic switch-over in *Isotoma nivea* Schäffer 1896. A new case of cyclomorphosis in Collembola (Isotomidae). *Norw. J. Entomol.*, **25**, 221–222.

Fjellberg, A. 1978b. New species of the genus *Isotoma* Bourlet, 1839 from North America (Collembola: Isotomidae). *Entomol. Scand.*, **9**, 93–110.

Fjellberg, A., and Greve, L. 1968. Notes on the genus *Boreus* in Norway. *Norsk Entomol. Tiddskr.*, **15**(1), 33–34.

Flatz, S. 1979. *Winteraktivität epigäischer Arthropoden (ibs. Aranei, Carabidae) im Bereich der Landesanstalt für Pflanzenzucht und Samenprüfung Rinn (Nordtirol, 900 m NN)*. Mag. thesis, Universität Innsbruck, Austria.

Flatz, S., and Thaler, K. 1980. Winteraktivität epigäischer Aranei und Carabidae des Innsbrucker Mittelgebirges (900 m NN, Tirol, Österreich). *Anz. Schädlingskde. Pflanzenschutz, Umweltschutz*, **53**, 40–45.

Formosov, A.N. 1946. Snow cover as an integral factor of the environment and its importance in the ecology of mammals and birds. *Moscow Soc. Naturalists Mater. Fauna Flora USSR,*

N.S. (Zool.), **5**, 1–152 [English translation 1969: Boreal Institute, University of Alberta, Occasional Publication 1, 1–141].

Fox, A.D., and Stroud, D.A. 1986. Diurnal rhythyms in a snow-surface springtail (*Isotoma violacea*, Collembola) and its predators at Eqalungmint Nunaat, West Greenland. *Pedobiologia*, **29**, 405–412.

Fuller, W., Stebbins, L.L., and Dyke, D.R. 1969. Overwintering of small mammals near Great Slave Lake, Northern Canada. *Arctic*, **22**(1), 34–55.

Fuller, W.A., Martell, A.M., Smith, R.F.C., and Speller, S.W. 1975. High arctic lemmings (*Dicrostonyx groenlandicus*): I. Natural history observations. *Can. Field Nat.*, **89**, 223–233.

Geiger, R. 1965. *The Climate near the Ground*. Harvard University Press, Cambridge, MA [translated from German by Scripta Technica].

Goodman, D. 1971. *Ecological Investigations of Ice Worms on Casement Glacier, Southeastern Alaska*. Report no. 39, Institute of Polar Studies. Columbus, OH.

Grainger, J.P., and Fairley, J.S. 1978. Studies on the biology of the Pygmy shrew *Sorex minutus* in the west of Ireland. *J. Zool. London*, **186**, 109–141.

Granström, U. 1977. *Spindlar (Araneida) på Ängsmark i norra Sverige-Fauna-sammansättning, säsongs- och dygnsaktivitet*. Doctoral thesis, University of Umeå, Sweden.

Green, K. 1982. Notes on winter-active invertebrates beneath the snow. *Vict. Nat.*, **99** (July, August), 144–146.

Hågvar, S. 1971. Field observations on the ecology of a snow insect, *Chionea araneoides* Dalm. (Dipt., Tipulidae). *Norsk Entomol. Tiddskr.*, **18**, 33–37.

Hågvar, S. 1973. Ecological studies on a winter-active spider *Bolyphantes index* (Thorell) (Araneida, Linyphiidae). *Norsk Entomol. Tiddskr.*, **20**, 309–314.

Hågvar, S. 1976. Phenology of egg development and egg-laying in a winter-active insect, *Chionea araneoides* Dalm (Dipt., Tipulidae). *Norw. J. Entomol.*, **23**, 193–195.

Hågvar, S. 1995. Long distance, directional migration on snow in a forest collembolan, *Hypogastrura socialis* (Uzel). *Acta Zool. Fennica*, **196**, 200–205.

Hågvar, S., and Østbye, E. 1973. Notes on some winter-active Chironomidae. *Norsk Entomol. Tidsskr.*, **20**, 253–257.

Hamilton, W.J. Jr. 1933. The weasels of New York. *Am. Midl. Nat.*, **14**, 289–344.

Hanski, I. 1984. Food consumption, assimilation and metabolic rate in six species of shrew (*Sorex* and *Neomys*). *Ann. Zool. Fennici*, **21**, 157–165.

Heiniger, P.H. 1989. Arthropoden auf Schneefeldern und in schneefreien Habitaten im Jungfraugebiet. *Berner Oberland Schweiz. Mittel. der Schweiz. Ent. Gesell*, **62**, 375–386.

Heydemann, V. 1956. Untersuchungen über die Winteraktivität von Staphyliniden auf Feldern. *Entomol. Blätter*, **52**, 138–150.

Holmquist, A.M. 1926. Studies in arthropod hibernation. I. Ecological survey of hibernating species from forest environments of the Chicago region. *Ann. Entomol. Soc. Am.*, **19**, 395–426.

Howard, W.E. 1951. Relation between low temperatures and available food to survival of small rodents. *J. Mamm.*, **32**, 300–312.

Huhta, V. 1965. Ecology of spiders in the soil and litter of Finnish forests. *Ann. Zool. Fennici*, **2**, 260–308.

Huhta, V. 1971. Succession in the spider communities of the forest floor after clear-cutting and prescribed burning. *Ann. Zool. Fennici*, **8**, 485–542.

Huhta, V., and Viramo, J. 1979. Spiders active on snow in northern Finland. *Ann. Zool. Fennici*, **16**, 169–176.

Husby, J.A., and Zachariassen, K.E. 1980. Antifreeze agents in the body fluid of winter-active insects and spiders. *Experentia*, **36**, 963–964.

Hyvärinen, H. 1969. On the seasonal changes in the skeleton of the common shrew (*Sorex araneus* L.) and their physiological background. *Aquilo Ser. Zool.*, **7**, 1–32.

Hyvärinen, H. 1984. Wintering strategy of voles and shrews in Finland. In *Winter Ecology of Small Mammals*, J.F. Merritt (Ed.). Special Publication No. 10, Carnegie Museum of Natural History, Pittsburgh, pp. 139–148.

Itämies, J., and Lindgren, E. 1985. The ecology of *Chionea* species (Diptera, Tipulidae). *Notulae. Entomol.*, **65**, 29–31.

Janetschek, H. 1956. Das Problem der inneralpinen Eiszeitüberdauerung durch Tiere (ein Beitrag zur Geschichte der Nivalfauna). *Österr. Zool. Z.*, **6**, 421–506.

Janetschek, H. 1990. Als Zoologie am Dach der Welt. Faunistisch-ökologisch-biozonotische Ergenisse der 2. Expedition der Forschungs unternehmens Nepal Himalayas in den Khumbu Himal. *Ber. Nat.-med. Verein Innsbruck Suppl.*, **6**, 1–119.

Janetschek, H. 1993. Über Wirbellosen-Faunationen in Hochlagen der Zillertaler Alpen. *Ber. Nat.-med. Verein Innsbruck*, **80**, 121–165.

Jonsson, B., and Sandlund, O.T. 1975. Notes on winter activity of two *Diamesa* species (Dipt., Chironomidae) from Voss, Norway. *Norw. J. Entomol.*, **22**, 1–6.

Joosse, E.N.G., and Testerink, G.J. 1977. The role of food in the population dynamics of *Orchesella cincta* (Linné) (Collembola). *Oecologia*, **29**, 189–204.

Kaisila, J. 1952. Insects from arctic mountain snows. *Ann. Ent. Fennica*, **18**(1), 8–25.

Kaufmann, T. 1971. Hibernation in the arctic beetle *Pterostichus brevicornis* in Alaska. *J. Kansas Entomol. Soc.*, **44**(1), 81–92.

Kawakami, M. 1966. The feeding habits of alpine salamanders [In Japanese]. *Proc. Hokkaido Government High School Teachers' Association Science Research*, pp. 1–15.

Kelley, J.J., Weaver, O.F., and Smith, B.P. 1968. The variation of CO_2 under the snow in the Arctic. *Ecology*, **49**, 358–361.

Keränen, J. 1920. Uber die Temperatur des Bodens und der Schneedecke in Sondänkyla. Nach Beobachtungen mit Thermoelementen. *Ann. Acad. Sci. Fennicae*, **13**, 1–170.

Kevan, D.K.M. 1962. *Soil Animals*. Witherby, London.

Knight, C.B. 1976. Seasonal and microstratal dietary research on *Tomocerus* (Collembola) in two forested ecosystems of eastern North Carolina. *Rev. Écol. Biol. Sol.*, **13**, 595–610.

Kohshima, S. 1984. A novel cold-tolerant insect found in a Himalayan glacier. *Nature*, **310**, 225–227.

Kopeszki, H. 1988. Zur Biologie zweier hochalpiner Collembolen–*Isotomurus palliceps* (Uzel, 1891) und *Isotoma saltans* (Nicolet, 1841). *Zool. Jb. Syst.*, **115**, 405–439.

Koponen, S. 1983. The arthropod fauna on the snow surface in northern Lapland [In Finnish]. *Finnish Oulanka Rep.*, **4**, 58–61.

Koponen, S. 1989. Spiders (Araneae) on the snow surface in subarctic Lapland. *Aquilo Ser. Zool.*, **24**, 91–94.

Korhonen, K. 1980a. Ventilation in the subnivean tunnels of the voles *Microtus agrestris* and *M. oeconomus*. *Ann. Zool. Fennici*, **17**, 1–4.

Korhonen, K. 1980b. Microclimate in the snow burrows of willow grouse (*Lagopus lagopus*). *Ann. Zool. Fennici*, **17**, 5–9.

Kühnelt, W. 1969. Zur Ökologie der Schneerandfauna. *Zool. Anz. Suppl.*, **32**, 707–721.

Lee, R.E. Jr. 1991. Principles of insect low temperature tolerance. In *Insects at Low Temperature*, R.E. Lee Jr. and D.L. Denlinger (Eds.). Chapman & Hall, New York, Chap. 2, pp. 17–46.

Leinaas, H.P. 1981a. Activity of Arthropoda in snow within a coniferous forest, with special reference to Collembola. *Holarctic Ecol.*, **4**, 127–138.

Leinaas, H.P. 1981b. Cyclomorphosis in the furca of the winteractive Collembola *Hypogastrura socialis* (Uzel). *Entomol. Scand.*, **12**, 35–38.

Leinaas, H.P. 1983. Winter strategy of surface dwelling Collembola. *Pedobiologia*, **25**, 235–240.

Levander, K.M. 1913. Ett bidrag till kännedom on vår vinterfauna. *Medd. Fauna Flora Fennica*, **1913**, 95–104.

MacKinney, A.L. 1929. Effect of forest litter on soil temperature and soil freezing in autumn and winter. *Ecology*, **10**, 312–321.

MacLean, S.F. Jr. 1975. Ecological adaptations of tundra invertebrates. In *Proc. AIBS Symposium on Physiological Adaptation to the Environment*, J. Vernberg (Ed.). Intext Educational Publishers, Amhurst, MA, pp. 269–300.

MacNamara, C. 1924. The food of Collembola. *Can. Entomol.*, **56**, 99–105.

McNab, B.K. 1983. Energetics, body size, and the limits to endothermy. *J. Zool. Lond.*, **199**, 1–29.

Madison, D.M. 1984. Group nesting and its ecological and evolutionary significance in overwintering microtine rodents. In *Winter Ecology of Small Mammals*, J.F. Merritt (Ed.). Special Publication No. 10, Carnegie Museum of Natural History, Pittsburgh, pp. 267–274.

Mani, M.S. 1962. *Introduction to High Altitude Entomology*. Metheun, London.

Mani, M.S. 1968. *Ecology and Biogeography of High Altitude Insects*. Junk N.V. Publishers, The Hague.

Mann, D.H., Edwards, J.S., and Gara, R.I. 1980. Diel activity patterns in snowfield foraging invertebrates on Mt. Rainier, Washington. *Arctic Alpine Res.*, **12**, 359–368.

Masutti, L. 1978. Insetti e nevi stagionali. Riflessioni su reperti relativi alle Alpi Carniche e Guilie. *Boll. Entomol. Bologna*, **34**, 75–94.

Mendl, H., Müller, K., and Viramo, J. 1977. Vorkommen und Verbreitung von *Chionea araneoides* Dalm., *C. crassipes* Boh. und *C. lutescens* Lundstr. (Dipt., Tipulidae) in Nordeuropa. *Notulae Entomol.*, **57**, 85–90.

Merriam, G., Wegner, J., and Caldwell, D. 1983. Invertebrate activity under snow in deciduous woods. *Holarctic Ecol.*, **6**, 89–94.

Merritt, J.F. 1986. Winter survival adaptations of the short-tailed shrew (*Blarina brevicauda*) in an Appalachian montane forest. *J. Mamm.*, **67**(3), 450–464.

Meyer, E., and Thaler, K. 1995. Animal diversity at high altitudes in the Austrian Central Alps. In *Ecological Studies: Arctic and Alpine Biodiversity*, F.S. Chapin, III and C. Körner (Eds.). Springer-Verlag, Berlin, Vol. 113, pp. 97–108.

Mezhzherin, V.A. 1958. On the feeding habits of *Sorex araneus* L. and *Sorex minutus* L [In Russian]. *Zool. Zh.*, 37, 948–953.

Mezhzherin, V.A. 1964. Dehnel's Phenomenon and its possible explanation. *Acta Theriol.*, 8, 95–114.

Näsmark, O. 1964. Vinteraktivitet under snöu hos landlevande evertebrater. *Zoologisk Revy*, 26, 5–15.

Nero, R.W. 1969. The status of the Great Grey Owl in Manitoba, with special reference to the 1968–1969 influx. *The Blue Jay*, 27, 191–209.

Novikov, G. 1940. Observations on insects on the snow at the Arctic Circle. [In Russian] *Priroda*, **3**, 78.

Novikov, G.A. 1972. The use of under-snow refuges among small birds of the sparrow family. *Aquilo Ser. Zool.*, **13**, 95–97.

Ohymama, Y., and Asahina, E. 1972. Frost resistance in adult insects. *J. Insect Physiol.*, **18**, 267–282.

Olynyk, J., and Freitag, R. 1977. Collections of spiders beneath snow. *Can. Field Nat.*, **91**, 401–402.

Østbye, E. 1966. A spider which caught springtails in a web in winter time [In Norwegian]. *Fauna*, **19**, 43.

Østbye, E., and Sømme, L. 1972. The overwintering of *Pelophila borealis* Payk. I. Survival rates and cold-hardiness. *Norsk Entomol. Tiddskr.*, **19**, 165–168.

Oswald, E.T., and Minty, L.W. 1970. Acarine fauna in southeastern Manitoba. I. Forest soils. *Manitoba Entomol.*, **4**, 76–87.

Palmén, E. 1948. Felduntersuchungen und Experimente zur Kenntnis der Uberwinterung einiger Uferarthropoden. *Ann. Ent. Fennica*, **14**, 169–179.

Penny, C.E., and Pruitt, W.O. Jr. 1984. Subnivean accumulation of carbon dioxide and its effects on the winter distribution of small mammals. In *Winter Ecology of Small Mammals*, Special Publication No. 10, J.F. Merritt (Ed.). Carnegie Museum of Natural History, Pittsburgh, pp. 373–380.

Pernetta, J.C. 1976. Diets of the shrews *Sorex araneus* L. and *Sorex minutus* L. in Wytham grassland. *J. Anim. Ecol.*, **45**, 899–912.

Plassman, E. 1975. Zur Vorkommen imaginaler Pilzmucken (Diptera: Mycetophilidae) in Bodenfallen wahrend der Wintermonate im Messauregebiet. *Ent. Tidskr.*, **96**, 27–28.

Polenec, A. 1962. Winterspinnen aus Mala Hrastnica [In Slovenian]. *Loski Razgledi*, **9**, 66–70.

Pritchard, G., and Scholefield, P. 1978. Observations on the food, feeding behavior and associated sense organs of *Grylloblatta campodeiformis* (Grylloblattodea). *Can. Entomol.*, **110**, 205–212.

Pruitt, W.O. Jr. 1954. Ageing in the masked shrew, *Sorex cinereus cinereus* Kerr. *J. Mamm.*, **35**(1), 35–39.

Pruitt, W.O. Jr. 1970. Some ecological aspects of snow. In *Ecology of the Subarctic Regions*, UNESCO, Paris, pp. 83–100.

Pruitt, W.O. Jr. 1972. Synchronous fluctuations in small mammal biomass on both sides of a major geographic barrier. *Aquilo Ser. Zool.*, **13**, 40–44.

Pruitt, W.O. Jr. 1978. *Boreal Ecology*. Edward Arnold, London.

Pruitt, W.O. Jr. 1984. Snow and small mammals. In *Winter Ecology of Small Mammals*, Special Publication No. 10, J.F. Merritt (Ed.). Carnegie Museum of Natural History, Pittsburgh, pp. 1–8.

Pucek, Z. 1970. Seasonal and age changes in shrews as an adaptive process. *Symp. Zool. Soc. London*, **26**, 189–207.

Puntscher, S. 1979. *Verteilung und Jahresrhythmik von Spinnen im Zentralalpinen Hochgebirge (Obergurgl, Ötztaler Alpen)*. Ph.D. thesis, Universität Innsbruck, Austria.

Renken, W. 1956. Untersuchungen über Winterlager der Insekten. *Z. Morph. Ökol. Tiere*, **45**, 34–106.

Richardson, S.G., and Salisbury, F.B. 1977. Plant responses to the light penetrating snow. *Ecology*, **58**, 1152–1158.

Rudge, M.R. 1968. The food of the common shrew *Sorex araneus* L. (Insectivora: Soricidae) in Britain. *J. Anim. Ecol.*, **37**, 565–581.

Salt, R.W. 1953. The influence of food on cold-hardiness of insects. *Can. Entomol.*, **85**, 261–269.

Salt, R.W. 1961. Principles of insect cold-hardiness. *Annu. Rev. Entomol.*, **6**, 55–74.

Schaefer, M. 1976. Experimentelle Untersuchungen zum Jahreszyklus und zur Überwinterung von Spinnen (Araneida). *Zool. Jb. Syst.*, **103**, 127–289.

Schaefer, M. 1977a. Winter ecology of spiders (Araneida). *Z. Angew. Ent.*, **83**(2), 113–134.

Schaefer, M. 1977b. Zur Bedeutung der Winters fur die Populationdynamik von vier Spinnenarten (Araneida). *Zool. Anz., Jena*, **199**(1/2), 77–88.

Schaller, F. 1992. *Isotoma saltans* und *Cryptopygus antarcticus*, Lebenskunstler unter Extrembedingunger (Collembola: Isotomidae). *Entomol. Gener.*, **17**(3), 161–167.

Schmid, W.D. 1971. Modification of the subnivean microclimate by snowmobiles. In *Proc. Snow and Ice in Relation to Wildlife and Recreation Symposium*, A.O. Hanger (Ed.). Iowa Cooperative Research Unit, Iowa State University, Ames, IA, pp. 251–257.

Schmidt, P., and Lockwood, J.A. 1992. Subnivean arthropod fauna of southeastern Wyoming: habitat and seasonal effects on population density. *Am. Midl. Nat.*, **127**, 66–76.

Schmoller, R. 1970. Life histories of alpine tundra Arachnida in Colorado. *Am. Midl. Nat.*, **83**, 119–133.

Schwaller, W. 1980. Eine Pseudoskorpion Art, *Neobisium erythrodactylum* L. Koch 1873, in Süddeutschland aktiv auf Schnee (Arachnida: Pseudoscorpiones: Neobisiidae). *Ent. Zeit.*, **5**, 54–56.

Seton, E.T. 1909. *Life Histories of Northern Mammals*, Vol. II. Scribner's, New York.

Shorthouse, J.D. 1979. Observations of the snow scorpionfly *Boreus brumalis* Fitch (Boreidae: Mecoptera) in Sudbury, Ontario. *Quaest. Entomol.*, **15**, 341–344.

Shull, A.F. 1907. Habits of the short-tailed shrew *Blarina brevicauda* (Say). *Am. Nat.*, **41**, 459–522.

Simms, D.A. 1979. North American weasels: resource utilization and distribution. *Can. J. Zool.*, **57**, 504–520.

Sømme, L. 1982. Supercooling and winter survival in terrestrial arthropods. *Comp. Biochem. Physiol.*, **73A**, 519–543.

Sømme, L. 1989. Adaptations of terrestrial arthropods to the alpine environment. *Biol. Rev.*, **64**, 367–407.

Sømme, L. 1993. Living in the cold. *Biologist*, **40**(1), 14–17.

Sømme, L., and Østbye, E. 1969. Cold-hardiness in some winter active insects. *Norsk Entomol. Tiddskr.*, **16**(1), 45–48.

Soper, J.D. 1944. On the winter trapping of small mammals. *J. Mamm.*, **24**, 344–353.

Spencer, A.W. 1984. Food habits, grazing activities and reproductive development of long-tailed voles, *Microtus longicaudus* (Merriam) in relation to snow cover in the mountains of Colorado. In *Winter Ecology of Small Mammals*, Special Publication No. 10, J.F. Merritt (Ed.). Carnegie Museum of Natural History, Pittsburgh, pp. 67–90.

Steigen, A. 1975a. Respiratory rates and respiratory energy loss in terrestrial invertebrates from Hardangervidda. In *Fennoscandian Tundra Ecosystems*, F.E. Wielgolaski (Ed.). Springer-Verlag, Berlin, pp. 122–128.

Steigen, A. 1975b. Energetics in a population of *Pardosa palustris* (L.) (Araneae, Lycosidae) on Hardangervidda. In *Fennoscandian Tundra Ecosystems*, F.E. Wielgolaski (Ed.). Springer-Verlag, Berlin, pp. 129–144.

Storey, K.B. 1984. A metabolic approach to cold hardiness in animals. *Cryo-Lett.*, **5**, 147–161.

Strübing, H. 1958. *Schneeinsekten. Die Neue Brehm-Bücherei.* A. Ziemsen Verlag, Wittenburg, Lutherstadt.

Sutton, G.M., and W.J. Hamilton Jr. 1932. The mammals of Southampton Island. *Mem. Carnegie Mus.*, (part II, Section 1) **7**, 1–109.

Swan, L.W. 1961. The ecology of the high Himalayas. *Sci. Am.*, **205**, 68–78.

Tahvonen, E. 1942. Beobachtungen über Winterinsekten. *Ann. Ent. Fennica*, **8**, 203–214.

Thaler, K. 1988. Arealformen in der nivalen Spinnenfauna der Ostalpen (Arachnida, Aranei). *Zool. Anz.*, **220**(5/6), 233–244.

Thaler, K. 1989. Streufunde nivale Arthropoden in den mittleren Ostalpen. *Ber. Nat.-Med. Verein Innsbruck*, **76**, 99–106.

Thaler, K. 1992. Weitere Funde nivaler Spinnen (Aranei) in Nordtirol und Beifänge. *Ber. Nat.-Med. Verein Innsbruck*, **79**, 153–159.

Thaler, K., and Steiner, H.M. 1975. Winteraktive Spinnen auf einem Acker bei Grossenzerdorf (Niederösterreich). *Anz. Schädlingsk.*, **48**, 184–187.

Uchida, H., and Fujita, K. 1968. Mass occurrence and diurnal activity of *Dicyrtomina rufescens* (Collembola, Dicyrtomidae) in winter. *Sci. Rep. Hirosaki Univ.*, **15** (1–2), 36–48.

Viramo, J. 1983. Invertebrates active on snow in northern Finland. *Oulanka Rep.*, 3, 47–52.

Wallwork, J.A. 1970. *Ecology of Soil Animals*. McGraw-Hill, London.

Wanless, F.R. 1975. Spiders of the family Salticidae from the upper slopes of Everest and Makalu. *Bull. Br. Arach. Soc.*, 3(5), 132–136.

West, S.D. 1977. Midwinter aggregation in the northern red-backed vole, *Clethrionomys rutilis*. *Can. J. Zool.*, **55**, 1404–1409.

Whittaker, J.B. 1981. Feeding of *Onychiurus subtenuis* (Collembola) at snow melt in aspen litter in the Canadian Rocky Mountains. *Oikos*, **36**, 203–206.

Wigglesworth, V.B. 1964. *The Principles of Insect Physiology*. Metheun, London.

Willard, J.R. 1973. *Soil Invertebrates: II. Collembola and Minor Insects: Populations and Biomass*, Technical Report No. 19, Saskatoon, Canada; published by the Canadian Committee for the International Biological Programme, University of Saskatchewan, National Research Council of Canada.

Wojtusiak, H. 1951. The temperature preferendum of winter insects of the genus *Boreus* (Panorpatae) and *Chionea* (Diptera). *Bull. Acad. Polon. Sci. Lett. B*, **2**, 123–143.

Wolff, J.O. 1984. Overwintering behavioral strategies in taiga voles (*Microtus xanthognathus*). In *Winter Ecology of Small Mammals*, Special Publication No. 10, J.F. Merritt (Ed.). Carnegie Museum of Natural History, Pittsburgh, pp. 315–318.

Wolska, H. 1957. Preliminary investigations on the thermic preferendum of some insects and spiders encountered in snow [In Polish]. *Folia Biol. Krakow*, **5**, 195–208.

Ylimäki, A. 1962. The effect of snow cover on temperature conditions in the soil and overwintering of field crops. *Ann. Agric. Fenn.*, **1**, 192–216.

Yudin, B.S. 1962. Ecology of shrews (genus *Sorex*) in western Siberia [In Russian]. *Akad. Nauk SSSR, Siberian Section Trudy Inst. Biol.*, **8**, 33–134.

Yudin, B.S. 1964. The geographic distribution and interspecific taxonomy of *Sorex minutissimus* Zimmerman, 1780, in west Siberia [In Russian]. *Acta Theriol.*, **8**, 167–179.

Zacharda, M. 1980. Soil mites of the family Rhagidiidae (Actinedida: Eupodoidea). Morphology, systematics, ecology. *Acta Universitatis Carolinae-Biologia*, **1978**, 489–785.

Zacharda, M., and Daniel, M. 1987. The first record of the family Rhagidiidae (Acari: Prostigmata) from the Himalayan region. *Vest. cs. Spolec. Zool.*, **51**, 58–59.

Zettel, J. 1984. The significance of temperature and barometric pressure changes for the snow surface activity of *Isotoma hiemalis* (Collembola). *Experientia*, **40**, 1369–1372.

Zettel, J., and von Allmen, H. 1982. Jahresverlauf der Kälteresistenz zweier Collembolen-Arten in der Berner Voralpen. *Rev. Suisse Zool.*, **89**, 927–939.

Zinkler, D. 1966. Vergleichende Untersuchungen zur Atmungsphysiologie von Collembolen (Apterygota) und anderen Bodenkleinarthropoden. *Z. Vergl. Physiol.*, **52**, 99–144.

Zonov, G.B. 1982. Directions of ecological adaptations of birds and small mammals to winter conditions. *Soviet J. Ecol.*, **13**(5), 331–335.

6   Snow–Vegetation Interactions in Tundra Environments

D. A. WALKER, W. D. BILLINGS, AND J. G. DE MOLENAAR

6.1   Overview

Snow affects the plant species and ecosystem processes of numerous biomes. Pomeroy and Brun (Chapter 2) discuss the important role of intercepted snow in boreal forest biomes and its effect on patterns of soil moisture, depth of freezing, soil temperatures, and soil heat flux (Pomeroy et al., 1994; Pomeroy and Gray, 1995), while Tranter and Jones (Chapter 3) have described the role of snow in nutrient fluxes and ecosystem chemical budgets. Begin and Boivin (Chapter 7) document the effect of wind and heavy snow accumulations on the morphology and growth of trees. In prairie and tundra regions, topography and wind play more important roles in the distribution and physical properties of the snow cover. The importance of snow to agriculture in northern grasslands has been long recognized, and a wide variety of snow-management techniques have been used to both increase and decrease snow in selected portions of agricultural landscapes (Staple and Lehane, 1952, 1955; Staple, Lehane, and Wenhart, 1960; Stepphun, 1981). Of all biomes, tundra regions are the most strongly affected by snow. This chapter synthesizes the extensive existing information on tundra snow–vegetation interactions. It points toward a unified hierarchical understanding of species-, community-, landscape-, and biome-level responses to various snow regimes. We use three different approaches to examine snow–vegetation interactions. The first describes the influence of snow on the distribution of plant communities along topographic gradients. The second considers the effect of snow, or lack of snow, on plant physiology and growth. The third examines an experimental approach to study the effects of altered snow regimes on a variety of ecosystem properties. This is an attempt to interpret the ecosystem impacts of apprehended climate change (see Groisman and Davies, Chapter 1).

6.2   Introduction

Snow has long been recognized as the single most important variable affecting the patterns of vegetation of alpine regions (Vestergren, 1902). Early

266

descriptions of the influence of snow on alpine vegetation from the Swiss Alps mention "schneetälchen" (little snow valley) vegetation (Herr, 1836). These distinctive hollows were among the first vegetation units described in the vegetation literature of Europe, and many studies have since focused on descriptions of the vegetation in snowbeds and rocky windblown areas. Gjærevoll's (1956) classic monograph focused entirely on the snowbeds of Scandinavia, and Rønning (1965) described the vegetation of windblown areas of Svalbard. More recently, studies of plant ecophysiological response along snow gradients, experimental studies of plant response to altered snow regimes, and application of geographic information systems (GIS) have expanded our understanding of the extent and nature of snow-dominated arctic and alpine ecosystems.

6.3 Snow Gradients

Numerous studies have analyzed the more or less continuous changes in vegetation and soil properties that occur along snow gradients (Billings and Bliss, 1959; Billings, 1973; Flock, 1978; Hrapko and LaRoi, 1978; Bell and Bliss, 1979; Komárková, 1979; Burns, 1980; Alpert and Oechel, 1982; Burns and Tonkin, 1982; Miller, Mangan, and Kummerow, 1982; Ostler et al., 1982; Molenaar, 1987; Nams and Freedman, 1987a, 1987b). Billings (1973) developed the concept of a mesotopographic gradient to describe this variation; other similar approaches include the ecohydrological gradient of Molenaar (1987) and the synthetic alpine slope model of Burns and Tonkin (1982). The best-developed examples of snow gradients occur in windy temperate mountain ranges such as the east slope of the Colorado Front Range where deep snow drifts are interspersed with wind-scoured areas. A mesotopographic gradient for the Colorado Front Range is used throughout this chapter to discuss the effect of snow on site environmental factors and species response (Figure 6.1; Table 6.1).

6.3.1 Plant Species

Many arctic and alpine plants predictably occur in either windblown or snowbed habitats. In a study of plant species distribution on Niwot Ridge, Walker et al. (1993a,b) found that most species had an optimal range of snow depths where they were found most abundantly. For example, *Paronychia pulvinata*, a cushion plant, has its optimal distribution on stable, dry, windblown, rocky sites with less than 25 cm of mean maximum snow depth (Figure 6.2a). At the other end of the spectrum, the sedge *Carex pyrenaica* occurs only in deep snow areas with over 400 cm of mean maximum snow depth (Figure 6.2f). The species showing the clearest distributions

267

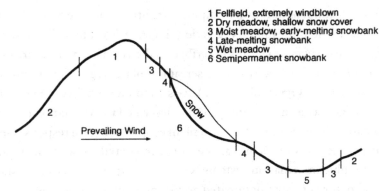

Figure 6.1. Mesotopographic gradient (Billings, 1973) adapted to the Niwot Ridge situation. Table 6.1 shows corresponding environmental information, associated vegetation, and soils. Summarized from Burns and Tonkin (1982), Komárková (1979), and May and Webber (1982).

with respect to the snow gradient are diagnostic taxa used to characterize the major plant associations of the Front Range (Komárková, 1979).

Wet areas beneath permanent snowpatches often contain rare plant species (e.g., the chionophytes *Saxifraga rivularis*, *Koenigia islandica*, *Phippsia algida* and *Haplomitrium hookeri* in the Front Range; *Stellaria umbellatus* in northern Alaska). *K. islandica* is rare in alpine areas of North America but occurs in the Colorado Front Range and again 700 km to the northwest in the Beartooth Mountains on the boundary between Wyoming and Montana. At the latter location, there are several scattered populations, all of which occur in wet mossy locations below long-lasting snowbanks (Johnson and Billings, 1962; Reynolds, 1984a, 1984b). Komárková (1979) notes that the restriction of many rare species to snowbed areas in the Colorado Front Range is typical of mountainous regions with strong winds (Jeník, 1959). Similar patterns are seen in the central European mountains of middle altitude, where snow redistribution into lee cirques has kept many of these areas free of forest vegetation during periods of treeline advance and allowed rare arctic–alpine plant species to survive warm postglacial periods (Jeník, 1959). Snow also appears to contribute to the high endemic ratio of forested areas subject to exceptionally high snowfall such as the Chubu District in central Japan (Uemura, 1989).

6.3.2   Plant Communities

Classification of snowbed communities has yielded much understanding of their environmental relationships. Braun-Blanquet (1949a, 1949b, 1950)

Table 6.1. *Site characteristics of the Niwot Ridge alpine snow gradient.*

Site (code in Figures 6.3 and 6.4)	Microsite description	Snow-free days	Typical plant associations (Komárková, 1979)	Soils, US soil taxonomy (Burns, 1980)
Fellfield, extremely windblown (1)	Xeric, extremely wind-exposed ridges and west-facing slopes	>200	*Sileno-Paronychietum, Potentillo-Careetum rupestris, Trifolietum dasyphylli*	Dystric Cryochrept, Typic Cryumbrept
Dry meadow, windblown (2)	Subxeric to mesic turfs on gentle wind-exposed west-facing slopes	150–200	*Selaginello densae-Kobresietum myosuroidis*	Dystric Cryochrept, Typic Cryumbrept
Moist meadow, early-melting snowbank (3)	Earlier melting snowpatches of the Front Range, subxeric to mesic snowpatches, leeward slopes and depressions	100–150	*Acomastylidetum rossii, Deschampsio caespitosae-Trifolietum parryi, Stellario laetae-Deschampsietum caespitosae*	Typic Cryumbrept, Pachic Cryumbrept, Dystric Cryochrept
Late-melting snowbank (4)	Later melting snowpatches of the Front Range, includes a wide variety of microhabitats from subxeric margins of late melting snow to subhygric, bryophyte-dominated, very late-melting snowpatches	50–100	*Toninio-Sibbaldietum, Caricetum pyrenaicae, Juncetum drummondii, Phleo commutati-Caricetum nigricantis, Poo arcticae-Caricetum hydenianae, Polytrichastro alpini-Anthelietum juratzkanae*	Dystric Cryochrept
Wet meadow (5)	Alpine fens, willow shrublands and springs; subhygric to hydric	>100	*Caricetum scopulorum, Rhodiolo integrifoliae-Salicetum planifoliae,*	Pergelic Cryaquept, Humic Pergelic Cryaquept, Histic Pergelic

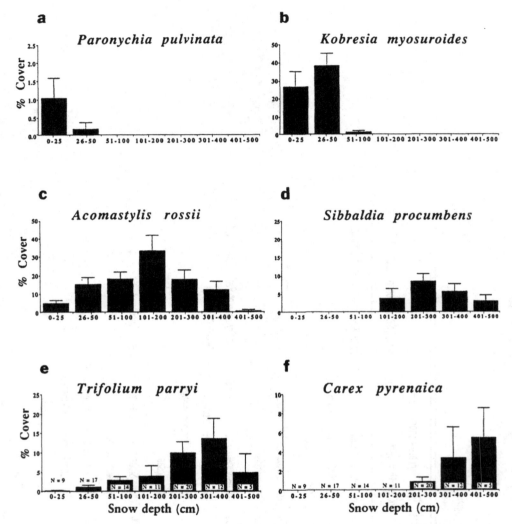

Figure 6.2. Distribution (plus or minus standard error) of six common species along the snow gradient. *Paronychia pulvinata* and *Kobresia myosuroides* are diagnostic taxa for the Alliance *Kobresio-Caricion rupestris*, which includes associations that are typical of broad, well-drained, stable windswept ridges in the Front Range. *S. procumbens*, *T. parryi*, and *C. pyrenaica* are diagnostic species for syntaxa within the snowpatch *Salicetae herbaceae*. *Acomastylis rossii* occurs across a broad range of snow-depth classes and is most abundant in areas with moderate snow cover. From Walker et al. (1993).

recognized a suite of nine alpine vegetation classes based on the characteristic alpine habitats in the Grison Mountains, European Alps. (The terminology used here for vegetation community names is that of the Braun-Blanquet approach [Westhoff and van der Maarel, 1978; Ellenberg, 1988]). Four of these classes are directly related to snow cover or the lack of it (*Salicetae herbaceae*, snowpatch swards; *Elyno-Seslerietea* and *Caricetea curvulae*, calcareous and acidic alpine swards; *Oxytropi-Elynion* and *Loiseleurio-Vaccinion*, calcareous and acidic windswept places; and *Vaccinio-Piceetea*, subalpine dwarf shrub heaths) and two others are related to meltwater from snow-banks (*Scheuchzerio-Caricetea*, fens; and *Montio-Cardaminetea*, spring communities). Tundra snowbed vegetation has been described worldwide including studies from the central European Alps (Lüdi, 1921; Braun-Blanquet, 1932, 1949a, 1949b, 1950; Dierssen, 1984); Appenines (Tomaselli, 1991; Ferrari and Rossi, 1995); Scandinavia (Vestergren, 1902; Fries, 1913; Nordhagen, 1928, 1936; Kalliola, 1939; Gjærevoll, 1950; Hedberg, Martensson, and Rudberg, 1952; Gjærevoll, 1954, 1956; Dahl, 1956; Gjærevoll and Bringer, 1965; Wielgolaski, 1997); Iceland (Hadac, 1971); Greenland (Böcher, 1954, 1959, 1963; Molenaar, 1976; Daniëls, 1982; Herk, Knaapen, and Daniëls, 1988); Svalbard (Elvebakk 1984a, 1984b, 1985, 1997); Canada (Lambert, 1968; Barrett, 1972; Bliss, Kerik, and Peterson, 1977; Hrapko and LaRoi, 1978; Bliss and Svoboda, 1984; Bliss, Svoboda, and Bliss, 1984; Nams and Freedman, 1987a; Nams and Freedman, 1987b; Bliss, 1997); the Presidential Range, New Hampshire (Bliss, 1963); the Colorado Rocky Mountains (Kiener, 1967; Komárková, 1979, 1980; Willard, 1979; Haase, 1987); Washington North Cascades (Kuramoto and Bliss, 1970; Douglas and Bliss, 1977; Evans and Fonda, 1990); Alaska (Gjærevoll, 1954; Cooper, 1986; Komárková and McKendrick, 1988; Walker, 1985, 1990; Walker, Walker, and Auerbach, 1994a; Walker and Walker, 1996); Chukotka, Russia (Razzhivin, 1994a); the central Himalayas (Miehe, 1997); Japan (Ohba, 1974); Australia (Williams, 1987; Williams and Ashton, 1987); and New Zealand (Mark, 1965, 1970, 1975; Mark and Bliss, 1970; Mark and Dickinson, 1997).

Gjærevoll's descriptions of the snowbed communities in Scandinavia remain the most thorough descriptions of snowbed plant communities (Gjærevoll, 1950, 1956). He described 65 snowbed plant communities based on a combination of lateness of snowmelt, soil pH, and site moisture. In the United States, Komárková (1979, 1980) used the Braun-Blanquet approach to classify the vegetation in the Indian Peaks, Front Range Colorado. She recognized a total of 52 alpine plant associations. Of these, 20 occurred in snowbeds, 7 in windblown areas, 6 in rock crevices, 12 in wet meadows, 8 in shrublands, and 4 in spring areas. Figure 6.1 shows the relationship of several of Komárková's associations to the conceptual mesotopographic gradient on Niwot

Ridge, Colorado. It includes a suite of six vegetated microenvironments: (1) extremely windblown fellfields found on ridge crests and knolls and dominated by cushion-plant communities (e.g., association *Sileno-Paronychietum*), (2) windblown dry sedge meadows found mostly on gentle windward slopes, (e.g., association *Selaginello densae-Kobresietum myosuroidis*), (3) moist meadows that are covered by shallow snowpatches with grass and forb communities (primarily association *Acomastylidetum rossii* and association *Stellario laetae-Deschampsietum caespitosae*), (4) late-melting snowpatches that include a wide variety of plant communities from association *Toninio-Sibbaldietum* in the relatively early melting, well-drained portions of snowbanks to well-drained late-lying snowbed areas dominated by sedges (association *Caricetum pyrenaicae*) to hydrophilous moss communities at the base of late-lying snowpatches (e.g., association *Polytrichastro alpini-Anthelietum juratzkanae*), (5) wet sedge meadows at the base of snowpatch runoff areas with fen communities, willow shrublands, and spring communities (e.g., association *Caricetum scopulorum*, association *Rhodiolo integrifoliae-Salicetum planifoliae*, and association *Clementsio rhodanthae-Calthetum leptosepalae*), and (6) semipermanent snowbeds that are for the most part unvegetated (e.g., *Oxyrio digynae-Poetum arcticae*) (Table 6.1).

Komárková (1979) commented on the large number of associations found in the snowpatch class (*Salicetae herbaceae*) compared with other alpine vegetation classes. She attributed this phenomenon to the steep environmental gradients within snowbed areas. Many studies have noted major floristic differences between early-melting and late-melting snowbed communities (Nordhagen, 1943; Gjærevoll, 1950, 1956; Molenaar, 1976; Komárková, 1979; Walker, 1985; Walker et al., 1993, 1994a; Razzhivin, 1994; Walker and Walker, 1996). Other descriptions have noted the differences between communities on acidic and nonacidic substrates (Gjærevoll, 1950, 1956; Elvebakk, 1984a, 1984b; Ellenberg, 1988; Razzhivin, 1994; Walker et al., 1994). Stability of the site and position with respect to summer meltwaters are other contributing factors. Komárková (1993) also noted that compared with alpine areas, snowbed vegetation types in the Arctic at all hierarchy levels are more numerous, less well defined, and less clearly distributed along the controlling environmental gradients. She attributes this to the more numerous surface disturbances in the Arctic associated with, for example, the thaw–lake cycle, extensive reworking by rivers, cryoturbation, and wind.

6.3.3   Snow Flush Areas, Ribbon Forests, and Krummholz

Of special note are the communities that occur along gradients at the extremes of tundra regions. For example, within the northern polar-desert regions, most landscapes are nearly totally barren because of a lack of soil moisture through

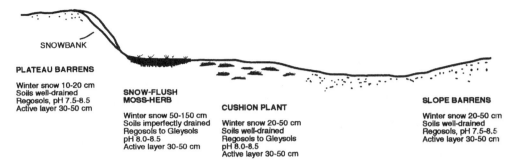

Figure 6.3. Generalized diagram to show the relationship of snow-flush, cushion-plant, and polar-barren plant communities within a polar-desert landscape. Modified from Bliss et al. (1984).

much of the growing season. Distinctive snow flush communities occur downslope of late-melting snowbeds and occupy about 3–5 percent of the polar desert areas (Figure 6.2f). Bliss et al. (1984) sampled twelve of these communities on three islands in the Canadian Archipelago, and reported species richness (9–14 spp.), and standing crop (120–740 g m^{-2}), compared to the polar barrens of the surrounding terrain (5–7 spp., and 5–60 g m^{-2}). A mesotopographic gradient for a polar-desert landscape (Figure 6.3) shows the landscape relationship of the snow-flush communities with respect to polar-barren and cushion-plant communities.

Billings (1969) has described a different wind–snow–vegetation gradient at the forest boundary that he terms the "Ribbon Forest–Snow Glade-Alternating Series" (Figure 6.4). This phenomenon occurs near alpine timberline in the Rocky Mountains from Canada to New Mexico on high, relatively level, windswept plateaus. In this series, snow blows laterally for considerable distances from the alpine tundra (either natural or fire caused) down into spruce-fir subalpine forest where it piles up within the forest. Parallel "ribbon forests" act as snow fences accumulating long-lasting snowdrifts that result in the grass- and sedge-dominated "snow glades" between the "forest ribbons" (Figure 6.4). These ribbon forests have been studied by Buckner (1977) and Earle (1993), and a similar wave pattern occurs in *Abies balsamea* forests in the Adirondack Mountains of northern New York State (Sprugel, 1976). The snowdrifts in the forests do not melt until late in the growing season. The result is the death of the older trees in parallel patterns where the snow is deepest and the soil most saturated with meltwater. The dead trees, over the years, are replaced by elongated parallel wet meadows ("snow glades").

Wind, topography, and snow also control patterns of krummholz (wind-shaped tree islands) at the forest-tundra ecotone on Niwot Ridge (Wardle, 1968; Marr, 1977;

Figure 6.4. Ribbon forest on Buffalo Pass, Colorado. Deep snowdrifts form in the open areas between strings of trees. Photo by David Buckner.

Benedict, 1984; Holtmeier and Broll, 1992; Pauker and Seastedt, 1996). These features form in shallow depressions on windswept ridges where site conditions are relatively favorable for seedling establishment (Figure 6.5). As the seedling grows it begins to form a snowdrift downwind, which changes the winter microclimate on the leeward side of the island. At maturity, the tree island totally controls the microsite. The island slowly dies off on the windward side because of physiological drought stress and frost damage, whereas the leeward side of the island is able to advance in the protection of the drift through the process of layering (Marr, 1977). Benedict (1984) estimated the leeward movement of krummholz to be about 2 cm year^{-1} over the past 500 years. In places the islands actually migrate downwind, leaving trails of scattered woody root fragments up to 15 m long (Marr, 1977). Vegetation on the windward side of the islands is generally typical fellfields or dry meadows. Inside the tree island, the dense evergreen foliage prevents much growth except for a few meager individual plants such as *Polemonium delicatum* and *Ribes montigenum*. Downwind, the vegetation is much more luxuriant and is typical of many subalpine meadows (e.g., *Vaccinium scoparium*, *Vaccinium myrtilus*, *Deschampsia*

274

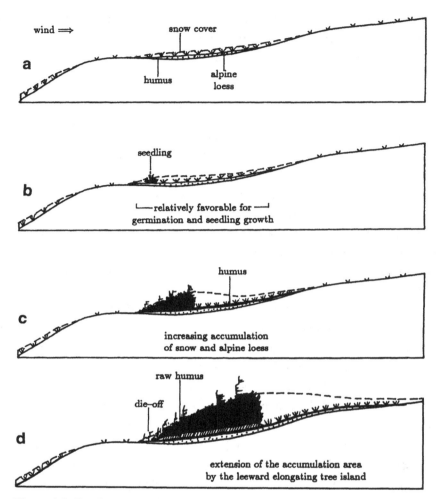

Figure 6.5. Development and migration of tree islands on Niwot Ridge. (a) Open alpine fellfield with slight depression and accumulation of fine-grained loess and humus. (b) Seedling establishment in area relatively favorable for germination and seedling growth and formation of incipient drift downwind of the seedling. (c) Tree island shaped by wind with modified habitat within and downwind of the island. (d) Tree island controls microsite, dieback occurs on windward side, and extension occurs on the leeward side through layering of the island. From Holtmeier and Broll (1992).

caespitosa). Species such as *Kobresia myosuroides* that are intolerant of deep snow are not present in the drift area (Bell and Bliss, 1979). There are also distinct trends of soil organic matter, nutrients (N, P, K), C/N ratios, cation exchange capacity, and pH associated with the island (Holtmeier and Broll, 1992; Pauker and Seastedt, 1996).

Within the tree island, soil formation is strongly influenced by the accumulation of evergreen leaf litter with thick organic horizons, relatively high soil moisture, and incipient podzolization (Holtmeier and Broll, 1992).

6.3.4   Soils

Soil development in arctic and alpine areas is slow and affected by snow cover and related conditions of drainage supply of meltwater, occurrence of permafrost, cryoturbation, and various forms of solifluction and mass wasting. The rootable soil is generally shallow, and rhizosphere biological activity is limited by the soil's instability, low temperatures, and wetness (Dierschke, 1977). Soils in many snowbed areas tend to be acidic and nutrient-poor because of leaching caused by large amounts of meltwater (Burns, 1980; Tomaselli, 1991; Stanton, Rejmánek, and Galen, 1994). However, snowbeds also trap windblown mineral and organic material, and the lower portions of snowbeds often accumulate this aeolian material. Regions with large annual inputs of loess, such as Prudhoe Bay, Alaska, have nonacidic snowbeds and support relatively rich plant communities (Walker, 1985, 1990). Soil fertility is also strongly correlated with the percentage of organic matter, and in very-late-lying snowbeds fertility declines sharply with later snowmelt dates (Stanton et al., 1994).

Snowbeds normally lie on hills with strong gradients of soil nutrients and water availability. Burns (1980) described a series of soils associated with the mesotopographic gradient on Niwot Ridge (see Figure 6.3). The chemical composition of snowbed soils is influenced by the local geology, the chemistry of the snowpack (Tranter and Jones, Chapter 3), nival biological activity (Hoham and Duval, Chapter 4), the breakdown of organic materials beneath the snowpack (Jones, 1991), and subsequent biological interactions with vegetation. Snowdrifts collect plant debris and mineral fines blown from the exposed places; part of this is carried off by meltwater, and some is held by the vegetation and adds to the fertility of the soil. Miller (1982) described a nitrogen and phosphorus gradient along an arctic–alpine hill-slope unit at Eagle Creek, Alaska. The upper portions of the snowbed were nitrogen-poor and composed of dwarf evergreen ericaceous shrub species (e.g., *Cassiope, Ledum, Vaccinium, Empetrum*). Nitrogen uptake was highest in lower snowbed areas receiving nitrogen and phosphorus from upslope and aeolian sources. Here, graminoid, forb, and shrub species with greater nitrogen demands were dominant (e.g., *Salix, Anemone, Artemisia, Pedicularis*) (Miller, 1982). The most nutrient-rich sites occurred near the bottom of slopes, whereas the sites poorest in nutrients were the heath communities near the top of the gradient.

The exchange of nutrients between communities was examined along a hill-slope gradient associated with a river bluff in northern Alaska (Giblin et al., 1991;

276

Nadelhoffer et al., 1991). The flow of nutrients was measured through the soils and plants over permafrost from a hilltop *Dryas* heath community, across a late-lying snowbed community of dwarf shrubs, lupine (*Lupinus*), and horsetails (*Equisetum*) to wet sedge tundra and riverside willows beyond the base of the slope. The hilltop heath site was nearly snow free in winter because of wind. A large amount of dissolved inorganic nitrogen and phosphorus was found in the 2- to 3-m-deep snowbank over the hillslope shrub-lupine and footslope *Equisetum* communities and contributed up to 10 percent of the nitrogen and phosphorus requirements of the hillslope and footslope ecosystems. The snowbank is an important mechanism of lateral nutrient transport down the topographic gradient, supplementing the downslope flow of water and soil solution through the thawed layer above the permafrost, which acts as an impermeable "basement." The redistribution of snow by wind is partially responsible for a heterogeneous pattern of nitrogen deposition in alpine and arctic environments. In an alpine system at Niwot Ridge, Bowman (1992) measured the amount of NO_3^- and NH_4^+ in snow and calculated the snow input as a percentage of the total atmospheric input in 1990–1991. Snowmelt nitrogen ranged from 5 to 20 percent of atmospheric input in dry meadows to 43 to 152 percent in the snowbed communities.

There are also important temporal nutrient patterns. Williams et al. (1995) developed a conceptual model of the mass balance of nitrogen pools and fluxes for the Emerald Lake watershed in the southern Sierras, which has a relatively pristine winter snowpack. They found that up to 90 percent of the annual wet deposition of nitrogen was stored in the seasonal snowpack and that about 80 percent of this was released in an ion-rich pulse with the first 20 percent of spring meltwater. This spring ionic pulse has been noted in numerous snow-dominated ecosystems (Tranter and Jones, Chapter 3). Mullen and Schmidt (1993) noted that the spring flush of nutrients occurs during times of low soil temperatures, and plants require special adaptations to take advantage of these early season conditions. For example, *Ranunculus adoneus*, the snow buttercup, is one of the few plants that flowers during the snowmelt. This plant accumulates phosphorus after seed set while the soil temperatures are relatively high and stores phosphorus over the winter to remobilize the following spring when the soil temperatures are close to $0°C$. Phosphorus uptake appears to coincide with the development of vesicular-arbuscular-mycorrhizal (VAM) fungi in the roots of *R. adoneus*, indicating that these fungi are important to the phosphorus nutrition of the plants (Mullen and Schmidt, 1993). In contrast to phosphorus uptake, nitrogen uptake occurs very early in the growing season before new roots and active VAM fungal structures are formed. At this time NH_4^+ is the main form of inorganic nitrogen. *R. adoneus* takes advantage of the early season flush by using a preexisting root system that is heavily infected with a "dark septate" fungi. These fungi are able to grow both saprophytically and symbiotically and thus may be an important sink for the early season flush of nitrogen.

6.3.5   Subnivean Animals

Subnivean activity by small mammals (Aitchison, Chapter 5) affects the distribution, biomass, and nutrients of associated vegetation communities. The presence of these animals causes considerable disturbance to the vegetation and soils. For example, gophers (*Thomomys talpoides*) are abundant in moderately deep early-melting alpine snowbeds of Colorado (Stoecker, 1976; Thorn, 1978; Willard, 1979; Burns, 1980; Thorn, 1982). Approximately 90 percent of the gopher activity on the Niwot Ridge Saddle is restricted to areas with moderately deep winter snow cover (Burns, 1980). Presumably, these are areas with sufficient unfrozen winter soils to permit winter digging activities and sufficient vegetation cover as food for the gophers. Thorn (1982) found that gopher activity was strongly associated with snowbed vegetation types on Niwot Ridge. The most heavily impacted plant associations have up to 80 percent of their surfaces disturbed by gopher activity. Soils in these areas have a very thick A horizon (>40 cm) formed by homogenized A and upper B horizons (Pachic Cryumbrepts). Gophers maintain plant species diversity by creating gaps in the plant canopy, redistributing nutrients and soil, and suppressing species that otherwise would dominate (Halfpenny and Southwick, 1982; Tilman, 1983; Andersen, 1987; Huntly and Inouye, 1987; Inouye et al., 1987). Gophers strongly affect the nutrients and soil organic carbon content of alpine soils. Chronic frequent disturbance by gophers appears to homogenize the soil particle-size distribution, bulk density, and available nutrients of the upper soil horizons (Litaor, Mancinelli, and Halfpenny, 1996). Gopher-affected areas have lower carbon content and total concentrations of carbon, nitrogen, and exchangeable calcium and potassium but higher available nitrate and nitrogen fluxes (Cortinas and Seastedt, 1996; Litaor et al., 1996).

In the Arctic, subnivean animal activities are partially responsible for heterogeneous spatial and temporal patterns of nutrients within arctic landscapes. At Barrow, Alaska, population pulses of brown lemmings (*Lemmus sibiricus*) are associated with major pulses of nutrients that are released at snowmelt from stocks of urine, feces, and litter that accumulate beneath the snow during peaks in the lemming population cycle (Pitelka, 1964; Schultz, 1964, 1969; Batzli et al., 1980). The lemmings can drastically reduce the standing crops of live and dead aboveground biomass, consuming up to 50 percent of the annual aboveground production and 20 percent of the total net production. Approximately 70 percent of this is returned to the soil as highly soluble feces and urine. Other animals that conceivably could play important roles that affect

278

nutrient regimes and vegetation patterns in arctic snowbeds include voles (*Microtus* spp.) (Batzli and Henttonen, 1990), collared lemmings (*Dicrostonyx rubricatus*), arctic ground squirrels (*Spermophilus parryi*), and willow ptarmigan (*Lagopus lagopus*) (Walker, 1990) (also see Aitchison, Chapter 5).

6.3.6   Other Factors Affecting Plant Composition of Snowbeds

Many snowbed plant species are found in relatively narrow ranges of habitats, particularly in alpine areas (Komárková, 1979; Walker et al., 1993). A good example is the snow buttercup *R. adoneus*, which occurs almost exclusively in late-melting snowbed areas and is nearly absent from adjacent turf and grassy meadow habitats, although habitats outside snowbeds appear to be favorable for *R. adoneus* growth (Scherf, Galen, and Stanton, 1994). One factor that restricts the movement of *Ranunculus* from snowbeds is that its seeds are rarely dispersed more than a few centimeters beyond the maternal plants. Secondary seed movement by water, gravity, and animals also appears to be limited. Transplanted *R. adoneus* seedling survival is statistically the same in meadow and snowbed sites, but causes of mortality are quite different. Seedling desiccation during the summer is much higher in the snowbed areas, whereas predation by small mammals occurs primarily in the meadow sites. This implies that genotypes appropriate for regeneration in the snowbed are probably poorly suited for colonization of adjacent plant communities. The sites most suitable for *R. adoneus* growth are those that offer opportunities for rapid infection by mycorrhizal fungi in the absence of neighboring vegetation. Scherf et al. (1994) conclude that such combinations of environmental factors are probably rare and ephemeral.

Ecotypic variation has been suspected in plants that occur across large portions of the snow gradients, such as *Dryas octopetala* and *Acomastylis rossii*, both of which show considerable morphological variation from fellfields to deep snow habitats (Billings and Bliss, 1959; May, Webber, and May, 1982; McGraw and Antonovics, 1983; McGraw, 1985a, 1985b, 1987). Billings and Bliss (1959) found that plants of *A. rossii* transplanted from a snowbed to a greenhouse maintained field morphological characteristics for at least 2 years, indicating possible ecoclinal or ecotypic status. Plants of *A. rossii* above the snowbank and toward the ridge were much smaller than the large, tall plants in the wet meadow. A similar pattern has been found in field reciprocal transplant experiments of *A. rossii* on Niwot Ridge (May, 1976; May and Webber, 1982; May et al., 1982), although reciprocal transplant studies with *R. adoneus* demonstrate that most phenotypic variation for this species across the snowmelt gradient is due to environmental, rather than genetic, variation.

Other factors that affect the composition of snowbed plant communities that have received some scientific study include different reproductive strategies (Marchand and Roach, 1980; Heide, 1992), pathogenic fungi (Sturges, 1989), wintertime herbivory by small mammals (Fox, 1981), seed germinability (Bock, 1976), and characteristics of the seed bank (Chambers, 1991, 1993; Chambers, MacMahon, and Brown, 1990).

6.3.7   Landscape and Regional Vegetation Patterns

The concept of snow gradients can be extended to three-dimensional landscapes with maps and GIS. In the Colorado Rocky Mountains, Walker et al. (1993a, 1993b) used a hierarchy of GIS databases to quantify the spatial correspondence between topography, snow, plant species, plant associations, and regional patterns of vegetation greenness. At the plant community level, 78 percent of Niwot Ridge is covered by vegetation that is typical of either windblown areas (class *Elyno-Seslerietea*) or snowpatches (class *Salicetea herbaceae*), an indication of the important role that wind and snow have in defining vegetation patterns at this site (Bell and Bliss, 1979; Komárková and Webber 1978). Correlations between snow depths and vegetation communities using a GIS analysis revealed that most plant associations, like plant species (see Section 6.3.1), are found in distinct portions of the snow gradient (Walker et al., 1993). These patterns reoccur in other portions of the Colorado Front Range. For example, Bell (1974) found that fellfields in Rocky Mountain National Park were totally snow free for 38 percent of February and March, whereas *Kobresia* meadows were snow free about 22 percent of the time, and snowbed communities were continuously covered during this period. Similar analyses have been done in the Alaskan Arctic foothills (Evans et al., 1989) and the Mosquito Range, Colorado (Stanton et al., 1994).

At regional scales, Walker et al. (1993) used SPOT satellite imagery and a digital terrain model to examine the patterns of the normalized difference vegetation index (NDVI) along elevation gradients on various slope aspects in the Front Range (Walker et al., 1993a, 1993b). They found that NDVI on east-, north-, and south-facing slopes decreased with elevation in response to the temperature gradient but that NDVI was low at all elevations on west-facing slopes because of the overriding influence of strong westerly winter winds that scour the west-facing slopes and constrain plant production at all elevations, particularly on the eastern slope of the Front Range. Measurements of NDVI at the 88 Niwot Ridge Saddle grid points and comparison with snow depth and soil moisture measurements at the same points confirmed an underlying relationship between soil moisture and vegetation greenness as inferred by NDVI (Walker et al., 1993). Duguay and Walker (1996) have shown that there can be high local variation in the NDVI–elevation relationship because of local climatic and topographic effects.

Table 6.2. *Comparison of wind, snow, and soil properties in moist nonacidic tundra and moist acidic tundra at the MNT–MAT boundary in northern Alaska.*

Ecosystem property	MNT	MAT
Wind	+	−
Snow depth	−	+
Depth hoar	−	+
Wind slab	+	−
Vegetation canopy height	−	+
Summer heat flux into the soil	+	−
Permafrost temperature	−	+
Active layer depth	+	−
Upward freezing from the base of the active layer in early winter	+	−
Downward freezing from the soil surface in early winter	+	−
Time for freeze-up in fall	+	−
Cryoturbation	+	−
Movement of subsurface calcareous soils to the surface	+	−

Note: Summarized from Walker et al. (1998).

In the Arctic, snow and its effect on soil thermal properties are thought to be at least partially responsible for the boundary between northern moist nonacidic tundra (MNT) and southern moist acidic tundra (MAT) ecosystems (Walker et al., 1998). The boundary stretches across all of northern Alaska near the northern edge of the Arctic foothills of the Brooks Range. The boundary is clearly visible on false color-infrared satellite images. The spectral boundary is caused by differences in greenness primarily because of greater shrub abundance (*Betula*, *Salix*, *Ledum*, and others) in MAT. Similar boundaries separate acidic Low Arctic from nonacidic High Arctic ecosystems in other parts of the Arctic (Alexandrova, 1980), but the boundary is enhanced in northern Alaska because of the strong climatic boundary at the northern front of the east-west-west trending Brooks Range. Differences in winter climate, primarily wind regimes, are thought to be one of the causes of the boundary (Table 6.2). Stronger more persistent winds on the northern coastal plain create thinner, more dense snowpack and colder soil surface termperature, which in turn promote more active cryoturbation (Sturm and Johnson, 1992; Benson et al., 1996; Zhang, Osterkamp, and Stamnes,

281

1996a, 1996b, 1997; Bockheim et al., 1998; Nelson et al., 1997a, 1997b; Sturm et al., 1997). The continual frost stirring of the soils tends to bring the nonacidic alluvial and aeolian subsurface deposits to the surface, which promotes the growth of nonacidic plant communities. Many of the shrub species dominant in acidic tundra are absent in nonacidic tundra; the shorter plant canopies also promote development of a thinner, less dense snowpack (Table 6.2).

Other integrated studies of snow–vegetation interactions involving mapping of snow and vegetation patterns have been done in the Apennines, Italy (Ferrari and Rossi, 1995) and Obergurgl, Austria (Hampel, 1963). An interdisciplinary ecographic approach used vegetation maps to analyze snow regimes and microclimatic conditions at the upper treeline (approximately 2,225 m) in the Ötz Valley near Obergurgl. Reforestation of the upper treeline is of great concern in the highly populated regions of the Alps. Much of the forest has been replaced by grazing lands, resulting in enhanced avalanche activity. Severe avalanche activity in 1951 and 1954 gave rise to studies that included site and vegetation description, ecophysiological studies of key species, microbiology, and investigation of engineered structures for avalanche prevention. The studies concluded that detailed maps of plant communities provided fundamental insights to habitat, particularly soil temperature, which controls root respiration, development of mycorrhizal mycelium, and other processes important for the establishment of trees at treeline.

Pomeroy, Marsh, and Gray (1997) applied a distributed snow model to analyze the interactions between vegetation patterns and snow hydrology in a 68-km^2 catchment near Inuvik, Northwest Territories, Canada. The model uses a Landsat-derived vegetation classification and a digital elevation model to segregate the basin into snow sources and sinks. The model relocates snow from the sources to sinks and calculates in-transit sublimation loss. The resulting annual snow accumulation in specific vegetation-landscape types was compared with the results of intensive field surveys of snow depth and density. On an annual basis, 28 percent of annual snowfall sublimated from tundra surfaces and 18 percent was transported to sink areas. Annual blowing snow transport to sink areas amounted to an additional 16 percent of annual snowfall to shrub tundra and an additional 182 percent to drifts. For the total budget, 86.5 percent of the total snowfall accumulated on the ground, 19.5 percent was lost to sublimation, and 5.8 percent was transported into the catchment.

Knowledge of snow–vegetation relationships has practical relevance for interpreting winter snowpack conditions during the summer months. During winter, the snow cover is often uniformly white on aerial photographs and other remotely sensed imagery, giving few clues about the depth of the snow. During summer one can estimate

282

rather closely what the relative winter snowpack, or lack of it, is by inspection of the topography along with the floristic structure and height of the plant communities (Pomeroy and Gray, 1995). Gjærevoll (1952) used vegetation information to help identify snowbed areas in the engineering of mountain roads in Norway. In northern Alaska, such information is profitably used in routing wintertime seismic operations (Felix and Raynolds, 1989a, 1989b; Emers, Jorgenson, and Raynolds, 1995).

Numerous remote-sensing techniques have been developed to aid in estimating snow water equivalent (Hall and Martinec, 1985), including satellite-mounted microwave sensors (Bales and Harrington, 1995), airborne γ-ray sensors (Carrol et al., 1995), and approaches using time series of Landsat images, digital terrain models, and distributed energy balance models. In some instances, remote sensing of the vegetation cover during the summer can give a better clue about average wintertime snowpack conditions than remote sensing of the snow surface itself, and could be used to provide independent verification of drift and windblown areas.

6.4   Plant Physiological Responses

At the plant level, the presence of a winter snow cover offers plants protection against frost damage, dehydration, and physical damage from wind and windblown particles (Wardle, 1968; Tranquillini, 1979; Chapter 7). It limits intensive and deep freezing of the soil and suppresses soil instability caused by frost action and weathering (Pomeroy and Brun, Chapter 2). However, the protection that snow provides against winter climate extremes is bought at the price of a belated, short growing season. Conversely, snow-free places are exposed to severe winds, and plants in these places are subject to high rates of evapotranspiration (LeDrew, 1975; Bell and Bliss, 1979; Isard, 1986; Isard and Belding, 1989).

It is useful to consider the factors that influence the aboveground plant environment (foliosphere) separately from those that influence the rooting environment (rhizosphere) (Figure 6.6). The most important site factors influencing the foliosphere are light, temperature, and humidity, whereas the most important factors in the rhizosphere are temperature, soil stability, and availability of moisture, nutrients, and oxygen. All these site factors are strongly influenced by snow, through its effect on either the microclimate or the local hydrology (Gjessing and Øvstedal, 1975; Molenaar, 1987).

The microclimate within the foliosphere of tundra vegetation is quite different from the climate measured at nearby climate stations. There are larger fluctuations in air temperature and air saturation deficit, and wind velocity is much decreased (Geiger,

283

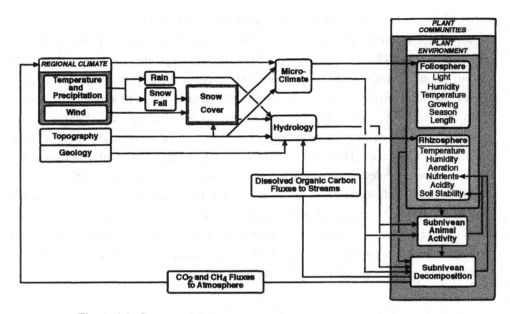

Figure 6.6. Conceptual diagram of snow–vegetation interactions. Regional climate, geology, and topography control snow distribution, which affects microclimates and hydrological regimes. The subnivean microclimate mainly influences the aboveground plant environment (foliosphere), whereas the hydrological regime mainly influences the belowground environment (rhizosphere). Subnivean animal activity is influenced by the subnivean microclimate and hydrology as well as availability of food; in turn it affects the stability and nutrient regime of the microsite. Subnivean decomposition is affected by animal activity, subnivean temperatures, and water availability; in turn it affects plant nutrient status as well as fluxes of trace gases to the atmosphere and dissolved organic matter to streams.

1965; Monteith, 1973). The intensity of irradiation and of heat loss, and thus the extremes of the microclimate, is especially great in high altitude sites but is also related to latitude, aspect, and slope of the site as well as the soil moisture conditions (Cline, 1995). The contrast between the microclimate of the foliosphere and that of the rhizosphere is particularly great in snow margin areas and can produce stressful conditions for many plants. The high surface temperatures promote photosynthesis at the same time the roots of the plant are near freezing, which limits uptake of nutrients and water.

6.4.1 Characteristics of Arctic and Alpine Snow Cover

The properties of the snowpack itself are important to the subnivean environment through control of heat flux into and from the soil surface (Pomeroy and

284

Brun, Chapter 2). Tundra snow cover is quite different from that in forested and more temperate regions (Sturm, Holmgren, and Liston, 1995; Chapter 2). Arctic tundra snow cover occurs north of the boreal treeline. It generally consists of wind-hardened, thin (10–75 cm), cold snowpack with a few centimeters of loose granular depth hoar at the base of the snowpack, overlain by multiple wind slabs. Also, there are often hard surface drift forms and an absence of ice lenses and other winter melt features.

In alpine areas, the snow cover is generally warmer and more heterogeneous because of variable topographic and climatic influences related to latitude, altitude, and continentality (Barry, 1992; Sturm and Holmgren, 1995). Generally, alpine snowpacks are deeper (75–250 cm), with alternating layers of wind-packed and soft snow and often with basal depth hoar (Sturm and Holmgren, 1995). An important contrast between arctic and alpine snowpacks is that deeper snowdrifts in the alpine may remain unfrozen at the base all winter, whereas most arctic snowpacks remain below freezing, thus limiting most biological activity. The snow cover is also affected locally by variation in the height of the tundra plant canopy (Pomeroy and Gray, 1995; Tabler and Schmidt, 1986).

The amount of snow (snow-water equivalent) on a site is the single most important factor governing the length of the growing season and hence the total amount of warmth available for plant development and growth (Billings and Bliss, 1959; Holway and Ward, 1963, 1965; Geiger, 1965; Miller, 1982; Walker et al., 1993; Stanton et al., 1994). Very deep snowpatches melt out in late summer or not at all in some years. In these extreme sites, the short growing season combined with other negative influences, such as wet, humus poor, and unstable soils, results in areas completely void of vegetation. In contrast, early-melting areas have the advantage of a relatively warm and protected winter environment combined with a long growing season, adequate soil moisture, and a relatively moderate summer microclimate (Billings and Bliss, 1959; Stanton et al., 1994).

The previously discussed snow gradient on Niwot Ridge, Colorado (Figure 6.1), is used here to illustrate the effect of snow cover on growing season length and other site factors (Figure 6.7). Species composition, biomass, phenology, and site factors have been monitored in five vegetation types representing the portion of the snow gradient from exposed ridge tops to shallow snowbeds (May, 1976; Komárková and Webber, 1978; Komárková, 1979, 1980; Burns and Tonkin, 1982; May and Webber, 1982; Walker et al., 1994b; Walker, Ingersoll, and Webber, 1995). Snow depths were monitored year round in 1972–1974. The mean maximum depths ranged from less than 10 cm in the fellfield community (association *Sileno-Paronychietum*) to 120 cm in the shallow snowbeds (association *Toninio-Sibbaldietum*) (Figure 6.7a). The corresponding

285

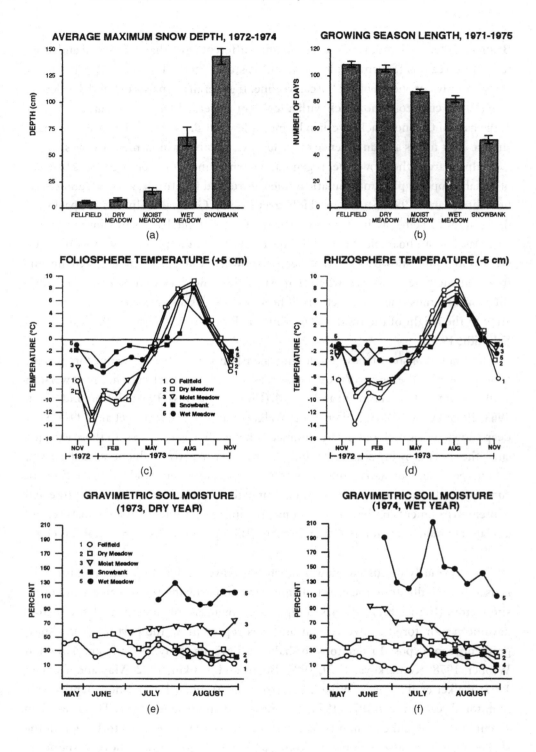

AVERAGE MAXIMUM SNOW DEPTH, 1972-1974

(a)

GROWING SEASON LENGTH, 1971-1975

(b)

FOLIOSPHERE TEMPERATURE (+5 cm)

1 O Fellfield
2 □ Dry Meadow
3 ▽ Moist Meadow
4 ■ Snowbank
5 ● Wet Meadow

(c)

RHIZOSPHERE TEMPERATURE (-5 cm)

(d)

GRAVIMETRIC SOIL MOISTURE
(1973, DRY YEAR)

1 O Fellfield
2 □ Dry Meadow
3 ▽ Moist Meadow
4 ■ Snowbank
5 ● Wet Meadow

(e)

GRAVIMETRIC SOIL MOISTURE
(1974, WET YEAR)

(f)

286

growing season length varied from 109 days in the fellfield community to 52 days in the snowbed (Figure 6.7b). The length of the growing season and the effects on vegetation have also been monitored at other sites (Billings and Bliss, 1959; Holway and Ward, 1965; Bliss, 1966; Svoboda, 1977; Miller, 1982; Williams, 1987; Stanton et al., 1994).

6.4.2   Phenology

In the already short growing season of arctic and alpine regions, delayed snowmelt strongly affects patterns of greenup, flowering, seed set, and total primary production. It is not uncommon to see the flowering of snow-margin species, such as the snow buttercup *R. adoneus*, occurring along the melting edge of snowpatches and moving as a wave across the tundra through the growing season. Delayed greenup occurs in small depressions where the snow melts late, and, conversely, delayed senescence may occur in the fall in snowbeds that are wetter than the surrounding dry tundra.

In some alpine and arctic plants, shoot growth and even flowering may commence from overwintering buds near the soil surface under snow before melt. Good examples of such early season growth under snow are *Polygonum davisiae* (Beyers, 1983), *Erythronium grandiflorum* (Caldwell, 1969; Salisbury, 1985), *Caltha leptosepala* (Rochow, 1969), *Oxyria digyna* (Mooney and Billings, 1961), *R. adoneus* (Caldwell, 1968; Salisbury, 1985; Galen and Stanton, 1991, 1993, 1995; Mullen and Schmidt, 1993; Galen, Dawson, and Stanton, 1993; Scherf et al., 1994; Mullen, 1995; Mullen, Schmidt, and Jaeger, 1998), *Saxifraga oppositifolia* (Moser, 1967), and *A. rossii* (Spomer and Salisbury, 1968; Chambers, 1991). Reradiation of heat from standing dead shoots from the previous season often melts slender holes in snow cover as deep as 20–30 cm. All the species mentioned above and many others can leaf out and flower quickly because their leaf primordia and flower buds preformed during the previous growing season (Sørenson, 1941; Billings and Mooney, 1968; Mark, 1970; Billings, 1974). They are also metabolically active with relatively high dark respiration rates at low temperatures under the snow.

Figure 6.7. Variation in site factors for five plant communities along an alpine snow gradient, Niwot Ridge, Colorado, 1972–73. (a) Average snow depths. (b) Length of the growing season. (c) Foliosphere temperatures at 5 cm above the soil surface for the same five plant communities. (d) Rhizosphere temperatures at 5 cm below the soil surface. (e) Gravimetric soil moisture in 1973 (a dry year). (f) Gravimetric soil moisture in 1974 (a wet year). Adapted from May (1976).

The effects of interannual climate variation on phenology of the alpine avens *A. rossii* and bistort *Bistorta bistortoides* have been studied over a 9-year period in the Colorado alpine (Walker et al., 1995). *B. bistortoides* showed a strong response to variation in the timing of snowmelt. During the heavy snowfall year of 1983, a major El Niño Southern Oscillation year in the southern Rocky Mountains, leaf lengths in *B. bistortoides* increased approximately 10 percent and the average number of leaves nearly doubled despite a much shorter growing season. Walker et al. (1995) concluded that the greater leaf lengths and number of leaves are related to soil moisture during the period of maximum growth, and this is controlled by the amount and timing of snowmelt. The positive role of deeper winter snowpack on plant growth is also supported in a subalpine study by Inouye and McGuire (1991), who showed a positive correlation between the number of flowers and cumulative snowfall at the Rocky Mountain Biological Laboratory, Crested Butte, Colorado, for the period 1973–1989. They found that, during years of lower snow accumulation, *Delphinium nelsonii* plants experienced lower temperatures between the period of snowmelt and flowering, which resulted in delayed and reduced floral production.

6.4.3 Production

Usually there is a gradient of plant production from the latest melting areas to the early-melting margins (Sørenson, 1941; Billings and Bliss, 1959; Holway and Ward, 1965; Bliss, 1966; Canaday and Fonda, 1974; May, 1976; Douglas and Bliss, 1977; Webber, 1978; Murray and Miller, 1982; Wijk, 1986a, 1986b; Nams and Freedman, 1987a, 1987b; Walker et al., 1994b). Billings and Bliss (1959) studied the effects of an alpine snowbank on soil moisture, plant growth, and productivity across a large snowbank at an elevation of approximately 3,350 m in the Medicine Bow Mountains of Wyoming at 41°20′N. Mean daily primary productivity and total production were much greater along the part of the transect below the snowbank (moist meadow) than along the transect above the snowbank (dry meadow). For the late-melting portions of the snowbank, daily primary production was $6.0 \text{ g m}^{-2} \text{ day}^{-1}$ and total production was 128 g m^{-2}, compared with $2.5 \text{ g m}^{-2} \text{ day}^{-1}$ and 36.2 g m^{-2} on the upper transect. Billings and Bliss concluded that soil moisture appears to be the most critical factor affecting vegetation differences both above and below the snowbank.

Measurements of biomass along a snow gradient on Niwot Ridge, Colorado were made for 7 consecutive years (1982–1989) in the five Niwot Ridge plant communities discussed previously (Figure 6.1) (May, 1976; Walker et al., 1994b). The least biomass was in the snowbed, $113 \pm 15 \text{ g m}^{-2}$, compared with $164 \pm 12 \text{ g m}^{-2}$ in the fellfield, $197 \pm 18 \text{ g m}^{-2}$ in the dry meadow, $218 \pm 23 \text{ g m}^{-2}$ in the moist meadow, and

214 ± 21 g m^{-2} in the wet meadow. Interannual climate variation accounted for 15–40 percent of the variation in phytomass in the five communities along the gradient. The biomass of fellfields and dry meadows was most sensitive to the previous year's precipitation, and the moist and wet meadow communities were most sensitive to the current growing season soil moisture. The only communities sensitive to growing season length were the snowbed communities. Surprisingly, none of the communities was sensitive to variation in the annual accumulated sum of daily mean temperatures above 0°C (thawing degree days). Soil moisture was overwhelmingly important to biomass production, and this was most strongly linked to the amount of spring precipitation, which mostly falls as snow. The amounts of winter and summer precipitation were less important.

6.4.4   Growth Strategies of Snowbed Plants

Several authors have commented on the growth strategies of plants in response to poor nutrient conditions found in snowbeds (Svoboda, 1977; Miller et al., 1982; Maesson et al., 1983; Nams and Freedman, 1987a; Williams, 1987; Williams and Ashton, 1987). In the High Arctic, evergreen *Cassiope tetragona* heaths are common in snowbeds. Above-ground to below-ground biomass ratios are high compared with other arctic plant communities. For example, biomass ratios of the *Cassiope* snowbeds on Ellesmere Island are nearly 1:1 compared with 1:3.8 to 1:6.7 in wet sedge-moss meadows (Miller et al., 1982; Nams and Freedman, 1987b). Evergreen species are long lived and slow growing and tend to have small below-ground biomass because of a storage of nutrients and photosynthates in leaves and stems, whereas deciduous shrubs and graminoids use large below-ground biomass as storage sites. Abundant attached dead tissue, common in snowbed plants such as *Cassiope* and *Dryas integrifolia*, also contributes to high aboveground:below-ground biomass ratios. Over 90 percent of the aboveground standing biomass in these heaths is composed of nonphotosynthetic tissues and 74 percent is dead tissue attached to live plants. These stress-tolerant species allocate a small proportion of total biomass to current-year production, and a large proportion of the total biomass is attached dead tissue. The accumulation of attached dead leaves also enhances snow accumulation in winter and creates a relatively favorable thermal regime in both summer and winter. The attached litter also may serve as a reservoir of nutrients (Nams and Freedman, 1987b).

Photosynthate is stored underground in the roots, rhizomes, or stem bases of these herbaceous perennials as carbohydrates in the form of sugars and starches (Mooney and Billings, 1960; Fonda and Bliss, 1966; Rochow, 1969; Shaver and Billings, 1976; Wallace and Harrison, 1978). Some of these plants, notably dwarf evergreen shrubs

289

and evergreen rosette herbs, also store a considerable amount of lipids in their leaves (Billings and Mark, 1961; Hadley and Bliss, 1964). For example, the alpine *Celmisia viscosa* in New Zealand stores lipids and oils in such quantity in their silvery evergreen leaves that they are easily flammable. This species occurs primarily under the shallower parts of semipermanent snowbanks in the lee of solifluction terraces on the crests of the Old Man Range of Central Otago in the stormy westerlies of the Southern Hemisphere at 45°S (Billings and Mark, 1961).

6.4.5   Wind

Wind is a primary factor in the dominance of cushion-plant, evergreen dwarf-shrub, and tussock-graminoid growth forms in fellfields and exposed tundra turfs (Norman, 1894; Vestergren, 1902; Gelting, 1934; Dahl, 1956; Gjærevoll, 1956; Bliss, 1956, 1962, 1969; Billings and Bliss, 1959; Hadley and Bliss, 1964). The tightly packed stems and leaves of these growth forms minimize winter abrasion from wind-transported particles and reduce drought stress during the summer. In contrast, many snowbed and "snow cranny" plants that are protected by snow during winter have erect growth forms and soft leaves and are not drought resistant (Mooney and Billings, 1968).

Perhaps the most thorough winter study of plant physiology during severe winter conditions was that of Katherine Bell, who monitored *Kobresia myosuroides* growth during winter in Rocky Mountain National Park, Colorado (Bell, 1974; Bell and Bliss, 1979). Although *K. myosuroides* is the dominant plant in large areas of the Colorado Front Range, it occurs only in a narrow range of snow accumulation regimes. It is a tussock-forming sedge that forms dense turfs in areas that are largely snow free during much of the winter except for microdrifts that form leeward of *Kobresia* tussocks. *Kobresia* does not occur in extreme windblown fellfields or in areas of even shallow snow accumulation. Bell was interested in the reasons for this limited distribution. She compared behavior of undisturbed *Kobresia* with that of transplants into habitats with more and less winter snow accumulation. She found that *Kobresia*'s success in snow-free meadows is related to rapid summer growth and to its use of an extended period for development, from about 1 April, well before snowmelt, to 20 October, after the beginning of drift development in snowbeds. Wintergreen *Kobresia* leaves can even elongate during warm periods (warmer than −4°C) in midwinter, an apparently unique phenomenon in tundra plants. New leaves begin elongation in the autumn and complete growth the following summer. Most carbohydrates are stored aboveground in leaves, primarily as oligosaccharides, sugars that likely contribute to frost hardiness of the evergreen leaves. Storage of carbohydrates in the leaves obviates the need for translocation from the roots in frozen soils in winter. Transplanted

290

Kobresia do not survive in fellfields because of mechanical damage by windblown snow and sand. Also, low soil water potentials create drought stress. In its preferred habitat, *Kobresia* takes advantage of shallow snow cover (approximately 15 cm) that melts in early spring, permitting leaf elongation in saturated soils. In sites of moderate to deep snow accumulation (>75 cm), autumn dieback is incomplete before drifts first form in September. A long snow-free period after early September is apparently necessary for proper onset of normal winter carbohydrate status in the leaf shoots. Winter freezing destroys the apparently unhardened leaf tissues and meristems, resulting in loss of carbohydrate reserves.

6.4.6   Radiation

Two aspects of the shortwave radiation environment are particularly important to plant growth: (1) the penetration of visible light through snow, which affects many plant processes including seed germination, emergence from underground organs, and photosynthesis; and (2) reflection of harmful ultraviolet B (UV-B) radiation from the surface of the snow (Pomeroy and Brun, Chapter 2).

Several investigators have noted the ability of plants to germinate or emerge from organs beneath snow. Kimball, Bennett, and Salisbury (1973) found an inverse relationship between snow depth and the chlorophyll content of spring beauty *Claytonia lanceolata* and baby blue eyes *Nemophila breviflora*, suggesting that plants may be able to photosynthesize beneath deep snowpacks. Curl, Hardy, and Ellermeier (1972) speculate that low levels of red and far-red light beneath shallow snow may be sufficient for germination of algal blooms in snowpack (see Hoham and Duval, Chapter 4). Any physiological activity that occurs beneath the snow could be very important, particularly for plants that grow in very-late-melting snowbanks and ephemerals, such as the snow buttercup *R. adoneus*, which complete their growth and senesce within a few weeks after emergence from the snow.

In early summer, sunlight may penetrate the snowpack and warm the dark ground beneath, causing thawing to begin from below. A kind of greenhouse may develop under the snow, with relatively high temperatures and sufficient light for photosynthesis to achieve a positive balance. In these conditions, the growing season also may start somewhat earlier than might be expected, resulting in the well-known phenomenon of early formation of green shoots, buds, and even flowers beneath shallow snowpacks (Sørenson, 1941; Holway and Ward, 1965; Remmert, 1965; Salisbury, 1985). Richardson and Salisbury (1977) found that camas *Camassia quamash* exposed to the light penetrating 173 cm for 2 months began turning green and had begun to unroll, but camas plants kept in darkness were white and noticeably etiolated. Several

291

subalpine and alpine herbaceous species either remain green all winter or turn green before meltout, including *A. rossii*, *R. adoneus*, and *E. grandiflorum* (Salisbury, 1985). Kimball et al. (1973) documented the synthesis of chlorophyll under snow. These observations are from a relatively warm subalpine environment at 2,300 m in the Bear River Mountains, Utah, and may not apply to the colder conditions in alpine and arctic regions. For example, Tieszen (1974, 1978), working in a much colder wet arctic tundra at 71°N latitude near Barrow, Alaska, before snowmelt in early and mid-June, measured net photosynthesis and carboxylation activity under snow cover in three species of tundra graminoids: *Eriophorum angustifolium*, *Carex aquatilis*, and *Dupontia fisheri*. Subnivean temperatures at the site remained near −7°C until meltwater percolated through the snow just before snow melting began. Only after the snow melted did plant temperatures rise above 0°C. Before the snow melted, the extinction of light through the snow resulted in <33 W m^{-2} reaching the plants even at solar noon on clear days. This, in combination with the low temperatures, did not allow much net photosynthesis in the plants. Additionally, during this subnivean period, the plant leaves were not photosynthetically competent and, therefore, were poorly capable of using the light that did penetrate to the plant surface. One reason for the lack of plant competency is the lack of a full complement of carboxylase activity in the leaves. Tieszen concluded that most plants in the wet coastal tundra do not photosynthesize appreciably until after snowmelt. However, temperatures beneath deeper alpine and arctic snowbeds are considerably warmer than in the open tundra, and conditions similar to those described by Salisbury may be present locally. Also, other light or dark reactions in plants under snow, including those related to growth hormones such as indoleacetic acid, photoperiodic effects, and the damage caused by UV-B irradiation, may still be operative on the chlorotic emerging shoots (Richardson and Salisbury, 1977).

At high altitudes, plant shoots emerging from under melting snowbanks are exposed to very high reflected UV-B radiation from surrounding areas of snow plus high downward UV-B flux from the sky and the sun. The effective UV dose could be doubled under these conditions (Caldwell, 1981). Another concern for high–altitude plants is the cumulative effects of long-term increases in UV-B due to alteration of the ozone layer in the atmosphere. Caldwell et al. (1980, 1982) measured solar UV-B radiation in tundra and alpine ecosystems from Atkasook, Alaska (71°N), through the Cordillera of the Rocky Mountains, to the Peruvian Andes at 14°S. Although maximum daily total shortwave irradiance along this great latitudinal gradient varies by a factor of only 1.6, maximum integrated effective UV-B irradiance varies by a full order of magnitude because of elevation. If reflectance of these wavelengths is

292

increased near snowbanks, the dose received by plants growing near the snow can be very great, especially in high middle-latitude and tropical mountains.

Caldwell (1968) found that extract absorbance values for *R. adoneus* both at 3,000 Å (UV-B) and 3,600 Å (UV-A) increased rapidly after being released from snow, whereas epidermal UV transmission decreased from approximately 13 percent under the snow to after the loss of the snow cover. This was accompanied by a sudden reddening of the stem tissue due to anthocyanin formation that effectively screened out UV-B irradiance. After a few days in the open sunlight, the stems turned green. The same phenomenon has been noted in the stems and leaves of *Polygonum davisiae* at the edge of snowbanks in the alpine zone of the Sierra Nevada, California, *B. bistortoides* on Niwot Ridge, and *A. rossii* in Colorado (Spomer, 1962). Caldwell (1968) found that *O. digyna* on Niwot Ridge, at an elevation of 3,700 m and 40°N latitude, screens out epidermally almost twice as much UV-B radiation as plants of the same species from Pitmegea River and Cape Thompson at 68°N in the Alaskan Arctic when grown together in growth chambers with normal visible irradiance plus UV-B. Within *Oxyria* there are, apparently, complex ecotypes in relation to light that involve at least UV-B inhibition of photosynthesis (30 percent) compared with that in arctic ecotypes (70 percent) (Caldwell et al., 1982).

6.4.7   Temperature

A surprising amount of biological activity occurs beneath deep snowpacks, and much of this can be attributed primarily to the relatively warm soil conditions (Salisbury, 1985; Pomeroy and Brun, Chapter 2). Winter soil temperature is intimately tied through various pathways to snow cover and hydrological conditions on the site and at sites upslope (see Figure 6.6). The effects of snow cover on rhizosphere and foliosphere temperatures are illustrated for five vegetation types along the alpine snow gradient on Niwot Ridge described earlier, where snow depth varied from less than 10 cm to about 150 cm and the growing season varied from 50 to 110 days (Figures 6.7a and b) (May et al., 1982). Mean monthly temperatures were monitored at 5 cm above the surface and 5 cm below the surface for 1 year (Figures 6.7c and d). During winter, the temperature contrast between the fellfields and snowbeds was the greatest. In December, the mean temperature of the fellfield foliosphere was −15.5°C, whereas the foliosphere of the snowbed community was a relatively warm −1°C. The mean monthly temperature of the snowbed foliosphere did not drop below −3.5° during the winter, but it also did not warm above the freezing point until mid-June, a full 6 weeks later than the fellfield, and remained cooler than the fellfield throughout the summer until September when senescence had begun in all the communities.

293

Foliosphere temperature of the dry and moist meadow communities was intermediate during both the winter and early summer (June–August) periods. Only during the late-summer-to-fall period (August–October) was the temperature comparable in all the communities. Temperatures in the rhizosphere were a few degrees warmer than the air temperature during the winter and few degrees cooler in the summer for all communities (Figure 6.7d). In winter, the greatest contrast between winter foliosphere and rhizosphere temperatures occurs in fellfields and the least contrast occurs in snowbeds. During the summer the situation is reversed, with the greatest contrast occurring in the snowbeds. The soil moisture of the snowbed sites was relatively low because they are early-melting snowbeds on slopes (Figures 6.7e and f).

The winter climate within the plant canopy is strongly affected by the depth of snow cover. Bell (1974) found that in the Colorado alpine, plant canopy temperatures beneath >50 cm of snow are very stable. In windblown alpine areas, soil-surface temperatures are less variable during winter than during summer because of stronger wind velocities that produce windpumping and mixing within the snowpack. Differences between the soil-surface temperature and the air temperature at 120 cm are rarely greater than 1°C to 3°C. Frequent storms with strong winds slow diurnal heating at the soil surface and accelerate soil cooling in the afternoon. Their net effect is reduced heating in the foliosphere and a shortened period of relative warmth during the day. At average wind speeds, the highest temperature in early afternoon in the winter is several centimeters below the snow surface. Microdrifts 1–5 cm deep in the lee of plants have little effect on temperature profiles, because they allow maximum temperature gradients of only 1°C to 3°C between the soil surface and the snow surface. During midwinter, crusts of snow and ice sheets a few centimeters above the ground allow only a slight warming of surface soil and trapped air (Bell, 1974). Auerbach and Halfpenny (1991) examined midwinter soil-surface temperatures on north, south, and valley bottom sites for the period 1981–1988 at the Teton Science School, Grand Teton National Park, Wyoming. They found that, despite colder air temperatures on the north-facing slopes (mean of −13.8°C versus −8.9°C on south-facing slopes), soil temperatures were warmer on the northern slopes (−1.7°C versus −2.4°C on south-facing slopes) because of deeper winter snowpack and higher thermal indices of the snowpack, and the differences were most pronounced in years of deep snow. The same pattern was also noted in the Arctic near Toolik Lake, Alaska, where the prevailing southerly winds out of the Brooks Range deposit deep snowdrifts on north-facing slopes. Arctic subnivean soil temperatures are somewhat colder than those noted at the Niwot alpine site (Figure 6.8). Minimum temperatures beneath a 400-cm-deep drift at −2 cm in the soil remained at about −7°C through most of the

294

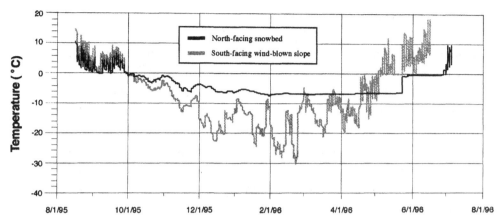

Figure 6.8. Soil temperature at 2-cm depth at two sites at Toolik Lake Alaska, August 1995–July 1996. (a) A windblown south-facing slope with minimal snow cover throughout the winter. Vegetation at this site is a dry *Dryas octopetela–Selaginella sibirica* community. (b) A deep north-facing snowdrift with over 500 cm of maximum snow accumulation. Vegetation at this site is a moist *Cassiope tetragona–Carex microchaeta* community.

winter, whereas soil temperatures dipped to about −30°C in a nearby windblown site.

In spring, wind speeds are much reduced, and the snow cover becomes the primary determinant of temperature profiles. Snow-free areas warm rapidly on sunny days. Bell (1974) found that fellfields and *Kobresia* meadows were 12°C to 15°C warmer than the air above on clear and overcast days. However, when snow falls in spring, the snow prevents any surface heating as long as it is present. Also, if the areas are inundated with water after the snow melts, the temperature around the plants remains close to 0°C as long as the water remains. Bell (1974) found that this common situation paradoxically leads to cooler plant temperatures in fellfields and *Kobresia* meadows during May after the melt begins than during the time in April when only a small part of the tundra is snow free. In arctic Alaska, Liston (1986) found bare tundra patches averaging 15°C to 24°C just before the general snowmelt and an extreme temperature of 42°C.

Snowdrifts also affect the summer air temperatures of sites marginal to the drifts. Cold air drainage, even off relatively small snowbanks, cools the area immediately downslope, resulting in cooler mean temperatures and shorter thaw seasons along snowbank margins. Billings and Bliss (1959) noted an average of 1.5°C cooling from the top to the bottom of a 100-m-long snowdrift in the Medicine Bow Mountains. The combined effect of the cool air drainage and cold water from the drift can create an exceptionally cold environment at the base of very late-lying snowdrifts, and

295

rhizosphere temperatures can be considerably colder than the foliosphere temperature in areas marginal to snowbeds (Holway and Ward, 1965). This can cause delayed phenology and reduced water and nutrient absorption at a time when photosynthetic activity is at a maximum. This can be important for phenological development. For example, Holway and Ward (1965) noted that alpine avens *A. rossii*, the most abundant alpine plant in the Colorado Front Range, consistently flowers only after the soil temperature at 3-cm depth exceeds 10°C, regardless of the aboveground temperatures.

Without a knowledge of the history of drift development during the winter and other influences of the site, it is not possible to predict soil temperatures on the basis of snow depth alone. Much variation in soil temperatures beneath snowpacks is related to the timing of drift development. A typical pattern for large late-lying alpine snowbeds is that the upper end of the snowbed forms early as snowdrifts into the snowbed depression (Dahl, 1956). These sites of early snow accumulation are sheltered from extreme temperatures throughout the winter, and soil temperatures are likely to remain relatively high (close to freezing) throughout the winter. As the drift builds, the lower ends of the snowbeds fill later after the soil temperatures have already been depressed. These sites may have deep snow cover and relatively cold soils.

Some decomposition occurs beneath deep snowpacks because of relatively warm winter soil temperatures (Bleak, 1970; O'Lear and Seastedt, 1994). Decomposition in soils, like many ecosystem processes, generally has a curvilinear response to snow depth (Webber et al., 1976; O'Lear and Seastedt, 1994). Decomposition is greatest in areas of intermediate snowpack and moderate soil moisture. A litterbag experiment on Niwot Ridge found that approximately 35 percent of green *Acomastylis* leaf litter and 70 percent of *Salix* leaf litter decomposed within 24 months in moderately wet snowbed areas, whereas 25 and 50 percent, respectively, decomposed in dry sites (Webber et al., 1976). O'Lear and Seastedt (1994) attribute the higher decomposition rates in the intermediate snow areas to relatively high microbial and invertebrate activities. Wintertime CO_2 and N_2O fluxes are evidence of subnivean microbial activity (Kelley, Weaver, and Smith, 1968; Sommerfeld, Mosier, and Musselman, 1993; Brooks et al., 1994, 1995, 1996, 1997; Oechel, Vourliltis, and Hastings, 1997; Fahnstock et al., 1997). On Niwot Ridge, CO_2 production occurs with ground surface temperatures as low as −5°C and N_2O production at a surface temperature of −1.5°C. The seasonal flux of CO_2 from snow-covered soils was related to both the severity of the freeze and the duration of snow cover. Whereas early-developing snowpacks resulted in warmer minimum soil temperature allowing production to continue for most of the winter, the highest CO_2 fluxes were recorded at sites that experienced a hard freeze before a consistent snowpack developed, suggesting that plant-cell freezing and lysis promotes

CO_2 flux. In contrast to CO_2, a hard freeze early in the winter did not result in greater N_2O loss. N_2O flux was related mainly to the length of time that soils were covered by a consistent snowpack. Brooks et al. (1997) concluded that subnivean microbial activity has the potential to mineralize from less than 1 percent to greater than 25 percent of the carbon fixed by aboveground net primary production, while the overwinter N_2O fluxes range from less than half to an order of magnitude higher than the growing season fluxes (Brooks et al., 1997). Simulations by global terrestrial biogeochemical models consistently underestimate the concentration of atmospheric carbon dioxide at high latitudes during the non-growing season, but consideration of the subnivean soil respiration improves these simulations considerably (McGuire et al., 2000).

6.4.8   **Soil Moisture and Drought Stress**

The local groundwater regime is a function of the site's position with respect to snow and slope gradients (Molenaar, 1987). In little-irrigated sites situated above snowdrifts, the soil dries and warms rapidly. In contrast, other sites with an equally long persisting snow cover, but situated below the snowdrift, are irrigated and exposed to leaching and remain moist and cold for a longer period. In temperate alpine areas, water is released over a prolonged period of time because of cool nighttime temperatures and deep snowdrifts (Caine, 1992; Caine and Thurman, 1990). Abundant well-defined snowbed plant communities, cold seeps, spring communities, and small wetlands are consequences of late-melting snow in alpine areas (Komárková, 1979). The relationship of snow depth to soil moisture conditions is illustrated by the five plant communities along the Niwot Ridge snow gradient (Figure 6.7, e and f). Soil moisture values are lowest in the fellfield and highest in the wet meadow at the lower margin of the Niwot Ridge snowdrift. The central part of the snowdrift has relatively low gravimetric soil moisture, comparable to that of the fellfield, because of rocky course-grained soils and a tendency to drain rapidly once the snow melts. The summer soil moisture in the fellfield and dry meadow generally declines throughout the summer but responds somewhat to rainfall events, whereas the moist meadow and wet meadow respond more to the supply of meltwater from the snowbed, remaining high in 1973 (Figure 6.7e), which had a relatively late meltout date, and declining during the low snow year of 1974 (Figure 6.7f). The lower margins of snowbanks melt early and are provided with a continuous supply of water as long as the snowbank persists.

A very different hydrological regime occurs in the Arctic, where the accumulated winter precipitation is released relatively quickly during the brief May–June snowmelt season because of continuous daylight and relatively warm nocturnal temperatures (Dingman et al., 1980; Kane et al., 1989, 1992; Everett, Kane, and Hinzman, 1996;

Hinzman et al., 1996). This water pools in low areas and persists on the soil surface throughout the summer because of the presence of permafrost. Snow is thus largely responsible for the wetland vegetation types and watertrack plant communities covering vast areas of the Low Arctic (Walker et al., 1989). Springs are relatively uncommon in the Arctic because of the lack of deep percolation of meltwater due to permafrost.

Soil moisture, in combination with timing of release from snow cover, governs plant water relations and primary production (Oberbauer and Billings, 1981; Beyers, 1983). Oberbauer and Billings (1981) did an intensive summer-long study of drought tolerance and water use in the Medicine Bow Mountains, Wyoming. They measured leaf water potentials and leaf stomatal conductances (predawn to sunset) in an array of species along a 420-m gradient from a windward slope across an alpine ridge at a 3,300-m elevation to a lee slope and wet meadow below a snowbank. The highest leaf water potentials and stomatal conductances occurred in the wet meadow and were lowest on the upper windward slope. Each of the 29 species measured at frequent intervals during the summer had its own individualistic distribution along the transect gradient. Rooting depth had considerable effect on leaf water potentials. For example, the deep-rooted *Trifolium parryi* always maintained steady midday leaf water potentials of approximately −2.00 MPa at each site, whereas its more shallowly rooted neighbors in stressed areas had very low potentials of near −3.5 to −4.0 MPa at midday.

Beyers (1983) did research on leaf water potential and stomatal conductance in relation to photosynthesis of three species of alpine perennial plants, *Polygonum davisiae*, *Lupinus lepidus* var. *lobbii*, and *Eriogonum incanum*, across a mesotopographic snowbank gradient at 2,750 m on Meiss Ridge, northern Sierra Nevada, California. This site has heavy winter snowfall and many snowbanks that persist throughout the dry summers. All three species showed reductions in gross photosynthetic rates during the growing season that were correlated with a decrease in soil moisture at 10- to 15-cm soil depths. Photosynthetic rates of all three species were more independent of measures of plant water status at the moist lower snowbank site than at the site above the snowbank. Photosynthesis decreased most steeply per decrease in leaf water potential in *Polygonum* (a deeply rooted species) and least steeply in *Eriogonum* (a shallowly rooted species), with *Lupinus* intermediate to the others. *Polygonum* maintained higher water potentials throughout most of the season and, therefore, carried on photosynthesis at rates closer to the maximum for more of the growing season than did *Lupinus* or *Eriogonum* (Beyers, 1983).

During the winter, snow protects plants from drought stress. Tranquillini (1964) noted that European alpine communities are seldom subjected to drought stress during the summer but that frozen soils and extreme winds often cause drought stress in

298

evergreen species during winter. The effects of winter drought stress have been most thoroughly studied in evergreen trees, shrubs, and krummholz near treeline (Lindsey, 1971; Tranquillini, 1979). On sunny winter days such plants lose a great deal of water by cuticular transpiration in spite of closed stomates; the stress can become more severe as winter progresses as the water cannot be replaced because of frozen soils. The leaves lose so much water that osmotic values can rise to over 40 atmospheres (Tranquillini, 1979). Shrubs covered with snow, however, actually gain water by absorption so that the osmotic values drop during the winter.

Evergreen species vary widely in their winter drought resistance. For example, *Rhododendron ferrugineum*, a European snowbed evergreen shrub, has low drought resistance and loses water rapidly if denuded during midwinter and may be killed. On the other hand, *Loiseleuria procumbens*, a prostrate evergreen shrub found in exposed areas, is very drought resistant. This plant has shallow adventitious roots that allow it take up surface meltwater late in winter when the soils are frozen (Larcher, 1963). Bell (1974) found that wintergreen plants in the Colorado *Kobresia* meadows and fellfields needed at least a small amount of protection provided by microdrifts. Wintergreen *Kobresia* leaf shoots elongated in winter only on warm days when the soil water was available and water potentials rose above −2.0 MPa.

6.5   Experimental Studies

If the snowpack is changed either regionally via climate change or locally via alteration of the wind patterns caused by construction of roads or buildings, the vegetation will respond. The effects of past climate changes are perhaps most evident in long-lived rock lichen communities. Large areas of the Canadian Arctic that were snow covered during the Little Ice Age (350–100 years before present) are still noticeably barren of lichens (Andrews, Davis, and Wright, 1976; Locke and Locke, 1977). The relatively lichen-free areas are light-colored on satellite images and contrast strongly with the heavily lichen-covered rocks that were not snow-covered. Locke and Locke (1977) used satellite images of Baffin Island to estimate that perennial snow cover in the vicinity of the Barnes Ice Cap increased about 35 percent during the Little Ice Age. In a boulder transplant experiment in the Colorado alpine, Benedict (1990) noted that rock lichens were killed after 5 to 8 years of increased summer snow cover. The ecosystem response of more complex tundra plant communities to changes in snow regime is currently unpredictable and depends on changes that occur at several different levels of ecosystem organization. Tundra vegetation communities do not

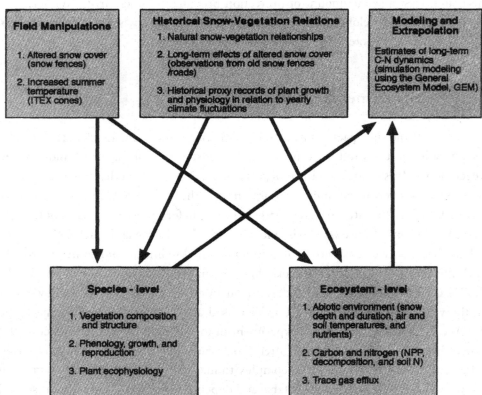

Field Manipulations

1. Altered snow cover (snow fences)

2. Increased summer temperature (ITEX cones)

Historical Snow-Vegetation Relations

1. Natural snow-vegetation relationships

2. Long-term effects of altered snow cover (observations from old snow fences /roads)

3. Historical proxy records of plant growth and physiology in relation to yearly climate fluctuations

Modeling and Extrapolation

Estimates of long-term C-N dynamics (simulation modeling using the General Ecosystem Model, GEM)

Species - level

1. Vegetation composition and structure

2. Phenology, growth, and reproduction

3. Plant ecophysiology

Ecosystem - level

1. Abiotic environment (snow depth and duration, air and soil temperatures, and nutrients)

2. Carbon and nitrogen (NPP, decomposition, and soil N)

3. Trace gas efflux

equilibrate quickly because changes to the below-ground resources and the substrate require long periods to adjust. A series of transient plant communities unlike existing ones are likely to occur (Galen and Stanton, 1995).

The effects of anthropogenically altered snow regimes have been studied next to roads (Willard and Marr, 1970, 1971; Bell, 1974) in experiments with infrared heaters (Harte and Shaw, 1995; Harte et al., 1995), shading with white reflective cloth to delay snowmelt (Galen and Stanton, 1993), removal of the snow to accelerate snowmelt (Galen and Stanton, 1993; Oberbauer, Starrm, and Pop, 1998), and snow fences (Bell, 1974; Weaver, 1974; Knight and Kyte, 1975; Outcalt et al., 1975; Slaughter et al., 1975; Emerick, 1976; Webber et al., 1976; Weaver and Collins, 1977; Bell and Bliss, 1979; Emerick and Webber, 1982; Sturges, 1989; Walker et al., 1993b, 1999). The importance of establishing long-term experiments to examine the effects of altered snow regimes was demonstrated by an early experiment on Niwot Ridge. The experiment used 1.2-m-high snow fences to examine the effects of enhanced snowpack in a dry *Kobresia myosuroides* meadow (Emerick and Webber, 1982). Early-season phenology of the plant canopy was delayed but soon caught up with the other vegetation during the late summer, and there were no other significant changes observed in the vegetation canopy during the 3-year experiment. In the years following the study, most of *Kobresia* died, and there were numerous other changes to the community composition. Other studies have since shown that snow accumulation strongly controls the distribution of *K. myosuroides* (Bell, 1974; Bell and Bliss, 1979). The delayed mortality observed after the Niwot Ridge experiments may have been due to exhaustion of the carbohydrate reserves in *Kobresia*'s wintergreen leaves after the experiment ended. Because of the short term of the experiment, these changes were not documented.

A new snow-fence experiment was started on Niwot Ridge in 1993 and at Toolik Lake, Alaska, in 1994 with the goal of determining the transient and long-term ecosystem responses of arctic and alpine tundras to altered snow regimes (Figure 6.9)

Figure 6.9. The Toolik Lake, Alaska, snow alteration experiment. (a) The arctic experiment on a dry site near Toolik Lake, Alaska. Vertical snow poles monitor snow depths. Small plastic cones are used to increase air temperatures above targeted plant species. Wooden fence is designed to lay flat on the ground during the summer to prevent alteration of the summer wind regimes. (b) Major components of the snow-alteration experiment. Experiment contains temperature and snow alterations, studies of natural snow gradients and long-term observations along existing fences and other areas of altered snow regimes, and a modeling component to estimate changes in carbon and nitrogen dynamics. Numerous abiotic variables are being monitored along with a variety of ecophysiological, phenological, plant community, and soil variables.

(Walker et al., 1993b, 1999). The experiment uses tall (2.6 m), long (60 m) fences that can be removed during summer to prevent alteration of the summer wind regimes. Species composition, vegetation structure, soil-carbon stores, geochemical composition of surface waters, decomposition, invertebrate and vertebrate populations, and the spectral reflectance of the vegetation are being monitored along the snow gradient leeward of the snow fence. The experiment is examining the combined effects of altered snow regimes and increased air temperature. Small open-topped chambers are being used to raise the air temperature surrounding targeted plant species (Figure 6.9) (Marion et al., 1997; Henry and Molau, 1997). In the first season, the drift caused a 50 percent reduction in the length of the growing season and, in the deepest part of the drift, raised the soil temperatures at 15-cm depth 15°C during midwinter (Figure 6.10). The mean winter temperature in dry *Kobresia* meadows was raised about 5°C. Late-season soil moisture within the drift area was raised about 33 percent compared with *Kobresia* meadows outside the drift (Figure 6.10b). The difference in soil moisture is thought to be the primary factor causing an increase in leaf lengths in *A. rossii* (Figure 6.10c). The combination of warmer subnivean temperatures and greater soil moisture increased litter decomposition by about 13 percent in the drift area (Figure 6.10d). Fluxes of N_2O and CO_2 increased as a result of the warmer winter soil temperatures (Figure 6.10e) (Brooks et al., 1995, 1996, 1997). Early September 1994 NDVI values in the deepest parts of the snowdrift area were about 30 percent greater than outside the fence, indicating delayed senescence in the drift area due primarily to greater soil moisture (Figure 6.10f). The opposite pattern is seen in early summer. Future monitoring at the experimental sites will document the long-term changes in plant community composition and ecosystem processes.

6.6   Conclusion

The initial motivation for this chapter was a perceived need for an integrated hierarchical approach to understanding snow–vegetation interactions. Although there is a long history of studies of the adaptations of organisms to cold and snow (e.g., Formozov, 1964; Halfpenny and Ozanne, 1989; Marchand, 1987; Merritt, 1984; Remmert, 1980), there has not been a synthesis regarding the role of snow in influencing alpine and arctic plant communities. This chapter reviews a variety of topics, many of which have not been previously emphasized in winter ecology texts.

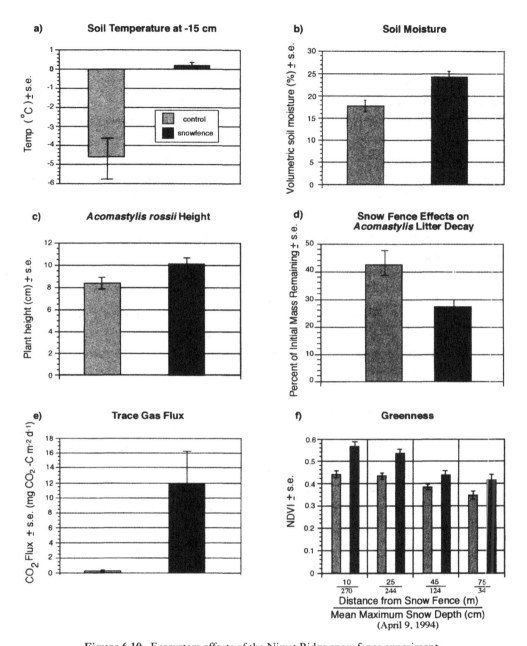

Figure 6.10. Ecosystem effects of the Niwot Ridge snow fence experiment.
(a–e) Comparison of effects in a control *Kobresia myosuroides* meadow and in the same community under a 270-cm drift. (a) Soil temperature at −15-cm depth. Temperatures were monitored periodically throughout the winter with buried thermisters.
(b) Volumetric soil moisture (August) determined by time-domain reflectometry.
(c) Average height of *Acomastylis rossii*. (d) Decomposition of *Acomastylis rossii* leaves after 678 days in the field. (e) Annual CO_2 flux (Brooks et al., 1996). (f) NDVI of areas inside and outside the fence at four points along the snowdrift gradient. NDVI was monitored with a handheld spectrometer.

303

The need for snow–vegetation research in especially important in the face of possible rapid climate change in northern latitudes (Oechel, Callaghan, and Gilmanov et al., 1996; Maxwell, 1997; Groisman and Davies, Chapter 1). The effects of altered snow regimes are of particular concern in the Arctic and Subarctic where changes in winter snow regimes and associated vegetation changes could dramatically change soil surface temperatures and permafrost conditions (Tyrtikov, 1976; Nelson et al., 1993; Zhang et al., 1997), the flux of nutrients, energy, and greenhouse gases from the soil (Oechel and Billings, 1992; Shaver et al., 1992; Fahnestock, Jones, and Brooks et al., 1998; Kolchugina and Vinson, 1993; Oechel et al., 1997; McGuire et al., 2000), and the way in which these systems are utilized by animals and man (Eastland, Boyer, and Fancy, 1989; Pomeroy and Gray, 1995). Some of the key research needs are:

1. **Species-level studies.** More detailed winter autecological studies of key tundra plants are needed to provide a better understanding of how these organisms cope with subnivean conditions and exposed windy conditions during the winter. The biogeography of snowbed plant communities has also not been studied in much detail. Much is still unknown regarding the worldwide floristic characteristics of snowbed vegetation and the relationship to site factors. The role of subnivean herbivores, for example, gophers, in snowbed plant communities is also an understudied topic.

2. **Snow–ecosystem studies.** The feedbacks between snow, vegetation, and climate are complex and occur at multiple scales. A combination of models, observations along natural snow gradients, and experimental studies are needed to help predict aboveground and below-ground plant production and other ecosystem processes resulting from altered snow regimes. In windy environments, small changes in the amount of snow or wind could dramatically change snowdrift locations and sizes. Currently available snow distribution models (e.g., Pomeroy, Marsh, and Gray, 1997) need to be linked to ecosystem models to help develop predictions of change. Simulated changes to landscape paterns could be explored by linking ecosystem models to maps of modern vegetation through the use of geographic information systems, as has been done to examine the effects of altered precipitation in the Great Plains (Burke et al., 1990). At watershed scales, models linking topography, snow, vegetation patterns, and spectral reflectance through hydrological models appear to be quite promising (e.g., Ostendorf et al., 1996). Such models predict vegetation, spectral reflectance, biomass,

and nutrient patterns solely on the basis of runoff volumes derived from digital elevation models. The addition of heterogeneous snow cover to these models will undoubtedly increase their predictive power. Also examination of regional patterns of the normalized difference vegetation index (NDVI) in relation to snow cover could provide insights regarding climate-biomass that would be useful to predict changes caused by altered climate (Walker et al., 1993a). Experimental research using subnivean access chambers (Salisbury, 1985; Williams et al., 1996) and alteration of snow regimes through enhancement (Walker et al., 1999) or removal (Galen and Stanton, 1993) is needed in a wider variety of snowy ecosystems. Larger scale experiments are needed to examine the ecosystem effects of increased snowpack in whole watersheds or larger ecosystems (Steinhoff and Ives, 1976).

3. **Global snow-ecosystem patterns.** As Groisman and Davies (Chapter 1) point out, there is much that needs to be done linking ecosystems with snow climate and ecosystems in GCMs. Thus far, snow-vegetation interactions have not been successfully incorporated into global biogeochemical models (McGuire et al., 2000). There is a large need to examine global vegetation and NDVI patterns in relation to snow regimes and not merely as a function of precipitation and temperature. In the Arctic, snow cover patterns may be largely responsible for some of the zonal vegetation patterns observed on satellite imagery (Walker et al., 1998), and these boundaries could shift with a change in snow regimes.

4. **Paleo-snow-ecosystem studies.** Past northern ecosystems probably had very different snow regimes than the present. For example, relatively dry, less-snowy conditions in northeastern Siberia and Alaska during the late-Pleistocene resulted in colder permafrost, deeper summer thaw, and drier grassier, less-mossy environments with many grazers, such as the steppe bison, horse, saiga antelope, and mammoth (Guthrie, 1990). Knowledge of these patterns would be extremely useful for understanding past paleoecological relationships and for predicting the possible consequences of future snow regime scenarios. Unfortunately, the past effects of changes in snow regimes caused by climate change are not presently apparent from the paleorecord except for the recent past where there is a record in the tree rings of some species (Begin and Boivin, Chapter 7). Knowledge of past snow–ecosystem relationships, possibly through studies of diatoms

and pollen from plants, with known snow-regime preferences could help in predictions of future environments.

6.7   Acknowledgments

This paper resulted from talks D. A. Walker and J. G. de Molenaar presented at the Snow Ecology Workshop, Québec City, June 3–7, 1993, convened by the Snow Ecology Working Group, a contribution of the International Commission on Snow and Ice to the International Geosphere-Biosphere Programme. Special thanks to Dr. Gerry Jones for organizing the Quebec workshop and urging us to push forward with these thoughts. Larry Bliss, Ron Hoham, Gerry Jones, John Pomeroy, and Maureen Stanton provided excellent review comments. Andrew Lillie drafted the figures and helped greatly in assembling the manuscript. This work was funded through the Arctic System Science Flux Study (NSF/OPP 9214959), the Niwot Long-Term Ecological Research project (NSF/BSR 9211776), and the U.S. International Tundra Experiment (NSF/OPP 9400083).

6.8   References

Alexandrova, V.D. 1980. *The Arctic and Antarctic: Their Division into Geobotanical Areas.* Cambridge University Press, Cambridge.

Alpert, P., and Oechel, W.C. 1982. Bryophyte vegetation and ecology along a topographic gradient in montane tundra in Alaska. *Holarctic Ecol.*, 5, 99–108.

Andersen, D.C. 1987. Below-ground herbivory in natural communities: a review emphasizing fossorial animals. *Q. Rev. Biol.*, 62, 261–286.

Andrews, J.T., Davis, P.T., and Wright, C. 1976. Little Ice Age permanent snowcover in the eastern Canadian Arctic: extent mapped from Landsat-1 satellite imagery. *Geografiska Annaler*, 58A, 71–81.

Auerbach, N.A., and Halfpenny, J.C. 1991. Snowpack and the subnivean environment for different aspects of an open meadow in Jackson Hole, Wyoming, USA. *Arctic Alpine Res.*, 23, 41–44.

Bales, R.C., and Harrington, R.F. 1995. Recent progress in snow hydrology. *Rev. Geophys.*, 33, 1011–1020.

Barrett, P.E. 1972. *Phytogeocoenoses of a Coastal Lowland Ecosystem, Devon Island, N.W.T.* Ph.D. thesis, University of British Columbia, Vancouver.

Barry, R.G. 1992. Mountain climatology and past and potential future climatic changes in mountain regions: a review. *Mountain Res. Dev.*, 12, 71–86.

Batzli, G.O., and Henttonen, H. 1990. Demography and resource use by microtine rodents near Toolik Lake, Alaska, USA. *Arctic Alpine Res.*, **22**, 51–64.

Batzli, G.O., White, R.G., MacLean, S.F. Jr., Pitelka, F.A., and Collier, B.D. 1980. The herbivore-based trophic system. In *An Arctic Ecosystem: The Coastal Tundra at Barrow, Alaska*, J. Brown, P.C. Miller, L.L. Tieszen, and F.L. Bunnell (Eds.). Dowden, Hutchinson and Ross, Stroudsburg, PA, pp. 335–410.

Bell, K.L. 1974. *Autecology of Kobresia bellardii: Why Winter Snow Accumulation Limits Local Distribution*. Ph.D. thesis, University of Alberta, Edmonton.

Bell, K.L., and Bliss, L.C. 1979. Autecology of *Kobresia bellardii*: why winter snow accumulation limits local distribution. *Ecol. Monogr.*, **49**, 377–402.

Benedict, J.B. 1984. Rates of tree-island migration, Colorado Rocky Mountains, U.S.A. *Ecology*, **65**, 820–823.

Benedict, J.B. 1990. Lichen mortality due to late-lying snow: results of a transplant study. *Arctic Alpine Res.*, **22**, 81–89.

Benson, C., Kane, D.L., Sturm, M., Holmgren, J., Liston, G., McNamara, J., and Hinzman, L. 1996. Snow distribution in the Alaskan Arctic. *EOS Trans.*, **77**, F185.

Beyers, J.L. 1983. *Physiological Ecology of Three Alpine Plant Species along a Snowbank Gradient in the Northern Sierra Nevada*. Ph.D. thesis, Duke University, Durham, NC.

Billings, W.D. 1969. Vegetational patterns near alpine timberline as affected by fire-snowdrift interactions. *Vegetatio*, **19**, 697–704.

Billings, W.D. 1973. Arctic and alpine vegetations: similarities, differences, and susceptibility to disturbances. *Bio. Sci.*, **23**, 697–704.

Billings, W.D. 1974. Arctic and alpine vegetation: plant adaptations to cold summer climates. In *Arctic and Alpine Environments*, R.G. Barry and J.D. Ives (Eds.). Methuen, London, pp. 403–443.

Billings, W.D., and Bliss, L.C. 1959. An alpine snowbank environment and its effects on vegetation, plant development, and productivity. *Ecology*, **40**, 388–397.

Billings, W.D., and Mark, A.F. 1961. Interactions between alpine tundra vegetation and patterned ground in the mountains of southern New Zealand. *Ecology*, **42**, 18–31.

Billings, W.D., and Mooney, H.A. 1968. The ecology of arctic and alpine plants. *Biol. Rev.*, **43**, 481–529.

Bleak, A.T. 1970. Disappearance of plant material under a winter snow cover. *Ecology*, **51**, 915–917.

Bliss, L.C. 1956. A comparison of plant development in microenvironments of arctic and alpine tundras. *Ecol. Monogr.*, **26**, 303–337.

Bliss, L.C. 1962. Adaptations of arctic and alpine plants to environmental conditions. *Arctic*, **15**, 117–144.

Bliss, L.C. 1963. Alpine plant communities of the Presidential Range, New Hampshire. *Ecology*, **44**, 678–697.

Bliss, L.C. 1966. Plant productivity in alpine microenvironments on Mt. Washington, New Hampshire. *Ecol. Monogr.*, **36**, 136–155.

Bliss, L.C. 1969. Alpine community pattern in relation to environmental parameters. In *Essays in Plant Geography and Ecology*, K.N.H. Greenidge (Ed.). Nova Scotia Museum, Halifax, Nova Scotia, pp. 167–184.

Bliss, L.C. 1997. Arctic ecosystems of North America. In *Polar and Alpine Tundra*, F.E. Wielgolaski (Ed.). Elsevier, Amsterdam, pp. 551–683.

Bliss, L.C., and Svoboda, J. 1984. Plant communities and plant production in the western Queen Elizabeth Islands. *Holarctic Ecol.*, 7, 325–344.

Bliss, L.C., Kerik, J., and Peterson, W. 1977. Primary production of dwarf shrub heath communities, Truelove Lowland. In *Truelove Lowland, Devon Island, Canada: A High Arctic Ecosystem*, L.C. Bliss (Ed.). University of Alberta Press, Edmonton, Canada, pp. 217–224.

Bliss, L.C., Svoboda, J., and Bliss, D.I. 1984. Polar deserts, their plant cover and production in the Canadian High Arctic. *Holarctic Ecol.*, 7, 305–324.

Böcher, T.W. 1954. Oceanic and continental vegetational complexes in southwest Greenland. *Meddelelser Grönland*, 148, 1–336.

Böcher, T.W. 1959. Floristic and ecological studies in middle west Greenland. *Meddelelser Grönland*, 156, 1–68.

Böcher, T.W. 1963. Phytogeography of middle west Greenland. *Meddelelser Grönland*, 148, 1–287.

Bock, J.H. 1976. The effects of increased snowpack on the phenology and seed germinability of selected alpine species. In *Ecological Impacts of Snowpack Augmentation in the San Juan Mountains, Colorado*, H.W. Steinhoff and J.D. Ives (Eds.). U.S. Department of the Interior, Bureau of Reclamation, Denver, CO, pp. 265–271.

Bockheim, J.G., Walker, D.A., Everett, L.R., Nelson, F.E., Shiklomanov, N.I., 1998. Soil and cryoturbation in moist nonacidic and acidic tundra in the Kuparuk River basin, Arctic Alaska, U.S.A. *Arctic Alpine Res.*, 30, 166–174.

Bowman, W.D. 1992. Inputs and storage of nitrogen in winter snowpack in an alpine ecosystem. *Arctic Alpine Res.*, 24, 211–215.

Braun-Blanquet, J. 1932. *Plant Sociology: The Study of Plant Communities*. McGraw-Hill, New York.

Braun-Blanquet, J. 1949a. Übersicht der pflanzengesellschaften ratiens (iv). *Station Intern. de Geobot. Medit. et Alp., Montpellier*, 24, 20–37.

Braun-Blanquet, J. 1949b. Übersicht der pflanzengesellschaften ratiens (v). *Station Intern. de Geobot. Medit. et Alp., Montpellier*, 11, 214–237.

Braun-Blanquet, J. 1950. Übersicht der pflanzengesellschaften ratiens (vi). *Station Intern. de Geobot. Medit. et Alp. Montpellier*, 29, 341–360.

Brooks, P.D., Schmidt, S.K., Sommerfeld, R., and Musselman, R. 1994. Distribution and abundance of microbial biomass in Rocky Mountain spring snowpacks. In *Proc. 50th Annual Eastern Snow Conf.*, pp. 301–306.

Brooks, P.D., Williams, M.K., Walker, D.A., and Schmidt, S.K. 1995. The Niwot Ridge snowfence experiment: biogeochemical responses to changes in the seasonal snowpack. In *Biogeochemistry of Seasonally Snow Covered Basins*, K.A. Tonnessen, M.W. Williams, and M. Tranter (Eds.). IAHS Publication No. 228, IAHS Press, Wallingford, U.K., pp. 293–302.

Brooks, P.D., Williams, M.W., and Schmidt, S.K. 1996. Microbial activity under alpine snowpacks, Niwot Ridge, Colorado. *Biogeochemistry*, **32**, 93–113.

Brooks, P.D., Schmidt, S.K., and Williams, M.W. 1997. Winter production of CO_2 and N_2O from alpine tundra: environmental controls and relationship to inter-system C and N fluxes. *Oecologia*, **110**, 403–413.

Buckner, D.L. 1977. *Ribbon Forest Development and Maintenance in the Central Rocky Mountains of Colorado*. Ph.D. thesis, University of Colorado, Boulder.

Burke, I.C., Schimel, D.S., Yonker, C.M., Parton, W.J., Joyce, L.A., and Lauenroth, W.K. 1990. Regional modeling of grassland biogeochemistry using GIS. *Landscape Ecol.*, **4**, 45–54.

Burns, S.F. 1980. *Alpine Soil Distribution and Development, Indian Peaks, Colorado Front Range*. Ph.D. thesis, University of Colorado, Boulder.

Burns, S.J., and Tonkin, P.J. 1982. Soil-geomorphic models and the spatial distribution and development of alpine soils. In *Space and Time in Geomorphology*, C.E. Thorn (Ed.). Allen and Unwin, London, pp. 25–43.

Caine, N. 1992. Modulation of the dirunal streamflow response by the seasonal snowcover of an alpine snowpatch. *Catena*, **16**, 153–162.

Caine, N., and Thurman, E.M. 1990. Temporal and spatial variations in the solute content of an alpine stream, Colorado Front Range. *Geomorphol. Abstr.*, **4**, 55–72.

Caldwell, M.L.H. 1969. *Erythronium*: comparative phenology of alpine and deciduous forest species in relation to environment. *Am. Midl. Nat.*, **82**, 543–558.

Caldwell, M.M. 1968. Solar ultraviolet radiation as an ecological factor for alpine plants. *Ecol. Monogr.*, **38**, 243–268.

Caldwell, M.M. 1981. Plant response to solar ultraviolet radiation. In *Encyclopedia of Plant Physiology*, O.L. Lange (Ed.). Vol. 12A, Springer-Verlag, Berlin, pp. 169–197.

Caldwell, M.M., Robberecht, R., and Billings, W.D. 1980. A steep latitudinal gradient of solar ultraviolet-B radiation in arctic-alpine life zone. *Ecology*, **61**, 600–611.

Caldwell, M.M., Robberecht, R., Nowak, R.S., and Billings, W.D. 1982. Differential photosynthetic inhibition by ultraviolet radiation in species from the arctic-alpine life zone. *Arctic Alpine Res.*, **14**, 195–202.

Canaday, B.B., and Fonda, R.W. 1974. The influence of subalpine snowbanks on vegetation pattern, production, and phenology. *Bull. Torrey Botanical Club*, **101**, 340–350.

Carrol, S.S., Day, G.N., Cressie, N., and Carroll, T.R. 1995. Spatial modelling of snow water equivalent using airborne and ground-based snow data. *Environmetrics*, **6**, 127–139.

Chambers, J.C. 1991. Patterns of growth and reproduction in a perennial tundra forb (*Geum rossii*): effects of clone area and neighborhood. *Can. J. Bot.*, **69**, 1977–1983.

309

Chambers, J.C. 1993. Seed and vegetation dynamics in an alpine herb field: effects of disturbance type. *Can. J. Bot.*, **71**, 471–485.

Chambers, J.C., MacMahon, J.A., and Brown, R.W. 1990. Alpine seedling establishment: the influence of disturbance type. *Ecology*, **71**, 1323–1341.

Cline, D. 1995. Snow surface energy exchanges and snowmelt at a continental alpine site. In *Biogeochemistry of Seasonally Snow Covered Basins*, K.A. Tonnessen, M.W. Williams, and M. Tranter (Eds.). IAHS Publication No. 228, IAHS Press, Wallingford, U.K., pp. 157–166.

Cooper, D.J. 1986. Arctic-alpine tundra vegetation of the Arrigetch Creek Valley, Brooks Range, Alaska. *Phytocoenologia*, **14**, 467–555.

Cortinas, M.R., and Seastedt, T.R. 1996. Short- and long-term effects of gophers (*Thomomys talpoides*) on soil organic matter dynamics in alpine tundra. *Pedobiologia*, **40**, 162–170.

Curl, H., Hardy, J.T., and Ellermeier, R. 1972. Spectral absorption of solar radiation in alpine snow fields. *Ecology*, **53**, 1189–1194.

Dahl, E. 1956. Rondane mountain vegetation in South Norway and its relation to the environment. *Norske Videnskaps Akademi*, **1956**, 1–394.

Daniëls, F.J.A. 1982. Vegetation of the Angmagssalik District, Southeast Greenland, IV. Shrub, dwarf shrub and terricolous lichens. *Meddelelser om Grønland, Bioscience*, **10**, 1–78.

Dierschke, H. 1977. *Vegetation und Klima*. J. Cramer, Vaduz.

Dierssen, K. 1984. Vergleichende vegetationakundliche Untersuchungen an Schneeböden (Zur Abgrenzung der Klasse *Salicetea herbacae*). *Ber. Deutsch. Bot. Ges.*, **97**, 359–382.

Dingman, S.L., Barry, R.G., Weller, G., Benson, C., LeDrew, E.F., and Goodwin, C.W. 1980. Climate, snow cover, micro-climate, and hydrology. In *An Arctic Ecosystem: The Coastal Tundra at Barrow, Alaska*, J. Brown, P.C. Miller, L.L. Tieszen, and F.L. Bunnell (Eds.). Dowden, Hutchinson and Ross, Stroudsburg, PA, pp. 30–65.

Douglas, G.W., and Bliss, L.C. 1977. Alpine and high subalpine plant communities of the North Cascades Range, Washington and British Columbia. *Ecol. Monogr.*, **47**, 113–150.

Duguay, C.R., and Walker, D.A. 1996. Environmental modeling and monitoring with GIS: Niwot Long-Term Ecological Research site. In *GIS and Environmental Modeling: Progress and Research Issues*, M.F. Goodchild, L.T. Steyaert, B.O. Parks, C.A. Johnston, D.R. Maidment, M. Crane, and S. Glendinning (Eds.). GIS World, Ft. Collins, CO, Chap. 41, pp. 219–223.

Earle, C.J. 1993. *Forest Dynamics in a Forest-Tundra Ecotone, Medicine Bow Mountains, Wyoming*. Ph.D. thesis, University of Washington, Seattle.

Eastland, W.G., Bowyer, R.T., and Fancy, S.G. 1989. Effects of snow cover on selection of calving sites by caribou. *J. Mamm.*, **70**, 824–828.

Ellenberg, H. 1988. *Vegetation Ecology of Central Europe*. Cambridge University Press, Cambridge.

Elvebakk, A. 1984a. Contributions to the lichen flora and ecology of Svalbard, Arctic Norway. *Bryologist*, **87**, 308–313.

Elvebakk, A. 1984b. Vegetation pattern and ecology of siliceous boulder snow beds on Svalbard. *Polarforschung*, **54**, 9–20.

Elvebakk, A. 1985. Higher phytosociological syntaxa on Svalbard and their use in subdivision of the Arctic. *Nordic J. Bot.*, **5**, 273–284.

Elvebakk, A. 1997. Tundra diversity and ecological characteristics of Svalbard. In *Polar and Alpine Tundra*, F.E. Wielgolaski (Ed.). Elsevier, Amsterdam, pp. 347–359.

Emerick, J.C. 1976. *Effects of Artificially Increased Winter Snow Cover on Plant Canopy Architecture and Primary Production in Selected Areas of Colorado Alpine Tundra*. Ph.D. thesis, University of Colorado, Boulder.

Emerick, J.C., and Webber, P.J. 1982. The effects of augmented winter snow cover on the canopy structure of alpine vegetation. In *Ecological Studies in the Colorado Alpine: A Festschrift for John W. Marr*, J.C. Halfpenny (Ed.). Occasional Paper No. 37, University of Colorado, Institute of Arctic and Alpine Research, Boulder, CO, pp. 63–72.

Emers, M., Jorgenson, J.C., and Raynolds, M.K. 1995. Response of arctic tundra plant communities to winter vehicle disturbance. *Can. J. Bot.*, **73**, 905–917.

Evans, R.D., and Fonda, R.W. 1990. The influence of snow on subalpine meadow community pattern, North Cascades, Washington. *Can. J. Bot.*, **68**, 212–220.

Evans, B.M., Walker, D.A., Benson, C.S., Nordstrand, E.A., and Petersen, G.W. 1989. Spatial interrelationships between terrain, snow distribution and vegetation patterns at an arctic foothills site in Alaska. *Holarctic Ecol.*, **12**, 270–278.

Everett, K.R., Kane, D.L., and Hinzman, L.D. 1996. Surface water chemistry and hydrology of a small arctic drainage basin. In *Landscape Function and Disturbance in Arctic Tundra*, J.F. Reynolds and J.D. Tenhunen (Eds.). Springer-Verlag, Berlin, pp. 185–201.

Fahnstock, J.T., Jones, M.H., Brooks, P.D., Walker, D.A., and Welker, J.M. 1997. Winter and early spring CO_2 flux from tundra communities of northern Alaska. *J. Geophys. Res*, 29023–29027.

Felix, N.A., and Raynolds, M.K. 1989a. The effects of winter seismic trails on tundra vegetation in northeastern Alaska, U.S.A. *Arctic Alpine Res.*, **21**, 188–202.

Felix, N.A., and Raynolds, M.K. 1989b. The role of snow cover in limiting surface disturbance caused by winter seismic exploration. *Arctic*, **42**, 62–68.

Ferrari, C., and Rossi, G. 1995. Relationships between plant communities and late snow melting on Mount Prado (Northern Apennines, Italy). *Vegetatio*, **120**, 49–58.

Flock, J.W. 1978. Lichen-bryophyte distribution along a snow-cover-soil-moisture gradient, Niwot Ridge, Colorado. *Arctic Alpine Res.*, **10**, 31–47.

Fonda, R.W., and Bliss, R.C. 1966. Annual carbohydrate cycle of alpine plants on Mt. Washington, New Hampshire. *Bull. Torrey Botanical Club*, **93**, 268–278.

Formozov, A.N. 1964. *Snow Cover as an Integral Factor of the Environment and Its Importance in the Ecology of Mammals and Birds*. Translated from Russian by W. Prychodko and W.O. Pruitt, Jr. Occasional Publication No. 1, University of Alberta, Boreal Institute for Northern Studies, Edmonton.

Fox, J.F. 1981. Intermediate levels of soil disturbance maximize alpine plant diversity. *Nature*, **293**, 564–565.

311

Fries, T.C.E. 1913. *Botanische Untersuchungen im nördlichsten Schweden*. Almquest and Wiksell, Uppsala.

Galen, C., and Stanton, M.L. 1991. Consequences of emergence phenology for reproductive success in *Ranunculus adoneus* (Ranunculaceae). *Am. J. Bot.*, **78**, 978–988.

Galen, C., and Stanton, M.L. 1993. Short-term responses of alpine buttercups to experimental manipulations of growing season length. *Ecology*, **75**, 1546–1557.

Galen, C., and Stanton, M.L. 1995. Responses of snowbed plant species to changes in growing-season length. *Ecology*, **74**, 1546–1557.

Galen, C., Dawson, T.E., and Stanton, M.L. 1993. Carpels as leaves: meeting the carbon cost of reproduction in an alpine buttercup. *Oecologia*, **95**, 187–193.

Geiger, R. 1965. *The Climate Near the Ground*. Harvard University Press, Cambridge.

Gelting, P. 1934. Studies in the vascular plants of East Greenland between Franz Joseph Fjord and Dove Bay. Meddleser om Grønland, **101**, 1–340.

Gerdel, R.W. 1948. Penetration of radiation into the snowpack. *Trans. Am. Geophys. Union*, **29**, 366–374.

Giblin, A.E., Nadelhoffer, K.J., Shaver, G.R., Laundre, J.A., and McKerrow, A.J. 1991. Biogeochemical diversity along a riverside toposequence in arctic Alaska. *Ecol. Monogr.*, **61**, 415–435.

Gjærevoll, O. 1950. The snow-bed vegetation in the surroundings of Lake Tornetrask, Swedish Lappland. *Svensk Botanisk Tidskrift*, **44**, 387–440.

Gjærevoll, O. 1952. Botanical assistance to roadmaking in the mountains (Eng. trans.). *Syn og Segn*, **1952**, 1–8.

Gjærevoll, O. 1954. Kobresieto-Dryadion in Alaska. *Nytt Magasih Botanikk*, **3**, 51–54.

Gjærevoll, O. 1956. The plant communities of the Scandinavian alpine snow-beds. *Det. Kgl. Norske Vidensk. Selsk. Skrift.*, **1**, 1–405.

Gjærevoll, O., and Bringer, K.-G. 1965. Plant cover of the alpine regions. *Acta Phytogeograph. Suec.*, **50**, 257–268.

Gjessing, Y.T., and Øvstedal, D.O. 1975. Energy budget and ecology of two vegetation types in Svalbard. *Astarte*, **8**, 83–92.

Guthrie, R.D. 1990. *Frozen Fauna of the Mammoth Steppe: The Story of Blue Babe*. University of Chicago Press, Chicago, 323 pp.

Haase, R. 1987. An alpine vegetation map of Caribou Lake Valley and Fourth of July Valley, Front Range, Colorado, U.S.A. *Arctic Alpine Res.*, **19**, 1–10.

Hadac, E. 1971. Snow-land communities of Reykjanes Peninsula, SW Iceland (Plant communities of Reykjanes Peninsula, Part 4). *Folia Geobot. Phytotaxon.*, **6**, 105–126.

Hadley, E.B., and Bliss, L.C. 1964. Energy relationships of alpine plants on Mt. Washington, New Hampshire. *Ecol. Monogr.*, **34**, 331–357.

Halfpenny, J.C., and Ozanne, R.D. 1989. *Winter: An Ecological Handbook*. Johnson Books, Boulder, CO, 273 pp.

Halfpenny, J.C., and Southwick, C.H. 1982. Small mammal herbivores of the Colorado alpine tundra. In *Ecological Studies in the Colorado Alpine: A Festschrift for John W. Marr*, J.C. Halfpenny (Ed.). Occasional Paper No. 37, University of Colorado, Institute of Arctic and Alpine Research, Boulder, CO, pp. 113–123.

Hall, D.K., and Martinec, J. 1985. *Remote Sensing of Ice and Snow*. Chapman and Hall, London.

Hampel, v. R. 1963. Ecological investigations in the sub-alpine zone for the purpose of reforestation at regions of high elevation (Part I and II). *Bewirtschaft. Forstl. Bundes-Versuchsanst Mariabrunn*, **60**, 865–879.

Harte, J., and Shaw, R. 1995. Shifting dominance within a montane vegetation community: results of a climate-warming experiment. *Science*, **267**, 876–880.

Harte, J., Torn, M.S., Chang, F.R., Feifarek, B., Kinzig, A.P., Shaw, R., and Shen, K. 1995. Global warming and soil microclimate: results from a meadow-warming experiment. *Ecol. Appl.*, **5**, 132–150.

Hedberg, O., Martensson, O., and Rudberg, S. 1952. Botanical investigations in the Paltsa region of northernmost Sweden. *Bot. Notiser Suppl.*, **3**, 1–209.

Heer, O. 1836. Die vegetationverhaltnisse des sudostlilchen Teils des Kantons Glarus. *Mitt. aus dem Gebiete der theoretischen Erdkunde*, J.J. Frobe und O. Heer, Zurich.

Heide, O.M. 1992. Flowering strategies of the high-arctic and high-alpine snow bed grass species *Phippsia algida*. *Physiol. Plant.*, **85**, 606–610.

Henry, G.H.R., and Molau, U. 1997. Tundra plants and climate change: the International Tundra Experiment (ITEX). *Global Change Biol.*, **3**, 1–9.

Herk, C.M.v., Knaapen, J.P., and Daniëls, F.J.A. 1988. Phytomass and duration of snow cover in a snowbed in southeast Greenland. In *Diversity and Pattern in Plant Communities*, H.J. During, M.J.A. Werger, and J.H. Williams (Eds.). SPB Academic, The Hague, Netherlands, pp. 67–75.

Hinzman, L.D., Kane, D.L., Benson, C.S., and Everett, K.R. 1996. Energy balance and hydrological processes in an arctic watershed. In *Landscape Function and Disturbance in Arctic Tundra*, J.F. Reynolds and J.D. Tenhunan (Eds.). Springer-Verlag, New York, pp. 131–154.

Holtmeier, F.-K., and Broll, G. 1992. The influence of tree islands and microtopography on pedoecological conditions in the forest-alpine tundra ecotone on Niwot Ridge, Colorado Front Range, U.S.A. *Arctic Alpine Res.*, **24**, 216–228.

Holway, J.G., and Ward, R.T. 1963. Snow and meltwater effects in an area of Colorado alpine. *Am. Midl. Nat.*, **69**, 189–197.

Holway, J.G., and Ward, R.T. 1965. Phenology of alpine plants in northern Colorado. *Ecology*, **46**, 73–83.

Hrapko, J.O., and LaRoi, G.H. 1978. The alpine tundra vegetation of Signal Mountain, Jasper National Park. *Can. J. Bot.*, **56**, 309–332.

Huntly, N., and Inouye, R.S. 1987. Small mammal populations of an old-field chronosequence: successional patterns and associations with vegetation. *J. Mamm.*, **68**, 739–745.

Inouye, D.W., and McGuire, A.D. 1991. Effects of snowpack on timing and abundance of flowering in *Delphinium nelsonii* (Ranunculaceae): implications for climate change. *Am. J. Bot.*, **78**, 997–1001.

Inouye, R.S., Huntly, N.J., Tilman, D., and Tester, J.R. 1987. Pocket gophers (*Geomys bursarius*), vegetation, and soil nitrogen along a successional sere in east central Minnesota. *Oecologia*, **72**, 178–184.

Isard, S.A. 1986. Factors influencing soil moisture and plant community distribution on Niwot Ridge, Front Range, Colorado, U.S.A. *Arctic Alpine Res.*, **18**, 83–96.

Isard, S.A., and Belding, M.J. 1989. Evapotranspiration from the alpine tundra of Colorado, U.S.A. *Arctic Alpine Res.*, **21**, 71–82.

Jeník, J. 1959. Kurzgefasste Übersicht der Theorie der anemoorographischen Systeme. *Preslia*, **31**, 337–357.

Johnson, P.L., and Billings, W.D. 1962. The alpine vegetation of the Beartooth Plateau in relation to cryopedogenic processes and patterns. *Ecol. Monogr.*, **32**, 105–135.

Jones, H.G. 1991. Snow chemistry and biological activity: a particular perspective on nutrient cycling. In *NATO ASI Series G: Ecol. Sci., Vol. 28, Seasonal Snowpacks, Processes of Compositional Change*. T.D. Davies, M. Tranter, and H.G. Jones (Eds.). Springer-Verlag, Berlin, pp. 173–227.

Kalliola, R. 1939. Über Weisen und wiesenartige Pflanzengesellschaften auf der Fischerhalbinsel in Petsamo Lappland. *Acta Forestalia*, **48**, 1–523.

Kane, D.L., Hinzman, L.D., Benson, C.S., and Everett, K.R. 1989. Hydrology of Imnavait Creek, an arctic watershed. *Holarctic Ecol.*, **12**, 262–269.

Kane, D.L., Hinzman, L.D., Woo, M., and Everett, K.R. 1992. Arctic hydrology and climate change. In *Arctic Ecosystems in a Changing Climate*, F.S. Chapin III, R.L. Jefferies, J.F. Reynolds, G.R. Shaver, and J. Svoboda (Eds.). Academic Press, New York, pp. 35–57.

Kelley, J.J., Weaver, D.F., and Smith, B.P. 1968. The variation of carbon dioxide under the snow in the Arctic. *Ecology*, **49**, 358–361.

Kiener, W. 1967. *Sociological Studies of the Alpine Vegetation on Longs Peak*. University of Nebraska Studies, New Series 34, Lincoln.

Kimball, S.L., Bennett, B.D., and Salisbury, F.B. 1973. The growth and development of montane species at near-freezing temperatures. *Ecology*, **54**, 168–173.

Knight, D.H., and Kyte, C.R. 1975. *The Effect of Snow Accumulation on Litter Decomposition and Nutrient Leaching*, pp. 215–224, in "The potential sensitivity of various ecosystem components to winter precipitation management." Final Report of the Medicine Bow Ecology Project, Prepared by University of Wyoming, Laramie, WY.

Kolchugina, T.P., and Vinson, T.S. 1993. Climate warming and the carbon cycle in the permafrost zone of the Soviet Union. *Permafrost Periglac. Processes*, **4**, 149–164.

Komárková, V. 1979. *Alpine Vegetation of the Indian Peaks Area, Front Range, Colorado Rocky Mountains*. J. Cramer, Vaduz.

Komárková, V. 1980. Classification and ordination in the Indian Peaks area, Colorado Rocky Mountains. *Vegetatio*, **42**, 149–163.

Komárková, V. 1993. Vegetation type hierarchies and landform disturbance in arctic Alaska and alpine Colorado with emphasis on snowpatches. *Vegetatio*, **106**, 155–181.

Komárková, V., and McKendrick, J.D. 1988. Patterns in vascular plant growth forms in arctic communities and environment at Atkasook, Alaska. In *Plant Form and Vegetation Structure*, M.J.A. Werger, P.J.M. van der Aart, H.J. During, and J.T.A. Verhoeven (Eds.). SPB Academic Publishing, The Hague, The Netherlands, pp. 45–70.

Komárková, V., and Webber, P.J. 1978. An alpine vegetation map of Niwot Ridge, Colorado. *Arctic Alpine Res.*, **10**, 1–29.

Kuramoto, R.T., and Bliss, L.C. 1970. Ecology of subalpine meadows in the Olympic Mountains, Washington. *Ecol. Monogr.*, **40**, 317–347.

Lambert, J.D.H. 1968. *The Ecology and Successional Trends in the Low Arctic Subalpine Zone of the Richardson and British Mountains of the Canadian Western Arctic*. Ph.D. thesis, University of British Columbia, Vancouver.

Larcher, W. 1963. Zur spätwinterlichen Erschwerung der Wasserbilanz vol Holzpflanzen au der Waldgrenze. *Ber. Naturwissen.-Med. Ver. Innsbruck*, **53**, 125–137.

Leadly, P.W., Li, H., Ostendorf, B., and Reynolds, J.F. 1996. Road-related disturbances in an arctic watershed: analyses by a spatially-explicit model of vegetation and ecosystem processes. In *Landscape Function and Disturbance in Arctic Tundra*, J.F. Reynolds and J.D. Tenhunan (Eds.). Springer-Verlag, New York, pp. 387–415.

LeDrew, E.F. 1975. The energy balance of mid-latitude alpine site during the growing season. *Arctic Alpine Res.*, **7**, 301–314.

Lindsey, J.H. 1971. Annual cycle of leaf water potential in *Picea engelmannii* and *Abies Lasiocarpa* at timberline Wyoming. *Arctic Alpine Res.*, **3**, 131–138.

Liston, G.E. 1986. *Seasonal Snowcover of the Foothills Region of Alaska's Arctic Slope: A Survey of Properties and Processes*. M.S. thesis, University of Alaska, Fairbanks.

Litaor, M.I., Mancinelli, R., and Halfpenny, J.C. 1996. The influence of pocket gophers on the status of nutrients in alpine soils. *Geoderma*, **70**, 37–48.

Locke, C.W., and Locke, W.W.I. 1977. Little ice age snow-cover extent and paleoglaciation thresholds: north-central Baffin Island, NWT, Canada. *Arctic Alpine Res.*, **9**, 291–300.

Lüdi, W. 1921. Die pflanzengesellschaften des Lauterbrunnentales und ihre Sukzession. *Beitr. Geobot. Landesafnamen*, **9**, 1–350.

Maesson, O., Freedman, B., Nams, M.L.N., and Svoboda, J. 1983. Resource allocation in high arctic vascular plants of differing growth form. *Can. J. Bot.*, **61**, 1680–1691.

Marchand, P.J. 1987. *Life in the Cold: An Introduction to Winter Ecology*. University Press of New England, Hanover, N.H.

Marchand, P.J., and Roach, D.A. 1980. Reproductive strategies of pioneering alpine species: seed production dispersal and germination. *Arctic Alpine Res.*, **12**, 137–146.

Marion, G.M., Henry, G.H.R., Freckman, D., Johnstone, J., Jones, G., Jones, M.H., Lévesque, E., Molau, U., Mølgaard, P., Parsons, A.N., Svoboda, J., Virginia, R.A. 1997. Open-top designs for manipulating field temperature in high-latitude ecosystems. *Global Change Biol.*, **3**, 20–32.

315

Mark, A.F. 1965. The environment and growth rate of narrow-leaved snow tussock, *Chionochloa rigida*, in Otago. *N. Zealand J. Bot.*, 3, 73–103.

Mark, A.F. 1970. Floral initiation and development in New Zealand alpine plants. *N. Zealand J. Bot.*, 8, 67–75.

Mark, A.F. 1975. Photosynthesis and dark respiration in three alpine snow tussocks (*Chionochloa* spp.) under controlled environments. *N. Zealand J. Bot.*, 13, 93–122.

Mark, A.F., and Bliss, L.C. 1970. The high-alpine vegetation of central Otago, New Zealand. *N. Zealand J. Bot.*, 8, 381–451.

Mark, A.F., and Dickénson, K.J.M. 1997. New Zealand alpine ecosystems. In *Polar and Alpine Tundra*, F.E. Wielgolaski (Ed.). Elsevier, Amsterdam, pp. 311–345.

Markon, C.J., Fleming, M.D., and Binnian, E.F. 1995. Characteristics of vegetation phenology over the Alaskan landscape using AVHRR time-series data. *Polar Record*, 31, 179–190.

Marr, J.W. 1977. The development and movement of tree islands near the upper limit of tree growth in the Southern Rocky Mountains. *Ecology*, 58, 1159–1164.

Maxwell, B. 1992. Arctic climate: potential for change under global warming. In *Arctic Ecosystems in a Changing Climate*, F.S. Chapin III, R.L. Jefferies, J.F. Reynolds, G.R. Shaver, and J. Svoboda (Eds.). Academic Press, San Diego, CA, pp. 11–34.

Maxwell, B. 1997. Recent climate patterns in the Arctic. In *Global Change and Arctic Terrestrial Ecosystems*, W.C. Oechel, T. Callaghan, T. Gilmanov, J.I. Holten, B. Maxwell, U. Molau, and B. Sveinbjörnsson (Eds.). Springer-Verlag, New York, pp. 21–46.

May, D.E. 1976. *The Response of Alpine Tundra Vegetation in Colorado to Environmental Modification*. Ph.D. thesis, University of Colorado, Boulder.

May, D.E., and Webber, P.J. 1982. Spatial and temporal variation of the vegetation and its productivity on Niwot Ridge, Colorado. In *Ecological Studies in the Colorado Alpine: A Festschrift for John W. Marr*, J.C. Halfpenny (Ed.). Occasional Paper No. 37, University of Colorado, Institute of Arctic and Alpine Research, Boulder, pp. 35–62.

May, D.E., Webber, P.J., and May, T.A. 1982. Success of transplanted alpine tundra plants on Niwot Ridge, Colorado. *J. Appl. Ecol.*, 19, 965–976.

McGraw, J.B. 1985a. Experimental ecology of *Dryas octopetala* ecotypes. III. Environmental factors and plant growth. *Arctic Alpine Res.*, 17, 229–239.

McGraw, J.B. 1985b. Experimental ecology of *Dryas octopetala* ecotypes: relative response to competitors. *New Phytol.*, 100, 233–241.

McGraw, J.B. 1987. Experimental ecology of *Dryas octopetala* ecotypes. IV. Fitness response to reciprocal transplanting in ecotypes with differing plasticity. *Oecologia*, 73, 465–468.

McGraw, J.B., and Antonovics, J. 1983. Experimental ecology of *Dryas octopetala* ecotypes. I. Ecotypic differentiation and life-cycle stages of selection. *J. Ecol.*, 71, 879–897.

McGuire, A.D., Melillo, J.M., Randerson, J.T., Parton, W.J., Heimann, M., Meier, R.A., Clein, J.S., Kicklighter, D.W., and Sauf, W. 2000. Modeling the effects of snowpack on heterotrophic respiration across northern temperate and high latitude regions: Comparison with measurements of atmospheric carbon dioxide in high latitudes. *Biogeochemistry*, 48, 91–114.

McKane, R.B., Rastetter, E.B., and Shaver, G.R. et al. 1997. Climatic effects on tundra carbon storage inferred from experimental data and a model. *Ecology*, 78, 1170–1187.

Merritt, J.F. 1984. *Winter Ecology of Small Mammals*. Special Publication No. 10. Pittsburgh: Carnegie Museum of Natural History.

Miehe, G. 1997. Alpine vegetation types of the central Himalayas. In *Polar and Alpine Tundra*, F.E. Wielgolaski (Ed.). Elsevier, Amsterdam, pp. 161–184.

Miller, P.C. 1982. Environmental and vegetational variation across a snow accumulation area in montane tundra in central Alaska. *Holarctic Ecol.*, 5, 85–98.

Miller, P.C., Mangan, R.M., and Kummerow, J. 1982. Vertical distribution of organic matter in eight vegetation types near Eagle Summit, Alaska. *Holarctic Ecol.*, 5, 117–124.

Molenaar, J.G.d. 1976. *Vegetation of the Angmagssalik District, Southeast Greenland II. Herb and Snow-Bed Vegetation*. Nyt Nordisk Forlag, Arnold Busck, København.

Molenaar, J.G.d. 1987. An ecohydrological approach to floral and vegetational patterns in arctic landscape ecology. *Arctic Alpine Res.*, 19, 414–424.

Monteith, J.L. 1973. *Principals of Environmental Physics*. Edward Arnold, London.

Mooney, H.A., and Billings, W.D. 1960. The annual carbohydrate cycle of alpine plants as related to growth. *Am. J. Bot.*, 47, 594–598.

Mooney, H.A., and Billings, W.D. 1961. Comparative physiological ecology of arctic and alpine populations of *Oxyria digyna*. *Ecol. Monogr.*, 31, 1–29.

Mooney, H.A., and Billings, W.D. 1968. The ecology of arctic and alpine plants. *Biol. Rev.*, 43, 481–529.

Moser, W. 1967. Einblicke in das Leben von Nivalpflanzen. *Jahrb. Ver. Schutze Alpenpfl. Tiere*, 32, 1–11.

Mullen, R.B. 1995. *Fungal Endophytes in the Alpine Buttercup Ranunculus adoneus: Implications for Nutrient Cycling in Alpine Systems*. Ph.D. thesis, University of Colorado, Boulder.

Mullen, R.B., and Schmidt, S.K. 1993. Mycorrhizal infection, phosphorus uptake, and phenology in *Ranunculus adoneus*: implications for the functioning of mycorrhizae in alpine systems. *Oecologia*, 94, 229–234.

Mullen, R.B., Schmidt, S.K., and Jaeger, C.H. III. 1998. Nitrogen uptake during snowmelt by the snow buttercup, *Ranunculus adoneus. Arctic Alpine Res.*, 30, 121–125.

Murray, D., and Miller, P.C. 1982. Phenological observations of major plant growth forms and species in montane and *Eriophorum vaginatum* tussock tundra in central Alaska. *Holarctic Ecol.*, 5, 109–116.

Nadelhoffer, K.J., Giblin, A.E., Shaver, G.R., and Laundre, J.A. 1991. Effects of temperature and substrate quality on element mineralization in six arctic soils. *Ecology*, 72, 242–253.

Nams, M.L.N., and Freedman, B. 1987a. Ecology of heath communities dominated by *Cassiope tetragona* at Alexandra Fiord, Ellesmere Island, Canada. *Holarctic Ecol.*, **10**, 22–32.

Nams, M.L.N., and Freedman, B. 1987b. Phenology and resource allocation in a high arctic evergreen dwarf shrub, *Cassiope tetragona*. *Holarctic Ecol.*, **10**, 128–136.

Nelson, F.E., Hinkel, K.M., Shiklomanov, N.I., Mueller, G.R., Miller, L.L., and Walker, D.A. 1997a. Active-layer thickness in northern Alaska: systematic sampling, scale, and spatial autocorrelation. *J. Geophys. Res.*, **103**, 28963–28973.

Nelson, F.E., Shiklomanov, N.I., Mueller, G.R., Hinkel, K.M., Walker, D.A., and Bockheim, J.G. 1997b. Estimating active-layer thickness over large regions: Kuparuk River Basin, Alaska. *Arctic Alpine Res.*, **29**, 367–378.

Nelson, F.E., Lachenbruch, A., Woo, M.K., Koster, E., Osterkamp, T., Gavrilova, M.K., and Cheng, G.D. 1993. Permafrost and changing climate. In *Sixth International Conference on Permafrost*, 2 (Wushan Guangzhou, China), South China University of Technology Press, pp. 987–1005.

Nordhagen, R. 1928. Die Vegetation und Flora des Sylenegebietes. I. Die Vegetation. *Skr. Nor. Vidensk. Akad. Oslo 1*, **1**, 1–612.

Nordhagen, R. 1936. Versuch einer neuen Einteilung der subalpinen-alpinen vegetation Norwegens. *Bergens Museums Arbok. Naturvidenskapelig*, **7**, 1–88.

Nordhagen, R. 1943. Sikilsdalen og Norges Fjellbeiter: en plantesosiologisk monografi. *Bergens Muse. Skr.*, **22**, 1–607.

Norman, J.M. 1894. *Norges Artiske Flora*. I. Kristiania, Oslo.

O'Lear, H.A., and Seastedt, T.R. 1994. Landscape patterns of litter decomposition in alpine tundra. *Oecologia*, **99**, 95–101.

Oberbauer, S.F., and Billings, W.D. 1981. Drought tolerance and water use by plants along an alpine topographic gradient. *Oecologia*, **50**, 325–331.

Oberbauer, S.F., Starr, G., and Pop, E.W. 1998. Effects of extended growing season and warming on carbon dioxide and methane exchange of tussock tundra in Alaska. *J. Geophys. Res.*, **103**, 29075–29082.

Oechel, W.C., and Billings, W.D. 1992. Effects of global change on the carbon balance of arctic plants and ecosystems. In *Arctic Ecosystems in a Changing Climate: An Ecophysiological Perspective*, F.S. Chapin III, R.L. Jefferies, J.F. Reynolds, G.R. Shaver, and J. Svoboda (Eds.). Academic Press, San Diego, CA, pp. 139–168.

Oechel, W.C., Callaghan, T., Gilmanov, T., Holtan, J.I., Maxwell, B., Molau, U., Svein-björnsson, B. (Eds.). 1996. *Global Change and Arctic Terrestrial Ecosystems*. Springer-Verlag, New York.

Oechel, W.C., Vourlitis, G., and Hastings, S.J. 1997. Cold season CO_2 emission from arctic soils. *Global Biogeochem. Cycles*, **11**, 163–172.

Ohba, T. 1974. Vergleichende Studien über die alpine Vegetation Japans. *Phytocoenologia*, **1**, 339–401.

Ostendorf, B., and Reynolds, J.F. 1993. Relationships between a terrain-based hydrologic model and patch-scale vegetation patterns in an arctic tundra landscape. *Landscape Ecol.*, 8, 229–237.

Ostendorf, B., Quinn, P., Bevan, K., and Tenhunen, J.D. 1996. Hydrological controls on ecosystem gas exchange in an Arctic landscape. In *Landscape Function and Disturbance in Arctic Tundra*, J.F. Reynolds and J.D. Tenhunen (Eds.). Springer-Verlag, New York, pp. 369–386.

Ostler, W.K., Harper, K.T., McKnight, K.B., and Anderson, D.C. 1982. The effects of increasing snowpack on a subalpine meadow in the Uinta Mountains, Utah, U.S.A. *Arctic Alpine Res.*, 14, 203–214.

Outcalt, S.I., Goodwin, C., Weller, G., and Brown, J. 1975. Computer simulation of the snowmelt and soil thermal regime at Barrow, Alaska. *Water Resources Res.*, 11, 709–715.

Pauker, S.J., and Seastedt, T.R. 1996. Effects of mobile tree islands on soil carbon storage in tundra ecosystems. *Ecology*, 77, 2563–2567.

Pitelka, F.A. 1964. The nutrient-recovery hypothesis for arctic microtine cycles. I. Introduction. In *Grazing in Terrestrial and Marine Environments: A Symposium of the British Ecological Society*, Bangor, 11-14 April 1962, D.J. Crisp (Ed.). Blackwell Scientific, Oxford, pp. 55–56.

Pomeroy, J.W., and Gray, D.M. 1995. *Snowcover: Accumulation, Relocation and Management*. NHRI Science Report No. 7, National Hydrology Research Institute, Saskatoon, Canada.

Pomeroy, J.W., Hedstrom, N., Dion, K., Elliot, J., and Granger, R.J. 1994. *Quantification of Hydrological Pathways in the Prince Albert Model Forest, 1993–1994 Annual Report*. Environment Canada Report to the Prince Albert Model Forest Association, National Hydrology Research Institute NHRI Contribution Series CS-94006, National Hydrology Research Institute, Saskatoon, Canada.

Pomeroy, J.W., Marsh, P., and Gray, D.M. 1997. Application of a distributed blowing snow model to the Arctic. *Hydrol. Processes*, 11, 1451–1464.

Razzhivin, V.Y. 1994. Snowbed vegetation of far northeastern Asia. *J. Vegetation Sci.*, 5, 829–842.

Remmert, H. 1965. Bilogische Periodik. In *Handuch der Biologie, Bd B*. Akad. Verlagsgesellschaft, Frankfurt/Konstanz, pp. 336-410.

Remmert, H. 1980. *Arctic Animal Ecology*. Springer-Verlag, New York, 250 pp.

Reynolds, D.N. 1984a. Alpine annual plants: phenology, germination, photosynthesis, and growth of three Rocky Mountain species. *Ecology*, 65, 759–766.

Reynolds, D.N. 1984b. Populational dynamics of three annual species of alpine plants in the Rocky Mountains. *Oecologia*, 62, 250–255.

Reynolds, J.F., Tenhunen, J.D., Leadley, P.W., Li, H., Moorhead, D.L., Ostendorf, B., and Chapin, F.S. III. 1996. Patch and landscape models of arctic tundra: potential and limitations. In *Landscape Function and Disturbance in Arctic Tundra*, J.F. Reynolds and J.D. Tenhunen (Eds.). Springer-Verlag, New York, pp. 293–324.

Richardson, S.G., and Salisbury, F.B. 1977. Plant responses to the light penetrating snow. *Ecology*, 58, 1152–1158.

319

Rochow, T.F. 1969. Growth, caloric content, and sugars in *Caltha leptosepala* in relation to alpine snowmelt. *Bull. Torrey Botanical Club*, **96**, 689–698.

Rønning, O.I. 1965. Studies on Dryadion on Svalbard. *Norsk Polarinst. Skr.*, **134**, 1–52.

Salisbury, F.S. 1985. Plant growth under snow. *Auilo Ser. Bot*, **23**, 1–7.

Scherf, E.J., Galen, C., and Stanton, M.L. 1994. Seed dispersal, seedling survival, and habitat affinity in a snowbed plant: limits to the distribution of the snow buttercup, *Ranunculus adoneus*. *Oikos*, **69**, 405–413.

Schultz, A.M. 1964. The nutrient recovery hypothesis for arctic microtine cycles. II. Ecosystem variables in relation to arctic microtine cycles. In *Grazing in Terrestrial and Marine Environments: A Symposium of the British Ecological Society*, Bangor, 11–14 April 1962, D.J. Crisp (Ed.). Blackwell Scientific, Oxford, pp. 57–68.

Schultz, A.M. 1969. A study of an ecosystem: the arctic tundra. In *The Ecosystem Concept in Natural Resource Management*, G.M. Van Dyne (Ed.). Academic Press, New York, pp. 77–93.

Seip, H.M. 1980. Acid snow- snowpack chemistry and snowmelt. In *Effects of Acid Precipitation on Terrestrial Ecosystems*, T.C. Hutchinson and M. Havas (Eds.). Plenum Press, New York, pp. 77–94.

Shaver, G.R., and Billings, W.D. 1976. Carbohydrate accumulation in tundra graminoid plants as a function of seasons and tissue ages. *Flora*, **165**, 247–267.

Shaver, G.R., Billings, W.D., Chapin, F.S. III, Giblin, A.E., Nadelhoffer, K.J., Oechel, W.C., and Rastetter, E.B. 1992. Global change and the carbon balance of arctic ecosystems. *BioSci.*, **42**, 433–441.

Slaughter, C.W., Mellor, M., Sellmann, P.V., Brown, J., and Brown, L. 1975. *Accumulating Snow to Augment the Fresh Water Supply at Barrow, Alaska*. U.S. Army Cold Regions Research and Engineering Laboratory CRREL Special Report 217, CRREL, Hanover, N.H.

Sommerfeld, R.A., Mosier, A.R., and Musselman, R.C. 1993. CO_2, CH_4, and N_2O flux through a Wyoming snowpack and implications for global budgets. *Nature*, **361**, 140–142.

Sørenson, T. 1941. Temperature relations and phenology of the Northeast Greenland flowering plants. *Meddelelser om Grønland*, **125**, 305.

Spomer, G.C. 1962. *Physiological Ecology of Alpine Plants*. Ph.D. thesis, Colorado State University, Boulder.

Spomer, G.G., and Salisbury, F.B. 1968. Eco-physiology of *Geum turbinatum* and implications concerning alpine environments. *Botan. Gazette*, **129**, 33–49.

Sprugel, D.G. 1976. Dynamic structure of wave-generated *Abies balsamea* forests in the northeastern United States. *J. Ecol.*, **64**, 889–911.

Stanton, M.L., Rejmánek, M., and Galen, C. 1994. Changes in vegetation and soil fertility along a predictable snowmelt gradient in the Mosquito Range, Colorado, U.S.A. *Arctic Alpine Res.*, **26**, 364–374.

Staple, W.J., and Lehane, J.J. 1952. The conservation of soil moisture in southern Saskatchewan. *Science Agric.*, **32**, 36–47.

Staple, W.J., and Lehanne, J.J. 1955. The influence of field shelterbelts on wind velocity, evaporation, soil moisture and crop yield. *Can. J. Agric. Sci.*, **35**, 440–453.

Staple, W.J., Lehane, J.J., and Wenhardt, A. 1960. Conservation of soil moisture from fall and winter precipitation. *Can. J. Soil Sci.*, **40**, 80–88.

Steinhoff, H.W., and Ives, J.D. (Eds.). 1976. *Ecological Impacts of Snowpack Augmentation in the San Juan Mountains, Colorado.* U.S. Department of the Interior, Bureau of Reclamation, Denver, CO.

Stepphun, H. 1981. Snow and agriculture. In *Handbook of Snow: Principles, Processes, Management and Use*, D.H. Gray and D.H. Male (Eds.). Pergamon Press, Toronto, pp. 60–125.

Stoecker, R. 1976. Pocket gopher distribution in relation to snow in the alpine tundra. In *Ecological Impacts of Snowpack Augmentation in the San Juan Mountains, Colorado*, H.W. Steinhoff and J.D. Ives (Eds.). U.S. Department of the Interior, Bureau of Reclamation, Denver, CO, pp. 281–287.

Sturges, D.L. 1989. Response of mountain big sagebrush to induced snow accumulation. *J. Appl. Ecol.*, **26**, 1035–1041.

Sturm, M., Holmgren, J., and Liston, G.E. 1995. A seasonal snow cover classification system for local to global applications. *J. Climate*, **8**, 1261–1283.

Sturm, M., and Johnson, J.B. 1992. Thermal conductivity measurements of depth hoar. *J. Geophys. Res.*, **97**(B2), 2129–2139.

Sturm, M., Holmgren, J., Konig, M., and Morris, K. 1997. The thermal conductivity of seasonal snow. *J. Glaciol.*, **43**, 26–41.

Svoboda, J. 1977. Ecology and primary production of raised beach communities, Truelove Lowland. In *Truelove Lowland, Devon Island, Canada: A High Arctic Ecosystem*, L.C. Bliss (Ed.). University of Alberta Press, Edmonton, pp. 185–216.

Tabler, R.D., and Schmidt, R.A. 1986. Snow erosion, transport and deposition. In *Proc. Symposium on Snow Management for Agriculture*, H. Stepphun and W. Nicholaichuk (Eds.). Great Plains Agricultural Council Publication, University of Nebraska, Lincoln, pp. 12–58.

Thorn, C.E. 1978. A preliminary assessment of the geomorphic role of pocket gophers in the alpine zone of the Colorado Front Range. *Geografiska Annaler*, **60A**, 181–187.

Thorn, C.E. 1982. Gopher disturbance: its variability by Braun-Blanquet vegetation units in the Niwot Ridge alpine tundra zone, Colorado Front Range, U.S.A. *Arctic Alpine Res.*, **14**, 45–51.

Tieszen, L.L. 1974. Photosynthetic competence of the subnivean vegetation of an arctic tundra. *Arctic Alpine Res.*, **6**, 253–256.

Tieszen, L.L. 1978. Photosynthesis in the principal Barrow, Alaska, species: a summary of field and laboratory responses. In *Vegetation and Production Ecology of an Alaskan Arctic Tundra*, L.L. Tieszen (Ed.). Springer-Verlag, New York, pp. 242–268.

Tilman, D. 1983. Plant succession and gopher disturbance along an experimental gradient. *Oecologia*, **60**, 285–292.

Tomaselli, M. 1991. The snow-bed vegetation in the Northern Apennines. *Vegetatio*, **94**, 177–189.

Tranquillini, W. 1964. The physiology of plants at high altitudes. *Annu. Rev. Plant Physiol.*, **15**, 345–362.

Tranquillini, W. 1979. *Physiological Ecology of the Alpine Timberline*. Springer-Verlag, Berlin.

Tyrtikov, A.P. 1976. *Effects of Vegetation on the Freezing and Thawing of Soils*. (English translation of 1969 publication for the U.S. Department of Agriculture, Forest Service, and the National Science Foundation) Amerind Publishing Co., New Delhi.

Uemura, S. 1989. Snowcover as a factor controlling the distribution and speciation of forest plants. *Vegetatio*, **82**, 127–137.

Vestergren, T. 1902. Om den olikformiga snöbetäokningens inflytande paa vegetationen in Sarekfjällen. *Bot. Notiser*, **1902**, 241–268.

Walker, D.A. 1985. *Vegetation and Environmental Gradients of the Prudhoe Bay Region, Alaska*. U.S. Army Cold Regions Research and Engineering Laboratory, Report 85-14, CRREL, Hanover, N.H.

Walker, D.A., and Walker, M.D. 1996. Terrain and vegetation of the Inavait Creek watershed. In *Landscape Function and Disturbance in Arctic Tundra*, J.F. Reynolds and J.D. Tenhunen (Eds.). Springer-Verlag, Berlin, pp. 73–108.

Walker, D.A., Binnian, E., Evans, B.M., Lederer, N.D., Nordstrand, E., and Webber, P.J. 1989. Terrain, vegetation and landscape evolution of the R4D research site, Brooks Range Foothills, Alaska. *Holarctic Ecol.*, **12**, 238–261.

Walker, D.A., Halfpenny, J.C., Walker, M.D., and Wessman, C. 1993a. Long-term studies of snow-vegetation interactions. *BioSci.*, **43**, 287–301.

Walker, D.A., Lewis, B.E., Krantz, W.B., Price, E.T., and Tabler, R.D. 1993b. Hierarchic studies of snow-ecosystem interactions: a 100-year snow-alteration experiment. In *Proc. 50th Annual Eastern Snow Conf.*, pp. 407–414.

Walker, D.A., Auerbach, N.A., Bockheim, J.G., Chapin, F.S. III, Eugster, W., King, J.Y., McFadden, J.P., Michaelson, G.J., Nelson, F.E., Oechel, W.C., Ping, C.L., Reeburg, W.S., Regli, S., Shiklomanov, N.I., Vourliltis, GL. 1998. Energy and trace-gas fluxes across a soil pH boundary in the Arctic. *Nature*, **394**, 469–472.

Walker, M.D. 1990. *Vegetation and Floristics of Pingos, Central Arctic Coastal Plain, Alaska*. Dissertationes Botanicae 149, J. Cramer, Stuttgart, pp. 1–283.

Walker, M.D., Walker, D.A., and Auerbach, N.A. 1994a. Plant communities of a tussock tundra landscape in the Brooks Range foothills, *Alaska. J. Vegetation Sci.*, **5**, 843–866.

Walker, M.D., Webber, P.J., Arnold, E.H., and Ebert-May, D. 1994b. Effects of interannual climate variation on aboveground phytomass in alpine vegetation. *Ecology*, **75**, 393–408.

Walker, M.D., Walker, D.A., Welker, J.M., Arft, A.M., Bordsley, T., Brooks, P.D., Fahnstick, J.T., Jones, M.H., Losleben, M., Davsons, A.N., Seastedt, T.R., and Turner, P.L. 1999. Long-term experimental manipulation of winter snow regime and summer temperature in arctic and alpine tundra. *Hydrol. Processes*, **13**, 2315–2330.

Walker, M.D., Ingersoll, R.C., and Webber, P.J. 1995. Effects of interannual climate variation on phenology and growth of two alpine forbs. *Ecology*, **76**, 1067–1083.

Wallace, L.L., and Harrison, A.T. 1978. Carbohydrate mobilization and movement in alpine plants. *Am. J. Bot.*, **65**, 1035–1040.

Wardle, P. 1968. Engelmann spruce (*Picea engelmannii* Engel.) at its upper limits on the Front Range, Colorado. *Ecology*, **49**, 483–495.

Weaver, T. 1974. Ecological effects of weather modification: effect of late snowmelt on *Festuca idahoensis* Elmer meadows. *Am. Midl. Nat.*, **92**, 346–356.

Weaver, T., and Collins, D. 1977. Possible effects of weather modification (increased snow-pack) on *Festuca idahoensis* meadows. *J. Range Manage.*, **30**, 451–456.

Webber, P.J. 1978. Spatial and temporal variation of the vegetation and its productivity, Barrow, Alaska. In *Vegetation and Production Ecology of an Alaskan Arctic Tundra*, L.L. Tieszen (Ed.). Springer-Verlag, New York, pp. 37–112.

Webber, P.J., Emerick, J.C., Ebert-May, D.C., and Komárková, V. 1976. The impact of increased snowfall on alpine vegetation. In *Ecological Impacts of Snowpack Augmentation in the San Juan Mountains, Colorado*, H.W. Steinhoff and J.D. Ives (Eds.). U.S. Department of the Interior, Bureau of Reclamation, Denver, CO, pp. 201–258.

Westhoff, V., and van der Maarel, E. 1978. The Braun-Blanquet approach. In *Classification of Plant Communities*, R.H. Whittaker (Ed.). Dr. W. Junk, Den Haag, pp. 287–399.

Wijk, S. 1986a. Performance of *Salix herbacea* in an alpine snow-bed gradient. *J. Ecol.*, **74**, 675–684.

Wijk, S. 1986b. Influence of climate and age on annual shoot increment in *Salix herbacea*. *J. Ecol.*, **74**, 685–692.

Wielgolaski, F.E. 1997. Fennoscandian tundra. In *Polar and Alpine Tundra*, F.E. Wielgolaski (Ed.). Elsevier, Amsterdam, pp. 27–83.

Willard, B.E. 1979. Plant sociology of alpine tundra, Trail Ridge, Rocky Mountain National Park, Colorado. *Colorado School Mines Q.*, **74**, 1–119.

Willard, B.E., and Marr, J.W. 1970. Effects of human activities on alpine tundra ecosystems in Rocky Mountain National Park, Colorado. *Biol. Conserv.*, **2**, 257–265.

Willard, B.E., and Marr, J.W. 1971. Recovery of alpine tundra under protection after damage by human activities in the Rocky Mountains of Colorado. *Biol. Conserv.*, **3**, 181–190.

Williams, M.W., Bales, R.C., Brown, A.D., and Melack, J.M. 1995. Fluxes and transformations of nitrogen in a high-elevation catchment, Sierra Nevada. *Biogeochemistry*, **28**, 1–31.

Williams, M.W., Brooks, P.D., Mosier, A., and Tonnessen, K.A. 1996. Mineral nitrogen transformations in and under seasonal snow in a high-elevation catchment in the Rocky Mountains, USA. *Water Resources Res.*, **32**, 3175–3185.

Williams, R.J. 1987. Patterns of air temperature and accumulation of snow in subalpine heathlands and grasslands on the Bogong High Plains, Victoria. *Austral. J. Ecol.*, **12**, 153–163.

Williams, R.J., and Ashton, D.H. 1987. The composition, structure and distribution of heath

323

and grassland communities in the subalpine tract of the Bogong High Plains, Victoria. *Austral. J. Ecol.*, **12**, 57–71.

Zhang, T., Osterkamp, T.E., and Stamnes, K. 1996a. Influence of the depth hoar layer of the seasonal snow cover on the ground thermal regime. *Water Resources Res.*, **32**, 2075–2086.

Zhang, T., Osterkamp, T.E., and Stamnes, K. 1996b. Some characteristics of the climate in Northern Alaska. *Arctic Alpine Res.*, **28**, 509–518.

Zhang, T., Osterkamp, T.E., and Stamnes, K. 1997. Effect of climate on the active layer and permafrost on the North Slope Alaska, U.S.A. *Permafrost Periglacial Processes*, **8**, 45–67.

7   Tree-Ring Dating of Past Snow Regimes

YVES BÉGIN AND SIMON BOIVIN

7.1   Introduction

Changes in the local distribution of snow or in the amount of regional snowfall can cause ecological changes on land. For example, an increase in snow depth may result in (1) peat inception as the water supply increases; (2) flooding of lakeshores, which can occur because of greater snowmelt (Bégin and Payette, 1988; Bégin, 2000); (3) the extension of summer snowpatches into the surrounding forests (Morin and Payette, 1986); or (4) the development of thermokarstic landscape in permafrost zones because of low heat loss under the insulation of a thick snow cover (Allard, Caron, and Bégin, 1996). Changes in snow accumulation regimes generally result in gradual environmental changes, but, because the snow cover is seasonal, the processes of change are complex, involving a series of events transcending the seasons.

Dendrochronology is one of the most useful methods for studying the details of past environmental changes (Schweingruber, 1996). It provides an accurate technique for dating along with substantial information on environmental conditions affecting the growth of trees. In the Subarctic, trees and tundra patches form a mosaic and are useful markers of processes related to the uneven distribution of snow (Payette, 1983). Winter exposure of plants to the wind is mainly controlled by the surface distribution of the snow cover (Frey, 1983; Friedland et al., 1984). Such features especially affect the evergreen species, whose photosynthetic potential may vary according to individual ability to withstand severe winter conditions. Needle loss associated with storm winds or with winter dessiccation, bud abrasion by snow crystals blown by the wind, and damage to cambium cells due to freezing of cell contents may result in alteration of the growth form of trees (Lavoie and Payette, 1992). Although exposure may be a determinant factor in the growth stages of young trees, once they develop to large sizes, the trees considerably affect the distribution of snow in their surroundings (Pomeroy and Brun, Chapter 2). Dendrochronology can be used to date the

325

development of irregular tree-growth forms and their reactions to injuries. Combined with analysis of the positions of the eroded and damaged organs along the tree stems, this method yields useful markers for reconstructing local past snow conditions.

This chapter presents evidence of changes in snow depth distribution through analysis of black spruce and white spruce on the coast of Hudson Bay, north of Whapmagustui-Kuujjuarapik, and on the central islands of Lake Bienville in northern Québec (Figure 7.1). Analysis of the development of tree-growth forms and tree-ring series is utilized to show the local influence of stand development on snow distribution

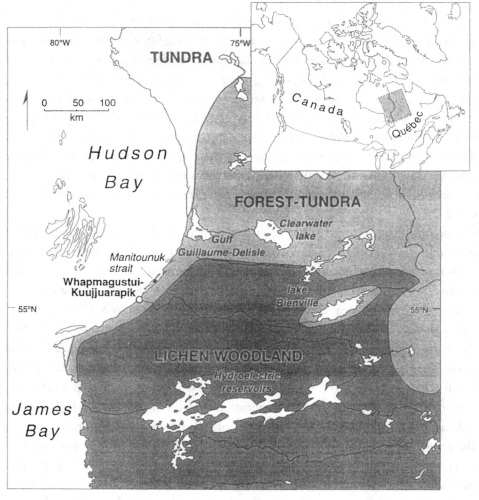

Figure 7.1. Ecological zones in the Hudson Bay area in northern Québec (after Payette, 1983).

and to support the hypothesis of increased regional snowfalls since 1880 as indicated by earlier studies throughout the north of the Québec-Labrador peninsula.

7.2   Dendrochronological Approaches and Basic Recognition of Past Snow Levels

Studies that use dendrochronology to reconstruct past precipitation records are concentrated in areas affected by water deficits – that is, where any precipitation event has a beneficial effect on tree growth. For example, in the western United States, rainfall has been reconstructed over the past three centuries on the basis of the relationships between regional master tree-ring chronologies and instrumental precipitation records for the past decades (Graumlich, 1987). Although such reconstructions are convincing in areas where drought is the main limiting factor of tree growth (Stockton and Meko, 1983), the application of dendrochronology to precipitation studies is much more difficult in northern areas where trees are also affected by temperature stresses. Snow plays an important role in tree development in the forest tundra, north of the boreal forest and the $-4°C$ isotherm (Payette, 1983). Windblown snow is unevenly distributed and relocated from aerodynamically swept surfaces to areas with obstacles (Filion and Payette, 1976; Payette, Filion, and Ouzilleau, 1973; Payette, Ouzilleau, and Filion, 1975; Payette and Lagarec, 1972; Thom and Grandberg, 1970; Chapter 2). The resulting variations in snow cover depths, partly controlled by vegetation structures and surface topography, give rise to protection against breakage of low plants and basal branches of trees by windblown snow crystals and against dessiccation due to extreme low temperatures and evaporation exacerbated by the wind. Living tissues in leaves, buds, branches, and stems are often exposed to stress, sometimes beyond the critical threshold, causing plant organ mortality (Kudo, 1992; Scott, Hansell, and Erickson, 1993; Valinger, Lundqvist, and Bondesson, 1993; Valinger, Lundqvist, and Brandel, 1994; Yoshino, 1973). Replacement of axes (branches and new leading stems) as a reaction to stress causes trees to adopt irregular shapes (Arseneault and Sirois, 1990; Bégin, 1991). Payette (1974, 1983) and Lavoie and Payette (1991, 1992) described the various tree-growth forms exhibited by both black and white spruce that dominate the northern forest tundra up to the latitudinal and maritime tree lines in northern Québec. The gradient in tree-growth forms (from erect normal-shaped trees to prostrate krummholz) expresses locally the contrasts of in-site exposure (Daly, 1984) and regionally the increasing climatic stress with latitude or proximity to the sea and large inland lakes (Payette, 1983). Furthermore, Lavoie and Payette (1992) showed that the

327

coexistence of generations of trees exhibiting contrasted growth habits may be indicative of past changes in local exposure conditions and, when occurring over a wide range of topographic situations, may provide evidence for past regional climate changes.

Cold, relatively dry, and windy winter conditions are also prerequisites for the ground surface redistribution of small, resistant, and densely crystalline snowpacks. In subarctic climates, the early onset of extremely cold conditions associated with the dominating arctic air masses fulfill these prerequisites soon in November and December. Snow rarely covers tree branches, as it is constantly swept away by the wind. Damage to branches occurs within 1–2 m above (erosional features) and below (loading injuries) the snow surface (Paatalo, Peltola, and Kellomaki, 1999; Peltola et al., 1999; Valinger and Fridman, 1999).

Through periodic and incremental radial growth, and the ability to adapt their growth form, most coniferous trees are useful indicators of past environmental and climatic changes. In climatically stressed environments, tree growth is generally sensitive and shows accordingly large variations in ring-width series (Travis, Meentemeyer, and Bélanger, 1990). In the forest tundra, tree organs can be damaged by a combination of snow, wind, and temperature conditions in the dormant season. The injuries can affect the photosynthetic potential during the following growing season. Growth conditions of the previous year – which determine the reserve of sugar available – and subsequent spring weather conditions for the onset of growth also contribute to wood production in a given year. The interpretation of ring-width series in an environment with multiple limiting factors, such as the forest tundra, is thus complex compared with areas where a single factor dominates, like semiarid areas. In the Subarctic, the role of snow may produce effects on tree growth combined with other environmental factors. Evidence of persistent snow levels producing erosion of buds, leaves, and shoots, and of tree organ breakage associated with heavy snow accumulation, are direct indicators of past events that can be studied by means of dendrochronology.

7.3   The Role of Snow in Tree Survival

In areas exposed to the wind, tree survival depends mainly on the degree of protection provided by the snow cover in winter. For example, on emergent shorelines in the Hudson Bay area, initial colonization is inhibited by waves and sea ice action. However, as the emerging shoreline moves above this zone (rates of emergence 1.2–1.5 m per century; Bégin, Bérubé, and Grégoire, 1993), the vegetation still suffers from extreme exposure to wind. Grégoire and Bégin (1993) showed that, in primary

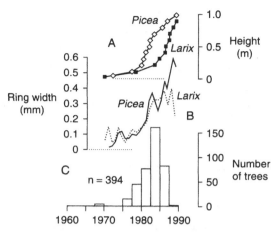

Figure 7.2. Growth of young white spruce (*Picea glauca*) and larch (*Larix laricina*) trees below the average level of snow. (A) Elongation of saplings in the early stages of growth. (B) Ring-width series from the basal section of trees. (C) Age structure of shrubs surrounding trees (adapted from Bégin et al., 1993). In both trees, elongation rates benefited from densification of the shrub thicket. Shrubs collect the wind-blowing snow and protect young trees against wind exposure.

succession, the relationship between the growing vegetation structures and the local accumulation of snow determines successful establishment of the subsequent survival of conifers along the coast. The early pioneer plants (mainly *Lathyrus japonicus* Willd., *Elymus arenarius* L., and *Arenaria peploides* L.) and shrubs (*Salix candida* Flügge, *Salix glauca* L., *Myrica gale* L., *Betula glandulosa* Michx.) help stabilize development of the vegetation cover (von Mörs and Bégin, 1993). For example, within a dense stand of *S. glauca* at 30 m from water's edge, the growth of two trees (*Picea mariana* and *Larix laricina* [DuRoi] K. Koch.) matched the densification of the shrub mat (Figure 7.2). As they grow denser and higher, shrubs retain large volumes of snow in the very early snowfalls in late October through early November. The freeze-up of nearshore waters in December creates a surface that collects the snow, which is soon blown landward by the westerly winds. Shoreline shrub mats catch this snow and reach saturation early in December (1 m deep in the case of Figure 7.2). Smaller plants, like conifer seedlings that are established in the shrub mats, thus benefit from the protection of snow before the onset of the severe cold period. Such protection during their dormant period allows them to keep their leaves, buds, and branches intact and to survive until

329

the next growing season. Shrubs have a genetically limited height (1–1.5 m for *S. glauca*, 1 m for *S. candida*). However, they rarely grow over 50–80 cm in exposed areas, a height that corresponds to the average snow depth. Shoots that emerge above the snow cover rarely survive because of severe exposure to frost and winds (Payette, 1983). As long as conifers are protected by snow, they develop a normal symmetrical growth form, but exposure above the snow will considerably modify their phenotype.

7.4   The Critical Period of the Snow Cover Breakthrough

The elongation of a conifer's stem above the average snow surface is the most critical stage of its life. This period determines the type of growth form (erect or prostrate) the tree will adopt. At this stage of growth, the harshness of winters (i.e., frequency of intensive cold and windy periods, wind strength) determines the probability of survival of above-snow tree structures, until the individual has grown beyond the snow saltation levels (usually 1–2 m in forest areas and 30–80 cm on tundra surfaces). The surviving exposed stems develop sufficiently resistant structures (thick bark, numerous branches) to allow the tree to withstand subsequent episodes of stress such as branch and foliage loss (Baig and Tranquillini, 1976, 1980; Hadley and Smith, 1983, 1986, 1987, 1989; Wardle, 1981; Péreg and Payette, 1998). Favorable summer conditions are also necessary for the tree to develop protective features that enhance its resistance to subsequent winter stress periods (leaf and bud cuticles, sugar storage) (Hadley and Smith, 1989; Payette et al., 1994).

Nevertheless, the successful breakthrough of the critical snow–air interface does not preclude later apex mortality. If the tree has kept abundant basal foliage, such apical death generally is soon compensated by development of a new leader – that is, a new shoot developing from a lateral bud or an existing branch that starts to grow vertically. An example of this process is given from the detailed analysis of the currently mature trees that established on the emerging Gulf Guillaume-Delisle coastline about 2 centuries ago (Figure 7.1). The black spruce shown in Figure 7.3 took 44 years to reach the present snow surface (basal part of the tree 0, period a–d, 1900–1943). The tree grew faster in diameter during the period 1922–1935 (b–c on the diagram), but no vertical growth improvement was observed. Could the stress at the position of the apex close below the average maximum snow level have limited the shoot elongation and favored the use of photosynthats in radial growth? The analysis shows that stem growth above the snow was very slow at first (10 cm in 8 years;

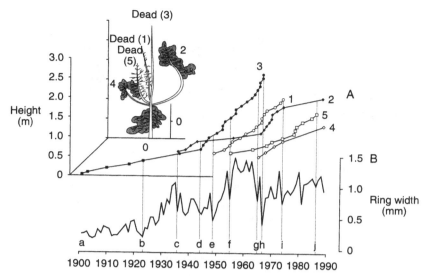

Figure 7.3. Development of a conifer stem above the average maximum snow level. (A) Longitudinal stem profiles in different parts of the tree indicated on the sketch diagram (left). (B) Ring-width series from basal section of tree. Lowercase letters refer to text.

c–d period), but the development of fast-growing branches (2 and later 1) near the snow–air interface stimulated subsequent vertical growth (c–i period). The development of branches generally coincides (pith years) with periods of slow radial growth, as shown by narrow rings in the basal section of the tree (e.g., at positions c and e–h). However, it often corresponds to an axis replacement (e.g., the mortality of the main stem [stem 3] was closely followed by the development of branch 4 and rapid vertical growth in branches 1 and 2). According to Bégin (1991) this process of axis replacement is very characteristic of the architecture of black spruce. This species has a true ability to change phenotypically and adopt growth forms that optimize its chances of survival in stressful conditions.

To study the process of stem development near the average and extreme snow levels and the resulting tree morphologies, a shore forest margin was studied in detail on a central island of Lake Bienville (Figure 7.1). The trees were mapped in a 20 × 40 m sampling quadrate and 27 trees were cut at various positions to date the different steps of their morphological development (Figure 7.4). To date the mortality of the dominant axis, we first sampled a cross section at the precise contact zone between the dead part of the stem and the level where cambial activity produced callous tissue to form a scar. The scars are dated by simply counting the rings (rings added since 1984 on the studied

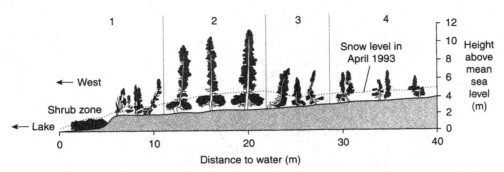

Figure 7.4. Transect showing gradient of black spruce growth forms at the western edge of a central island at Lake Bienville, northern Québec. Zones 1 and 2 are dominated by erosional tree-growth forms. Tree-growth forms in zones 3 and 4 result from recent past heavy snow accumulations.

example of Figure 7.5A). Second, the age of mortality of the main stem corresponds closely to that of the new leading stem that developed as a reaction. Such dating nonetheless entails some imprecision because the new leading stem can develop from an existing branch. The age of mortality is overestimated in this case. However, the branches have usually eccentric rings characterized by compression wood (Scurfield, 1973). Furthermore, branches have narrower rings than does the stem. By changing function, stem rings larger than earlier narrow branch rings will develop, increasing the accuracy of the tree-ring dating.

The combination of remnant dead apical shoots, adjacent new leading axis, and an underlying dense mat of branches makes it possible to identify the corresponding snow level. Dead apices along the current stem of a conifer are easy to identify because of the presence of a short dead wood segment with a small diameter (<2 cm) (Figure 7.5B). Resin is frequently found at the point of insertion of the dead shoot. Because the apical growth of conifers is orthotropic (vertical), the dead shoot, emerges vertically through the current stem, which makes it easy to distinguish from branches that insert horizontally (branch growth being plagiotropic [horizontal] in their first stages of development). New vertical axes, sprouting off a lateral bud after the death of the initial apical shoot, can be incorporated completely into newly formed wood. The bends in the new axis are often a good indicator of such an inclusion.

The black spruce in Figure 7.6 has two levels of apical loss that have affected the level of dense branch development. Approximately 80 cm above the ground, the apical dominance was taken by a branch already existing near the apex. The dead shoot is clearly visible (position of section 1). The radial growth curves in Figure 7.6 show that the death of the apex was not preceded by a period of growth supression. Moreover,

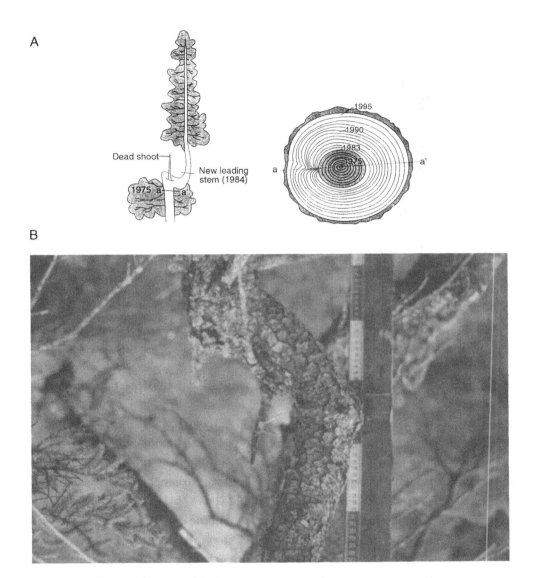

Figure 7.5. Dating a dead stem at the snow–air interface. (A) Dead stem can be dated from the resulting scar by subsequent rings beside the dead cambium. (B) Example of two dead apical remnants.

in 1821 it occurred after 2 years marked by large rings in the basal stem (section 0) and in the existing branch (section 1). This indicates that the mortality of the apex is not related to a period of decay but to a given year event. However, the dense mat of branches, the size of which suggests an upward progressive densification along the main stem, indicates a stagnation of the elongation of the stem between 30 and 60 cm

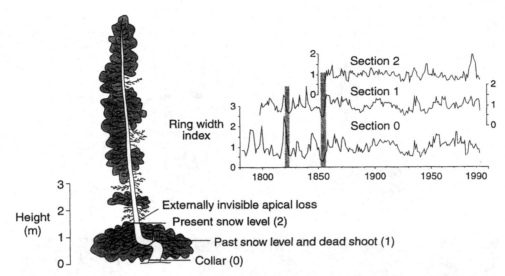

Figure 7.6. Tree shows morphological anomalies associated with average snow levels from different periods. Growth curves (right) pointed out on sketch of tree (left) indicate periods of stress (narrow rings) associated with the two subsequent snow levels (shaded areas centered in 1821 and 1853).

above the ground, which corresponds to the period 1821–1851 (average elongation 1 cm/year). At position 2, death of the apical leader occurred in 1851, as shown by a scar that is not externally visible. A new axis developed, but it is not possible to distinguish whether it came from a lateral bud or an existing branch. Nevertheless, the regular distribution of branches that developed along the straight stem above the snow indicates that no subsequent stress caused the tree to adopt any other anomalous morphological features. According to this example, the 1821–1851 period corresponds to the critical period of snow cover breakthrough, but the emergence of a supranivean regular shoot suggests that after this period of stress, favorable growth conditions occurred that allowed the apex to survive (no extremely narrow rings for a few decades following 1852; see basal section 0 in Figure 7.6). An explanation could be that densification of the forest stand caused an increase in local snow accumulation or simply created a mass effect to protect the tree against the wind. However, development of similar growth forms in many exposure conditions (Figure 7.4) brings forth evidence in favor of a climatic effect (increased snow accumulation or absence of extreme stress events; both can allow the apical part of the tree to survive) (see also Pereg, 1996).

The succession of dead apices along the stem (superimposed hook shapes) indicates past episodes of stress. Combined with a mat of underlying dense branches, the lower dead apices indicate the position of a persisting yearly maximum snow level

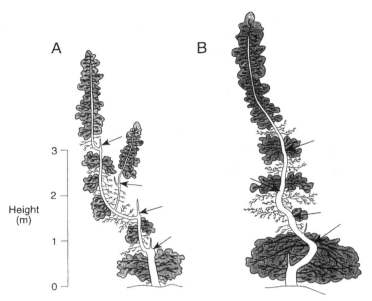

Figure 7.7. (A) Sketch of a tree showing multiple-axis mortality and new stems developed from lateral buds. This hook-like growth form is frequent in trees episodically exposed to wind stress. (B) The same phenomenon is presented on this sketch, but the dead apices are not externally visible (arrows). Occurrences can be identified by inflections in the main stem. This tree also shows a dense mat of branches below the average maximum snow level.

(Figure 7.7A). However, the dead apices may occur at any position above the snow level. Without the overramified section of the stem, they yield weak evidence of persisting past snow levels, but bends in the main axes could reveal the presence of included dead shoots, as mentioned earlier (Figure 7.7B).

7.5   The Effect of Snow Accumulation on Trees

Large accumulation of snow by drifting can have two different effects on trees. First, it generally delays snowmelt and reduces the growing period accordingly. The critical threshold for withstanding stress is also very low in this early stage of the tree's life cycle (Boivin and Bégin, 1997). Any mechanical damage to a seedling or intolerable growth conditions, such as a permanently saturated substrate or low soil temperatures, may end up killing the tree (Sakai, 1970). Second, changes in local snow accumulation may be reflected in tree growth forms. For example, a conifer that grew

335

normally exhibits a regular orthotropic main stem. Subsequent mechanical damage, such as apical or branch breakage, indicates that snow depth reached at least this level at the year of disturbance. The presence of new leading stems or the mats of branches as a reaction to mortality of tree axes gives the individual a "hybrid" shape, which is historically significant as indicating a rise in local snow level. The disturbances generally occur when the snow becomes denser because of increased water content. Injuries can be found at any position along the main axis of a seasonally buried tree. Because the snow is always blown off the branches, damages due to overloading occur within the snowpack. The effects of episodic heavy snow accumulation sometimes lead to intricate tree deformation (Figure 7.8). The annual layers of wood added each year

Figure 7.8. Trees subjected to heavy snow accumulation can develop bizarre shapes.

336

are unevenly distributed in distorted tree stems compared with normal upright trees, which have more regular ring widths everywhere on their main trunk. The xylem tends to develop at positions that compensate the mechanical stress within the tree structure (Mattheck, 1990, 1991; Petty and Worrell, 1981; Schmidtvogt et al., 1992). Stem and branch bending, tearing, and breakage are common in sites that accumulate windblown snow (Boivin and Bégin, 1997). Accurate dating of damage is possible with surviving trees by simply sectioning the stem at sites with external evidence of reaction. For example, it is suitable to sample the scars around the dead cambium following the tearing of a branch (Figure 7.9A), splitting along the stem's longitudinal axis (Figure 7.9B) or shoot bending (Figure 7.9C). The scar is formed by the wood added in the undamaged vicinity of the injury. A cross section of the stem including the callous margins of the scar generally shows relatively large annual rings, indicating the growth stimulus at the injury location that followed the disturbance event. Ring counts enable accurate dating of the damage at this position. Moreover, trees leaning under snow loading develop reaction wood in subsequent rings. In gymnosperms, the compression of cells by the overlignification on cell walls and the concentration of the xylem on the downward face of the tilted stem tend to compensate the loss of negative geotropism that characterizes nearly all living plants on Earth. Such "compression wood" (Scurfield, 1973) is clearly identifiable in cross section as eccentric rings with contrasting yellowish-brown color. The year of tree tilting is obtained by simply counting the number of reaction wood rings (Figure 7.9D). However, in most cases, the sequences of reaction wood are very complex. Most trees that intercept falling snow with their branches show short and intermittent sequences of reaction wood until the stem is sufficiently resistant to compensate for the mechanical tensions. To date the effect of significant heavy snow load episodes, one need concentrate only on prolonged and single orientation sequences of such eccentric growth rings. Such dating will make sense only on sites that are postulated to be affected by a prolonged or episodic shift away from "normal" snow accumulation conditions. Nevertheless, the interpretation of snow accumulation periods greatly benefits from a combination of several types of dendrochronological indicators.

7.6   Snow Levels and Past Climate

The literature contains many examples of tree-ring studies, which conclude that increasing snowfall accompanied the onset of warmer climatic conditions after 1880, which marked the end of the cold period known as the Little Ice Age. Jacoby,

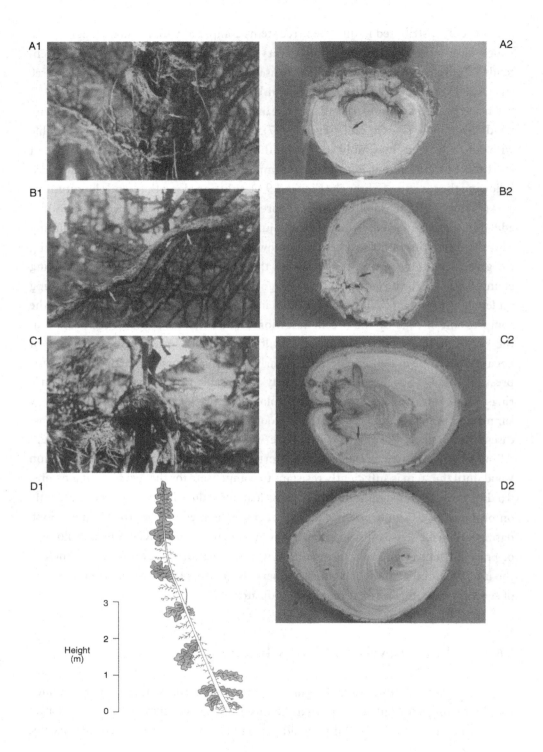

Ivanciu, and Ulan (1988) suggested that the warming started abruptly in the early 1880s, whereas other regional and northern studies tend to confirm that it took place gradually from 1880 to reach a thermal optimum in the 1940s (Jones et al., 1986; Lamb, 1977; Payette et al., 1985). This average temperature favored tree growth at the northern tree line and allowed the development of greater numbers of generally normal-shaped trees compared with those of colder earlier periods (Lavoie and Payette, 1992).

The forested shorelines of Lake Bienville (third-largest natural lake in Québec; 900 km^2) 350 km inland (Figure 7.1) exhibit an increase in snow damage at the same time during the onset of the warm period at the end of the nineteenth century (approximately 1880) (Boivin and Bégin, 1997). In these environments, greater snowfalls and local snow depths usually indicate less severe climatic conditions, favoring the densification of conifer stands. In a postfire stand that has developed since 1800–1820 from shoreline survivors, the growing forest achieved its maximum density in the early 1900s (Figure 7.10A). Between 1800–1820 and 1900, the stand progressively collected an increasing volume of snow swept by the wind over the adjacent lake-ice surface. At the exposed edge of the forest stand (Figure 7.4, zones 1 and 2), the level of apical shoot mortality rose irreversibly from the nineteenth century, indicating an increased snow depth even in exposed situations (Figure 7.10B). Apical loss usually occurs during the extreme cold of the midwinter period (February–March), while more than 60 percent of the snow has already fallen and while topographic saturation in snow has already been reached. In 1900, snow damage in the drifted snow accumulation zone had just begun (Figure 7.10C; trees in zones 3 and 4 on Figure 7.4) even though most of the trees currently alive were already 3–5 m tall. The concordance between the snow-depth increase in zones 1 and 2 and the recent injuries on trees located in

---

Figure 7.9. Anomalous tree-growth forms associated with excessive snow accumulation. (A) Torn-off branches produce a scar. On A1, resin is abundant (arrow). A cross section of the scar is shown in A2. Arrow indicates the first ring developed after damage. (B) Bending stems show open cracks along the longitudinal wood structures. The scar (arrows on B1) that develops at both ends of the crack allows the injury to be dated. (B2) Cross section of scar (arrow as in A2). (C) Stem bending creates intense mechanical stress along the longitudinal structure of the rings. Tension is often sufficient to damage the stem at the point of maximum flexure (C1). A scar develops around the damaged cambium (C2, arrow as in A2). (D) Trees leaning (D1) due to heavy snow load and showing sequences of compression wood (reaction wood) that can be used to date the tilting events (arrows on D2).

339

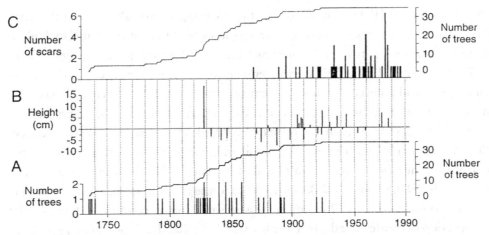

Figure 7.10. Chronology of tree-ring indicators of past snow levels at Lake Bienville. (A) Formation dates of branch tearing scars. (B) Position, above and below snow level of April 13, 1993, of dated dead shoots. (C) Age structure of overall black spruce stand. Top curves in A and C represent cumulative frequency of trees according to their dates of establishment (pith years in collar cross sections).

accumulation zones 3 and 4 of the forest stand indicates a general increase in snow depth that could be attributed to a secular augmentation in seasonal snowfalls.

Dendrochronological indications of snowier conditions since the beginning of this century have also been reported at many locations in northern Québec. In their study on the recent extension of snowpatches at the Guillaume-Delisle Gulf on the east coast of Hudson Bay (Figure 7.1), Morin and Payette (1986) provide good examples of tree-ring dating of different types of tree damage from snow overload. Furthermore, in the Guillaume-Delisle Gulf area, Laprise and Payette (1988) showed a rapid degradation of palsas by dating the periods of tree tilting due to substrate destabilization. They attributed the reduction in permafrost to climate warming and to a thicker snow cover, which prevented the winter heat loss that is necessary for permafrost to persist. Similar conclusions were reached with the detailed process studies of Allard et al. (1996) in the Manitounuk Strait 100 km south of Guillaume-Delisle Gulf. The recent extension of the northern and maritime range of porcupines has been determined by dating feeding scars on spruce stems (Payette, 1987). Because the younger winter feeding scars on spruce are at higher elevations compared with earlier periods, it can be assumed that there has been greater snow depth in the twentieth century.

Since the mid-1800s, higher frequency and magnitude of lake floods were dated by their effects on shoreline trees at the northern tree line (Payette and Delwaide,

1991) and in the southern part of the forest tundra (Bégin and Payette, 1988; Lepage and Bégin, 1996). The chronology of flood levels was obtained by dating the ice scars on stems formed when the lake level rises and allows the ice cover to reach the shoreline trees. Prolonged periods of high water levels in early summer (after the ice-breakup period) permits waves to destabilize shoreline trees. The dating of the reaction wood sequences in tilted trees leads to the chronology of early summer lake levels (Bégin, 1986). Since 1880 and especially during the 1930s and 1940s, the widespread occurrence of trees that were damaged by shore erosion or simply killed off by prolonged flooding around the large subarctic lakes indicates a shift in the flood regimes attributed to snowier winters.

Observations of the present regional climate lead to the hypothesis that snowfalls occur in response to the delayed southward seasonal movement of Arctic cold air masses. Subarctic climate over the Québec-Labrador peninsula varies considerably from season to season and from year to year (Bryson, 1966; Wilson, 1968). The rate of freeze-up of Hudson Bay has a considerable effect on the climate over the peninsula, which is very humid and close to maritime during the ice-free period and dry continental when the bay is completely ice covered (6 months/year; January–June). With the onset of Arctic cold conditions, the ice pack starts to expand southward from late October and covers the entire bay by mid-January. The cyclonic pathways of depressions follow the slow southward displacement of the Arctic front. The longer this displacement is, the later Hudson Bay freezes up. The frequent lows following the westerlies collect moisture over the bay and supply abundant snowfall on adjacent land (Barry, 1967, 1981; Dai, Trenberth, and Karl, 1998). Because close to half the annual snow falls before complete freeze-up of Hudson Bay, the length of this transition period determines regional snowfall regimes. After the onset of cold and dry winter conditions, snowfalls become less frequent and the snow cover is redistributed by the winds that prevail throughout winter.

7.7   Conclusion

Long-term reconstruction of interannual fluctuations of regional snowfall by means of dendrochronology faces the problem of distinguishing local variations in snow depth and average snowfalls. In essence, the tree-ring record of snow levels exists only where the snow is unequally distributed by being swept by the wind from open areas to accumulate in depressions and obstacles such as those created by vegetation. Suitable sites for dendrochronological studies are forests that border the margins of

341

ice-covered lakes and seas; the surroundings of burnt forest areas; and at the periphery of hills, plateaus, or terraces free from tall vegetation. In such forest stands, the collected snow can trigger ecological changes that can affect subsequent development of the vegetation and eventually limit its ability to collect large amounts of snow. Such a cyclic change may involve an initial increase in the snow cover duration and quantity that can reach limiting conditions for subsequent forest densification (short duration of the snow-free period, physical damage to the existing vegetation, changes in the soil hydric and thermal regime). Knowledge of exposure gradients asociated with the border effect is thus a fundamental skill in any tree-ring dating of snow-related processes.

To interpret changes in tree-growth forms and radial-growth variations in terms of fluctuations in winter precipitation, dendrochronological analysis greatly benefits from the insight of an event-based approach. Focusing on growth events (development of new stems, scars on stems, abrupt changes in radial growth rate) makes it possible to associate a sequence of dated tree-growth abnormalities to the cause of stress or disturbance. Compared with more classical dendroclimatic reconstructions (see Fritts, 1976; Schweingruber, 1988, 1996), such an event-based approach yields accurate dates of growth reactions allowing detailed site surveys of snow distribution over time. However, the discontinuous record of events does not allow a continuous reconstruction of precipitation. Tree-ring dating of given stem and branch sections enables one to interpret historical snow depths that leave their mark by affecting tree radial growth or by causing external injuries. The technique is based on the chronology of morphological changes that occurred near past snow–air interfaces. The concordance of various well-dated tree reactions yields evidence of the effects of changes in snow regime. This ecological approach is helpful in this type of dendroclimatological interpretation (e.g., data on the stages of tree development, status of the individual in the community, occurrence of any other site limitation or source of stress). However, because tree morphology records only discrete events, these results must be interpreted qualitatively. It is not possible to reconstruct a precipitation series by these methods as, for example, Graumlich (1987) has done. Persisting snow levels are marked by erosional features on the tree stem and discrete events associated with abnormal snow depths lead to mechanical injuries causing the tree to react morphologically.

Although this qualitative approach fails in the quantitative reconstruction of past precipitations, it provides potential for dating past extreme events before any instrumental register. In areas of short-term snowfall recording, such proxy data can also be valuable. With the omnipresence of sites affected by the uneven distribution of snow in the forest tundra, dendrochronology opens the perspective of spatial reconstruction of past snow covers. The year-to-year variations in seasonal snow covers are poorly documented by existing instrumental records. Dendrochronology represents a valuable future research skill that will considerably benefit our search for knowledge of the effective ecological role of snow in northern ecosystems.

7.8   Acknowledgments

This chapter was written to demonstrate possible applications of dendrochronology to subarctic environmental studies. Several of the basic principles were developed at Centre d'études nordiques (CEN) of Laval University in Québec. The idea of such a presentation would not be possible without the valuable contribution of Dr. Serge Payette (CEN) and his team of students. The authors are grateful to Dr. Brian Luckman (University of Western Ontario) and Dr. Fritz H. Schweingruber (Swiss Federal Institute for Forest, Snow and Landscape Research) for their useful comments on an earlier version of this chapter. Data gathering for part of this synopsis was financially supported by the Natural Sciences and Engineering Research Council of Canada (NSERC). S.B. received grants from NSERC and the Student Training Program of the Canadian Department of Indian Affairs and Northern Development for a Master's thesis that contributed to this chapter.

7.9   References

Allard, M., Caron, S., and Bégin, Y. 1996. Climatic and ecological controls on ice segregation and thermokarst: the case history of a permafrost plateau in northern Québec. *Permafrost Periglacial Processes*, 7, 207–227.

Arseneault, D., and Sirois, L. 1990. Forme et croissance de l'épinette noire (*Picea mariana* (Mill.) B.S.P.) avant-feu et après-feu en toundra forestière (Québec subarctique). *Naturaliste Can.*, 117, 1–7.

Baig, M.N., and Tranquillini, W. 1976. Studies on upper timberline: morphology and anatomy of Norway spruce (*Picea Abies*) and stone pine (*Pinus cembra*) needles from various habitat conditions. *Can. J. Bot.*, 54, 1622–1632.

Baig, M.N., and Tranquillini, W. 1980. The effects of wind and temperature on cuticular transpiration of *Picea abies* and *Pinus cembra* and their significance in dessication damage at alpine treeline. *Oecologia*, 47, 252–256.

Barry, R.G. 1967. Seasonal location of the arctic front over North America. *Geogr. Bull.*, 9, 79–95.

Barry, R.G. 1981. The nature and origin of climatic fluctuations in Northeastern North America. *Géogr. Phys. Q.*, 35, 41–47.

343

Bégin, C. 1991. *Analyse Architecturale et Dendroécologique d'une Pessière à Lichens à la Limite des Forêts*. Ph.D. thesis, Université Laval, Ste. Foy, Québec.

Bégin, C. 2000. Ice-push disturbances in High-Boreal and Subarctic lakeshore ecosystem since AD1830, Northern Québec, Canada. *The Holocene*, 10(2), 173–183.

Bégin, Y. 1986. *Dynamique de la Végétation Riveraine du Lac à l'Eau-Claire, Québec Subarctique*. Ph.D. thesis, Université Laval, Ste. Foy, Québec.

Bégin, Y., and Payette, S. 1988. Dendroecological evidence of lake level changes during the last three centuries in Subarctic Québec. *Q. Res.*, 30, 210–220.

Bégin, Y., Bérubé, D., and Grégoire, M. 1993. Downward migration of coastal conifers as a response to recent land emergence in Eastern Hudson Bay, Québec. *Q. Res.*, 40, 81–88.

Boivin, S., and Bégin, Y. 1997. Development of a black spruce *Picea mariana* (Mill) BSP shoreline stand in relation to snow level variations at Lake Bienville in Northern Quebec. *Can. J. Forest Res.*, 27, 295–303.

Bryson, R.A. 1966. Air masses, streamlines, and the Boreal forest. *Geogr. Bull.*, 8, 228–269.

Daly, C. 1984. Snow distribution patterns in the alpine krummholz zone. *Progr. Phys. Geogr.*, 8, 157–175.

Dai, A., Trenberth, K.E., and Karl, T.R. 1998. Global variations in droughts and wet spells: 1900–1995. *Geophys. Res. Lett.*, 25, 3367–3370.

Filion, L., and Payette, S. 1976. La dynamique de l'enneigement en région hémiarctique, Poste-de-la-Baleine, Nouveau-Québec. *Cahiers Géogr. Québec*, 20, 275–302.

Frey, W. 1983. The influence of snow on growth and survival on planted trees. *Arctic Alpine Res.*, 15, 241–251.

Friedland, A.J., Gregory, R.A., Kärenlampi, L., and Johnson, A.H. 1984. Winter damage to foliage as a factor in red spruce decline. *Can. J. Forest Res.*, 14, 963–965.

Fritts, H.C. 1976. *Tree Rings and Climate*. Academic Press, London.

Graumlich, L.J. 1987. Precipitation variation in the Pacific Northwest (1675–1975) as reconstructed from tree rings. *Ann. Assoc. Am. Geogr.*, 77, 19–29.

Grégoire, M., and Bégin, Y. 1993. The recent developement of a mixed shrub and conifer community on a rapidly emerging coast (Eastern Hudson Bay, Subarctic Québec, Canada). *J. Coastal Res.*, 9, 924–933.

Hadley, J.L., and Smith, W.K. 1983. Influence of wind exposure on needle dessication and mortality for timberline conifers in Wyoming, U.S.A. *Arctic Alpine Res.*, 15, 127–135.

Hadley, J.L., and Smith, W.K. 1986. Wind effects on needles of timberline conifers: seasonal influence on mortality. *Ecology*, 67, 12–19.

Hadley, J.L., and Smith, W.K. 1987. Influence of kruhmmolz mat microclimate on needle physiology and survival. *Oecologia*, 73, 82–90.

Hadley, J.L., and Smith, W.K. 1989. Wind erosion of leaf surface wax in alpine timberline conifers. *Arctic Alpine Res.*, 21, 392–398.

Jacoby, G.C., Ivanciu, I.S., and Ulan, L.D. 1988. A 263 year record of summer temperature for northern Quebec reconstructed from tree-ring data and evidence of major climatic shift in the early 1800's. *Paleogeogr. Paleoclimatol. Paleoecol.*, 64, 69–78.

Jones, P.D., Raper, S.C.B., Bradley, R.S., Diaz, H.F., Kelly, P.M., and Wigley, T.M.L. 1986.

Northern hemisphere surface air temperature variations: 1851–1984. *J. Am. Meteorol. Soc.*, **25**, 161–179.

Kirnbauer, R., Bloschl, G., and Gutknecht, D. 1994. Entering the era of distributed snow models. *Nordic Hydrol.*, **25**, 1–24.

Kudo, G. 1992. Effect of snow-free duration on leaf life-span of 4 alpine plant species. *Can. J. Bot.*, **70**, 1684–1688.

Lamb. H.H. 1977. *Climate, Present, Past and Future, Vol. 2: Climatic History and the Future*, Methuen, London.

Laprise, D., and Payette, S. 1988. Évolution récente d'une tourbière à palses (Québec subarctique): une analyse cartographique et dendrochronologique. *Can. J. Bot.*, **66**, 2217–2227.

Lavoie, C., and Payette, S. 1991. Les formes de croissance de l'épinette noire et les changements climatiques séculaires. *Natur. Monspeli.*, 222–227.

Lavoie, C., and Payette, S. 1992. Black spruce growth forms as a record of a changing winter environment at treeline, Québec, Canada. *Arctic Alpine Res.*, **24**, 40–49.

Lepage, H., and Bégin, Y. 1996. Tree-ring dating of extreme water level events at Lake Bienville, Subarctic Québec, Canada. *Arctic Alpine Res.*, **28**, 78–85.

Mattheck, C. 1990. Why they grow, how they grow: The mechanics of trees. *Arboriculture J.*, **14**, 1–17.

Mattheck, C. 1991. *Trees: The Mechanical Design*. Springer-Verlag, Berlin.

Morin, H., and Payette, S. 1986. La dynamique récente des combes à neige du golfe de Richmond (Québec nordique): une analyse dendrochronologique. *Can. J. Bot.*, **64**, 2113–2119.

Paatalo, M.L., Peltola, H., and Kellomaki, S. 1999: Modelling the risk of snow damage to forests under short-term snow loading. *Forest Ecol. Manage.*, **116**, 51–70.

Payette, S. 1974. Classification écologique des formes de croissance de *Picea glauca* (Moench.) Voss et de *Picea mariana* (Mill) BSP. En milieux subarctique et subalpin. *Naturaliste Can.*, **101**, 893–903.

Payette, S. 1983. The forest-tundra and present tree-lines of the northern Québec-Labrador Peninsula. Proceedings of the Northern Tree-Line Conference. *Nordicana*, **47**, 3–23.

Payette, S. 1987. Recent porcupine expansion at tree-line: a dendroecological analysis. *Can. J. Zool.*, **65**, 551–557.

Payette, S., and Delwaide, A. 1991. Variations séculaires du niveau d'eau dans le bassin de la rivière Boniface (Québec nordique): une analyse dendroécologique. *Géogr. Phys. Q.*, **45**, 59–67.

Payette, S., and Lagarec, D. 1972. Observations sur les conditions d'enneigement à Poste-de-la-Baleine, Nouveau-Québec. *Cahiers Géogr. Québec*, **16**, 469–481.

Payette, S., Filion, L., and Ouzilleau, J. 1973. Relations neige-végétation dans la toundra forestière du Nouveau-Québec, Baie d'Hudson. *Naturaliste Can.*, **100**, 493–508.

Payette, S., Ouzilleau, J., and Filion, L. 1975. Zonation des conditions d'enneigement en toundra forestière, Baie d'Hudson, Nouveau-Québec. *Can. J. Bot.*, **53**, 1021–1030.

Payette, S., Filion, L., Gauthier, L., and Boutin, Y. 1985. Secular climate change in old-growth tree-line vegetation of northern Québec. *Nature*, **314**, 135–138.

Payette, S., Delwaide, A., Morneau, C., and Lavoie, C. 1994. Stem analysis of a long-lived black spruce clone at treeline. *Arctic Alpine Res.*, **26**, 56–59.

Peltola, H., Kellomaki, S., Vaisanen, H., and Ikonen, V.P. 1999: A mechanistic model for assessing the risk of wind and damage to single trees and stands of Scots pine, Norway spruce, and birch. *Can. Forest Res.*, **29**, 647–661.

Pereg, D., 1996. *Les Formes de Croissance de l'Épinette Noire à la Limite des Arbres Comme Indicateurs de Changements Climatiques.* Master's thesis, Université Laval, Ste. Foy, Québec.

Péreg, D., and Payette, S., 1998. Development of black spruce growth forms at treeline. *Plant Ecol.*, **138**, 137–147.

Petty, J.A., and Worrell, R. 1981. Stability of coniferous tree stems in relation to damage by snow. *Forestry*, **54**, 115–128.

Sakai, A. 1970. Mechanism of dessiccation of conifers wintering in soil-frozen areas. *Ecology*, **51**, 657–664.

Schmidtvogt, H., Franke, A., Konnert, M., and Deichner, P. 1992. Identification, growth and resistance to snow break of spruce provenances from the black forest. *Allg. Forst Jagd.*, **163**, 145–154.

Schweingruber, F.H. 1988. *Tree Rings.* Kluwer, Dordrecht, The Netherlands.

Schweingruber, F.H. 1996. *Tree Rings and Environment: Dendroecology.* Swiss Federal Institute for Forest, Snow and Landscape Research at Birmensdorf, WSL/FNP, Berne, Stuttgart, Vienna, Haupt.

Scott, P.A., Hansell, R.I.C., and Erickson, W.R. 1993. Influences of wind and snow on northern tree-line environments at Churchill, Manitoba, Canada. *Arctic*, **46**, 316–323.

Scurfield, G. 1973. Reaction wood: Its structure and function. *Science*, **179**, 647–655.

Stockton, C.W., and Meko, D.M. 1983. Drought recurrence in the Great Plains as reconstructed from long-term tree-ring records. *J. Climate Appl. Meteorol.*, **22**, 17–29.

Thom, B.G., and Grandberg, H. 1970. Patterns of snow accumulation in a forest tundra environment, Central Labrador-Ungava. In *Proc. 27th Annual Eastern Snow Conf.*, pp. 76–86.

Travis, D.J., Meentemeyer, V., and Bélanger, R.P. 1990. Stressed trees produce a better climatic signal than healthy trees. *Tree-Ring Bull.*, **50**, 29–32.

Valinger, E., and Fridman, J. 1999. Models to assess the risk of snow and wind damage in pine, spruce, and birch forests in Sweden. *Environ. Manag.*, **24**, 209–217.

Valinger, E., Lundqvist, L., and Bondesson, L. 1993. Assessing the risk of snow and wind damage from tree physical characteristics. *Forestry*, **66**, 249–260.

Valinger, E., Lundqvist, L., and Brandel, G. 1994. Wind and snow damage in a thinning and fertilisation experiment in *Pinus sylvestris*. *Scand. J. Forest Res.*, **9**, 129–134.

von Mörs, I., and Bégin, Y. 1993. Shoreline shrub population extension in response to recent isostatic rebound in eastern Hudson Bay, Québec, Canada. *Arctic Alpine Res.*, **25**, 15–23.

Wardle, P. 1981. Winter dessication of conifer needles simulated by artificial freezing. *Arctic Alpine Res.*, **13**, 419–423.

Wilson, C. 1968. Notes on the climate of Poste-de-la-Baleine, Québec. *Nordicana*, **24**, 1–93.

Yoshino, M.M. 1973. Wind-shaped trees in the subalpine zone in Japan. *Arctic Alpine Res.*, **5**, 115–126.

346

Epilogue

This book reviews much of the basic science that relates to snow ecology; we hope the reader has found it satisfying in both the search for information and for stimulating interest in the science of snow. If the book has served as a catalyst for developing ideas and hypotheses on the functioning of snow and snow-covered ecosystems, then it has attained its purpose. The reader may find recommended directions for future research at the end of each chapter. However, there are broader concerns for the future of snow ecology, which are especially important in the face of two different aspects of human activity.

The first is the possible rapid climate change, with effects concentrated in the winter and spring periods (Groisman and Davies, Chapter 1). The effects of altered snow environments are of particular concern in high-latitude ecosystems where changes in winter snow regimes could dramatically alter land–atmosphere energy and water exchange (Pomeroy and Brun, Chapter 2). This, in turn, will modify weather systems, growing season, and the availability of water in soil and waterbodies. The dynamics of nutrient fluxes will also be subjected to change, particularly with regard to the emissions of greenhouse gases from the soil, deposition of anthropogenic nutrients to the ecosystem via snow (Tranter and Jones, Chapter 3), and in-pack microbiological activity (Hoham and Duval, Chapter 4) and associated invertebrate populations (Aitchison, Chapter 5) and vegetation (Walker, Billings, and de Molenaar, Chapter 6). The reading of past records by the use of dendochronology (Bégin and Boivin, Chapter 7) can help us ascertain the degree to which snow regimes have changed and how they may be linked to the overall changes in climate itself (Groisman and Davies, Chapter 1).

The second concerns the search for knowledge about the mechanisms that life can utilize in creating or finding energy conditions and liquid phases sufficient for survival in cold and "frozen" environments. If life can survive in cold regions dominated by the ice phase of water, then water-based life forms may have evolved on those planets and moons of the solar system – other than Earth – where phase change between ice and water is possible (e.g., Mars and Europa). In the current exploration of planets and their moons and the search for extraterrestrial life, one must be aware of the evolution and physiological adaptations of life in the extreme snow and ice environments here on Earth (Hoham and Duval, Chapter 4) and to the physical principles that govern phase change and energy balance in snow and ice systems (Pomeroy and Brun, Chapter 2).

We have found some focus for our lives in and with snow and have tried to express a small part of this fascination in describing the intricate principles and exotic phenomena and organisms that comprise snow ecosystems. Our understanding is not complete. If nothing else, we hope the reader is encouraged to go to the world of snow, study it, observe it, and describe it for the next generation of readers.

The Editors

Book Glossary

Abnormal mating pair: two gametes that fuse by means other than flagella.

Abrasion: erosion by friction.

Active layer: the layer of ground above the permafrost, which thaws and freezes annually.

Adiabatic effect: changes in temperature in ascending or descending air masses, the physical change occurring without the addition or loss of heat from the air mass.

Advanced very high resolution radiometer: instrument aboard National Oceanic and Atmospheric Administration satellites used to measure various spectral reflectance patterns on the Earth. The instrument has a 1.1-km unit of view (pixel size) at nadir and measures reflectance in five channels.

Advection: transport of heat from an external source into the air over a snowpack. During snowmelt this is usually due to warm air's pushing away colder air over the snow. Sources of advection are synoptic systems, warm forest canopies, bare fields, and buildings. The heat released by rain, falling onto cold snowpacks, and refreezing is sometimes termed advection.

Aeolian: wind-borne or transported materials.

Aerodynamic roughness height: the roughness of a surface as it affects wind flow above it, determined by the apparent intercept of the log-linear gradient of wind speed with height – i.e., the height at which the wind speed should reach zero.

Aerosol: a small particle ($<10\,\mu$m) with some liquid content, which is suspended in the atmosphere.

Albedo: degree of reflectance of visible light by a surface. An albedo equal to 1 indicates that the surface reflects light perfectly; an albedo equal to 0 indicates perfect light absorption.

349

Algae: mostly photosynthetic eukaryotes that lack complex reproductive structures typical of land plants. In some recent classification systems, algae may be classified into four Kingdoms.

Anaerobiosis: the ability to survive extended periods of time without oxygen at low temperatures – e.g., when encased in ice.

Anthocyanins: any of a number of water-soluble nitrogenous pigments, which contribute to the colors of the leaves of temperate-climate plants.

Aplanospore: an asexual cell without flagella that can develop into a vegetative cell that can further divide through mitosis to form more aplanospores.

Arachnids: a group of arthropod invertebrates with four pairs of legs.

Arthropods: invertebrates with a hardened exoskeleton, specialized body segments, and jointed appendages.

Asexual resting spore: a resistant cell with a thick wall that develops directly from a vegetative cell. There is no sexual process involved.

Association: 1) a stand, concrete community, or group of plants characterised by a definite floristic composition, presenting uniformity in physiognomy and structure, and growing under uniform habitat conditions. 2) In the Zurich-Montpellier (Braun-Blanquet) system, it is a unit of a hierarchical system consisting of associations, alliances, orders, and classes. This is the internationally accepted definition following the 1935 International Botanical Congress.

Autecology: the study of the individual plant, or members of a species collectively, in relation to environmental conditions.

Axenic: pure cultures of a single microbe.

Bacteria: mostly small unicells without nuclei and cellular organization.

Baroclinicity: the state of the atmosphere pertaining to, especially, temperature and pressure gradients. A strongly baroclinic atmosphere has strong gradients.

Basal ice: an ice layer that forms at the bottom of a snowpack due to meltwater ponding at the top of the soil and refreezing.

Biomass: the total amount of living material; often expressed per area of surface – e.g., $kg\ ha^{-1}$.

Biome: a large (continental-scale) area of the earth's surface with a broadly consistent vegetation type and mix (e.g., the boreal forest).

Blowing snow: transport and relocation of snow by the wind.

350

Callous tissue: a term used to designate the new wood formed around a damaged part of a tree stem. The tissues tend to cover the trunk and create a scar.

Cambium: The zone of cell division around a tree stem. The division produces wood (xylem) toward the inner part of the trunk and phloem on the outer part, just below the bark.

Canopy coverage: the area covered by canopy per unit area of the ground.

Carbohydrate: molecule of carbon, hydrogen and oxygen usually occurring in a ratio of 1:2:1.

Carboxylase: an enzyme that catalyses reactions in which a molecule of carbon dioxide is incorporated into an organic compound.

Carotenoids: accessory pigments comprised mostly of hydrocarbon chains that transfer absorbed energy to chlorophyll or screen out damaging UV light.

CCN (cloud condensation nuclei): small particulates around which ice crystals form in clouds.

Chemical flux: the amount of chemical species that is transported through an interface per unit time. In snowmelt studies the flux is often expressed as $mg\ m^{-2}\ day^{-1}$, $meq\ m^{-2}\ day^{-1}$, $kg\ ha^{-2}\ day^{-1}$, or $keq\ ha^{-2}\ day^{-1}$. The term eq is the equivalent of any chemical species; it is calculated as weight/equivalent weight – e.g., a flux of nitrate (NO_3^-) in meltwater of $124\ mg\ m^{-2}\ day^{-1}$ is equal to $2\ meq\ m^{-2}\ day^{-1}$.

Chemical load: the total amount of any chemical species per area of a storage milieu. Chemical loads in snow cover are usually expressed as $mg\ m^{-2}$, $meq\ m^{-2}$, $kg\ ha^{-1}$, or $keq\ ha^{-1}$.

Chemoautotroph: bacterium that lives in deep-sea hydrothermal vents.

Chill-coma temperature: a low temperature at and below which an invertebrate can no longer move.

Chionophile: an organism that can endure long-lasting snow cover during winter and spring or one that requires a snow cover in winter.

Chionophobe: an organism that cannot tolerate long-lasting snow cover during winter and spring, or one that can live with little or no snow cover in winter.

Chlorophyll: the generic name for the pigments in photosynthesis that absorbs primarily in the blue and red end of the spectrum. Five closely related chlorophylls, a through e, occur in higher plants and algae.

Chlorosis: a symptom of disease or disorder in plants that involves a reduction in, or

351

loss of, the normal green coloration of the plant due to the production of chlorophyll; consequently, the plants typically are pale green or yellow.

Chrysolaminarin: a carbohydrate food reserve of repeating units of β-glucose found in golden algae, diatoms and yellow-green algae.

Chytrids: a small group of fungal-like saprobic decomposers and parasites that are classified as protists.

Ciliates: mostly unicellular predators or parasites that are covered with locomotory cilia (structures like flagella, but are more numerous). These heterotrophs are classified as protists.

Class: the highest level in the Braun-Blanquet plant community classification approach, referring to very broad vegetation units.

Codon: a series of nitrogenous base triplets in a messenger RNA molecule that codes for a sequence of amino acids to form a protein.

Cohesion: the bonding of snow crystals to each other or to vegetation due to inter-crystal ice bonds or thin liquid layers surrounding ice particles.

Cold-hardiness: a physiological state of tolerating and surviving low temperatures.

Cold-hardy: being adapted physiologically to withstand low temperatures.

Cold-resistant: being able to survive low temperatures.

Cold stenothermy: adaptation to a narrow range of low temperatures.

Collection efficiency: the ratio of the amount of snowfall intercepted by a branch to that the branch received.

Collembola: also called springtails, are minute insects with chewing or piercing mouth parts that have a forked structure called a furcula used for jumping.

Compression wood: type of reaction wood developed in conifers. The cells that develop on the lower part of a tilted stem possess overlignified cell walls. Lignin is a very dense and resistant substance that gives the wood extreme rigidity to compensate for the mechanical stress developed in the tilted trunk.

Concentration: the amount of a component in a mixture either by volume or weight. The concentration of a chemical species in air is usually expressed by volume (e.g., ppmv, ppbv) and/or by weight (e.g., μg m^{-3}, μEq m^{-3}). The concentration of any chemical species in snow and meltwater is generally expressed by volume (e.g., μg L^{-1} or μEq L^{-1}).

352

Contractile vacuoles: structures fed by smaller vesicles that rhythmically expel water and solutes to the outside of the cell.

Cornice: an overhanging snowdrift, often at a mountaintop in alpine terrain.

Cryobiont: a snow or ice-inhabiting organism.

Cryoconite: an aggregation of particulate material from rock and soil found on the surface of snow and ice (see also *kryokonite*).

Cryophile (cryophilic): a cold tolerant organism (see also *psychrophile*).

Cryoprotectant: a substance protecting an invertebrate against low temperatures – e.g., a natural antifreeze such as sorbitol.

Cryoturbation: the process of stirring, heaving, and thrusting of the earth's mantle by frost action including frost heaving and differential mass movements like solifluction.

Cryptomonads: unicellular colorless or photosynthetic protistan flagellates with chlorophylls a and c and α-glucose food reserves called starch. Cells are surrounded by a protein periplast.

Cushion plant: a herbaceous or low woody-plant so densely branched that it forms a dense resilient mat or cushion.

Cuticle: a thin protective sheet (resin and wax) that develops in late summer on buds and leaves to insulate it from the cold in winter.

Cyanobacteria: bacteria with chlorophyll a, which produce oxygen, fix atmospheric nitrogen in many species, and cells have α-glucose food reserves called cyanophycean starch.

Cyclomorphosis: a unique morphological seasonal change seen in collembolans, with winter morphs acquiring winter-adaptive features and summer morphs being more normal.

Dark reaction: a photosynthetic reaction that involves the reduction of carbon dioxide to carbohydrates in the Calvin cycle, which can take place in darkness if there is sufficient ATP and NADPH.

Dehnel's phenomenon: a phenomenon in which small soricine shrews undergo morphological adaptations to winter, including reduction of body weight and shortened body length.

Dendrochronology: the study of annual tree rings.

Depth hoar: a layer of large (>2–3 mm) columnar or goblet-shaped crystals found usually just above the soil–snow interface. Depth hoar is formed when large vertical temperature gradients occur in the snowpack (see also *Temperature gradient metamorphism*).

Desiccation: the loss of water from living tissues because of drought events or sublimation of ice in extreme cold conditions.

Diapause: a period of suspended development and reduced physiological activity, usually an inactive resting stage in invertebrates during winter.

Diatoms: mostly unicellular photosynthetic protists with chlorophylls a and c and β-glucose food reserves called chrysolaminarin. Cells are surrounded by a silica shell.

Dinoflagellates: mostly unicellular colorless or photosynthetic protistan flagellates with chlorophylls a and c and α-glucose food reserves called starch. Cells are surrounded by an outer covering of cellulosic plates.

Diploid: having two sets of chromosomes ($2N$).

Dormant season: the period of metabolic inactivity in plants due to light availability and climatic conditions. Preparation for dormancy is a very complex physiologic feature involving the transfer of substances in the plant, especially carbohydrates, to cells, which can then avoid frost injury.

DTM (digital terrain model; also digital elevation model, DEM): a quantitative model of landform elevations in digital format.

Eccentric ring: asymmetric tree rings that develop due to uneven distribution of growth metabolic regulators in the stems (e.g., the rings that develop in compression wood).

Ecdysis: the act of moulting or skin shedding of an animal.

Ecophysiology: the study of the mechanics of living things within the framework of ecology (e.g., how living things interact with each other and with the nonliving constituents of their environment).

Ecotype: 1) a race within a species that is genetically adapted to a local habitat, which is different from the habitat of other races of that species. An ecotype usually consists of many biotypes, all of which have closely similar habitat requirements but differ in other respects. 2) An arbitrary segment of an ecoline. 3) Individuals, or groups of individuals, of a species that react differently to environmental factors than the rest of the individuals of that species.

354

Elasticity: the degree to which a material will deform when a force is applied.

El Niño: an extreme phase of the so-called Southern Oscillation. This occurs when a large part of the central Pacific Ocean has anomalous high sea-surface temperatures, associated with particular climate patterns in other parts of the Earth.

Elongation: the development of stems lengthwise.

Emergent shoreline: a feature due to isostasy (compensatory uplift of the Earth's crust over the inner plastic magma) created by the disappearance of the continental glaciers. In some parts of the world, the sea is still retreating slowly from marginal lands with the resulting formation of such shores.

Emissivity: the degree to which a material radiates long-wave radiation compared with an object (blackbody) that perfectly radiates long-wave radiation at the same surface temperature.

Endoevaporite: a microorganism that lives within salt crusts.

Endolith: a microorganism that lives within rocks.

Equitemperature metamorphism: metamorphism driven by the differences in saturation vapour pressure associated with the curvature of snow crystals. Ice sublimates from the sharp edges of crystals and small crystals, and the water vapour recrystallises on concave curvatures; often the ice bonds between crystals.

Erect form: the growth form of a tree with a normal upright main stem (see also *prostrate*).

Ericaceous: the heath family Ericacieae.

Eubacteria: true bacteria that exclude the archaebacteria.

Euglenoids: unicellular colorless or photosynthetic flagellates with chlorophylls a and b and β-glucose food reserves called paramylon. Cells are surrounded by a protein pellicle.

Eutrophic: an aquatic system with high nutrient levels.

Evaporation: the phase change from liquid to vapour.

Eyespot: a red- to orange-pigmented structure sensitive to light.

Fallow field: an agricultural field left bare of vegetation by periodic ploughing over the summer period.

Fatty acid: a long hydrocarbon chain bound to a carboxyl group ($-COOH$). If saturated, there are no double bounds; if monounsaturated, there is one double bond; if polyunsaturated, there are two or more double bonds in the hydrocarbon chain.

355

Feedback process: the interaction between two or more variables, which tends either to encourage further change in the same direction in the variable of interest (positive feedback) or to reverse the direction of change in the variable of interest (negative feedback).

Fellfield: from the Danish "fjoeld-mark" or rock desert. A type of tundra ecosystem characterised by rather flat relief, very stony soil, and low widely spaced vascular plants.

Flagella: locomotory structures associated with certain cells.

Flow fingers: irregular columns of coarse-grained, wet snow with large pore spaces or macropores, through which rapid and preferential flow of meltwater occurs.

Flux: a mass or energy flow rate, expressed as mass or energy per unit area per unit time (see also *chemical flux*).

Foliosphere: the above-ground environment of a plant in the vicinity of the foliage.

Food chain: straight line sequence of events of who eats whom in an ecosystem.

Food web: network of interlinked food chains composed of producers, consumers, decomposers and detrivores.

Forest tundra: the northern part of the Taiga made of scattered open conifer forests and patchy lichen-dwarf surfaces. Its southern limit is the continuous open boreal forest and its northern edge consists of the tree line.

Formation: in the oldest and widest (global) sense, a continental-scale vegetation unit comprising all plant communities that resemble each other in appearance and major features of their environment – e.g., northern coniferous forest or tropical rain forest defined through properties of vegetation cover.

Freezing resistant: the tolerance of intracellular freezing in body tissues by invertebrates.

Freezing sensitive (or frost sensitive): describes an invertebrate that avoids intracellular freezing but tolerates extracellular freezing in body tissue.

Fungi: heterotrophic decomposers and parasites with extracellular digestion.

Furcula: the forked "spring" on the ventral surface of a collembolan, tucked out of the way until disturbed when it is unfolded to propel the animal away.

Gamete: a $1N$ (haploid) sex cell, with or without flagella, that has the potential of uniting with another sex cell in a sexual process forming a $2N$ (diploid) zygote. This definition assumes the organism is not a polyploid.

GCM: general circulation model (or global climate model); a numerical model of the earth's climate.

Gene sequencing: the process of obtaining the order or sequence of the nucleotide bases in the genetic code.

GIS: geographical information system; a system of computer hardware and software used to store, retrieve, manipulate, analyse, and display spatial data.

GOES: geostationary satellite; a satellite that is stationary relative to a particular point on the earth's surface.

Golden algae: mostly unicellular and colonial photosynthetic or heterotrophic protistan flagellates with chlorophylls a and c and β-glucose food reserves called chrysolaminarin. Cells are often surrounded by a covering of overlapping silica scales.

Green algae: photosynthetic organisms with chlorophylls a and b and α-glucose food reserves called starch. Cells are usually surrounded by a cellulosic wall. Several groups are often classified as protists and one group is ancestral to land plants.

Greenhouse gas: an atmospheric gas that is relatively transparent to incoming shortwave solar radiation but absorbs relatively more of the outgoing long-wave terrestrial radiation. The increasing concentrations of such gases (e.g., carbon dioxide) are leading to global warming, and related climate change, because of the enhanced greenhouse gas effect.

Ground heat flux: the heat flow between the soil and the snowpack.

Growth form: the external appearance of a tree as expressed by its architecture.

Haemolymph: invertebrate blood.

Haploid: having one set of chromosomes ($1N$).

Hardness: an indicator of snow strength, measured as force per unit area necessary to cause a break or failure in snow structure. Ram hardness is measured by dropping a known mass a specified distance onto the snow; surface hardness is measured by applying a force with a spring-loaded plate until failure of the snow surface occurs.

Heterocyst: a thick-walled nitrogen-fixing cell found in cyanobacteria.

Heterotroph: an organism unable to make it own food, but feeds on other organisms or their wastes for organic compounds.

Hiemal threshold: the snow cover thickness at which the subnivean environment is insulated from diel fluctuations of ambient air temperature.

357

Hydrological cycle: the movement, and changes in phase, of water around the Earth.

Hydrophyte: a plant that is usually found growing in water or saturated soil.

Hyphae: the fine thread-like filaments from the main body of a fungus.

IAA: indoleacetic acid; a substance that acts as a growth hormone or auxin in plants, where it controls cell enlargement and, through interaction with other plant hormones, influences cell division.

Ice nucleators: any detritus or such agent in the gut of an invertebrate onto which ice crystals may form.

Ice worms: a group of segmented worms, or oligochaetes, belonging to the genus, *Mesenchytraeus*.

Infiltration: percolation of water into the soil.

Insects: a group of arthropods with three pairs of legs.

Interception: retention of snowfall by the leaves and branches of vegetation, most significantly by evergreen trees.

Interception efficiency: ratio of the amount of snow intercepted by a plant canopy to that which fell as snowfall.

Intranivean: within the snow cover.

Irreducible water content: the amount of water in snow that can be retained by capillary forces.

Isothermal: description of a snowpack, all of which is at the same temperature, usually 0°C.

Krummholz: from the German term for bent wood; eroded, mostly prostrated growth form of a tree. Krummholz is representative of trees that are subjected to climatic stress. In mountainous terrain the trees are often found in a characteristic zone at the limit of tree growth.

Kryokonite: an aggregation of fine micaceous particles with algae, protozoa, rotifers, tardigrades, pollen grains, and detritus found on alpine snow surfaces.

La Niña: a part of the so-called Southern Oscillation characterised by anomalous cold water off the Peruvian coast, associated with particular climate patterns in some other parts of the Earth.

Latent heat: the heat released or consumed upon phase change, expressed as energy per unit mass.

358

Latent (turbulent) heat transfer: the latent energy flux caused by a phase change at the snow surface and the subsequent or prior movement of water vapour via turbulent transfer in the atmosphere; associated with sublimation, evaporation, and condensation.

Layering: a means of plant reproduction by which a tree branch or a plant stalk may be covered by snow or soil, which then sprouts shoots through the material in which the parent plant is buried.

Leader stem: a stem that dominates all other shoots that develop in a tree.

Leaf-area index: the cumulative plan area of all needles, leaves, and stems of a canopy per unit area of ground.

Leaf stomatal conductance: the inverse of resistance to diffusion and directly proportional to transpiration. Calculation of conductance indicates how open or closed stomata are at a given time.

Leaf-water potential: the chemical potential of water in the leaf compared with the chemical potential of pure water at atmospheric pressure and at the same temperature. Solutes in the plant water raise the leaf-water potential, reducing the susceptibility to drought.

Lichens: a symbiotic association between fungi and photosynthetic algae or cyanobacteria.

Lignification: impregnation of lignin in wood cell walls.

Limiting factor: a substance or an environmental condition that is a prerequisite for plant growth.

Lipids: mainly nonpolar hydrocarbons, which may be associated with polar end groups.

Liquid-like layer: a very thin layer of liquid-like or quasi-liquid water that surrounds ice crystals even at temperatures well below freezing; its thickness increases with temperature.

Long-wave radiation: thermal infrared radiation emitted by all materials depending on their temperature; this is readily absorbed by snow.

Master chronology: the tree-ring series of the average population of trees that grow in the same conditions, thus expressing the common growth pattern for this population. Master chronologies mostly reflect the climatic signals recorded in tree rings.

359

Matrix flow: flow of meltwater through the fine-grained porous medium of the snowpack between flow fingers.

Mechanical damage: injury to a tree caused by an external physical agent.

Meiosis: a cell division in which the chromosome number is halved from $2N$ to $1N$ and four cells or products are usually produced.

Melt: the phase change from solid to liquid.

Mesophile (mesophilic): an organism that grows optimally at a temperature between a thermophile and a psychrophile.

Mesotopographic gradient: a conceptual gradient of soils, vegetation, and other factors associated with hill slopes.

Metamorphism: the process of change in snow crystal form that occurs within the snowpack under the influence of a variety of thermodynamic processes.

METEOSAT: an acronym for a meteorological satellite in a geostationary orbit relative to the earth's surface.

Microwave: radiation in the 1- to 10-cm range.

Mitosis: a cell division in which the chromosome number remains the same (either $1N$ or $2N$) and two cells are produced.

Mollisol: an order of soils with deep, dark, relatively fertile topsoil formed under grassland vegetation.

Monsoon: a strong seasonal reversal of windflow; in the case of the Indian monsoon, it is the summer rain-bearing southerly flow that is most often referred to as the monsoon.

Morphology (morphological): the shape or form of a inanimate object (e.g., snow crystal) or an organism.

Mutation: a change in the nucleotide sequence in the molecular structure of DNA, which may be hereditable.

Mycorrhizae: the joint or dual organs of absorption formed by the symbiotic association between the mycelium of a fungus and the roots of a plant.

NDVI: normalized difference vegetation index; a set of algorithms used to extract spectral information related to vegetation from remotely sensed data. NDVI is equal to the term $(NIR - R)/(NIR + R)$, where NIR is the spectral reflectance in the near-infrared band ($0.725-1.1$ μm), where light scattering from the canopy dominates, and R is the reflectance in the red chlorophyll-absorbing portion of the spectrum (0.58–0.68 μm). NDVI is related to the amount of illuminated chlorophyll in the plant canopy and is often used as an index of total green biomass or leaf-area index.

Negative geotropism: the tendency of plants to grow in the direction opposite to the force of gravity.

Nival: descibes an organism growing in snow or active on the snow surface (see also *intranivean, subnivean, supranivean*).

Nonobligate cryophile: an organism that is normally found in snow or ice but, to survive in soil or in water, has a greater temperature range for growth than a psychrophile.

Nonshivering thermogenesis: a physiological adaptation in small mammals that allows them to maintain their body temperature via the metabolism of brown adipose tissue.

Nucleic acids (DNA and RNA): a polymer composed of repeating nucleotide units (phosphates, sugars and bases).

Oligotrophic: an aquatic system with low nutrient levels.

Orographic effect: the uplift of air over high land reduces air mass temperature, leading to greater condensation of water vapour and, often, enhanced precipitation.

Orthotropic growth: vertical growth toward the sky.

Overramified stem: a tree stem that has been affected by terminal shoot death on repeated occasions at the same level (generally the snow–air interface). At this position, the lateral buds develop to compensate for the frequent mortality of the main stem and create a dense mat of branches.

Palaeoclimatic: state of the climate at some time in the past.

Paludification: conversion of previously dry terrain to a wetland through a process of plant succession and change in the hydrological regime.

Papilla: an extended area on a cell located between flagella; it is often used in the mating process.

Paramylon: a carbohydrate food reserve of repeating units of β-glucose found in euglenoids.

Parasite: an organism that obtains its nutrients directly from its host.

Partial correlation analysis: a technique that allows the correlation between a

361

dependent and an independent variable to be determined, while allowing for the correlation between the dependent variable and one or more other variable(s).

Pellicle: the protein outer covering of euglenoids.

Periplast: the protein outer covering of cryptomonads.

Permafrost: ground that remains frozen for at least one complete summer. Permafrost is defined on the basis of temperature ($<0°C$) and not on the basis of whether ice is present or not. The upper surface of permafrost is known as the permafrost table; the layer of ground above the table, which freezes and thaws each year, is known as the active layer.

pH: a measure of free hydrogen ions in solution.

Phenology: the study of the time of appearance of characteristic periodic events in the life cycles of organisms in nature and how these events are influenced by environmental factors, such as temperature, latitude, and altitude (e.g., flowering and leaf fall in plants).

Phenotype: the external visible shape of an organism corresponding to its architecture that can change because of external factors.

Phloem: the soft tissues located just below the tree bark. The phloem is the site of transport and redistribution of the sap enriched by the products of photosynthesis coming from the leaves.

Photosynthates: metabolic chemical species synthesised during photosynthesis.

Photosynthesis: the trapping of sunlight energy and its conversion to chemical energy to synthesise carbohydrates.

Phototaxis: movement in response to the direction of light.

Phylogeny: the history of evolutionary relationships.

Phytomass: total mass of plants, including dead attached parts per unit area (e.g., $g\ m^{-2}$; $kg\ ha^{-1}$) at a given time. It differs from biomass by including dead material.

Pitfall trap: a dry or fluid–filled cup-like trap set into the soil so that the upper lip is flush with the soil surface. It is used to collect small vertebrates and invertebrates.

Plagiotropic: horizontal growth.

Planozygote: a $2N$ (diploid) cell with flagella formed from the fusion of two gametes.

362

Podsolisation: displacement of iron and aluminium in soils, forming a distinctive leached soil horizon often above a reddish horizon of iron accumulation.

Polar anticyclone: a high sea-level pressure area, which often characterises polar latitudes, with slowly subsiding air.

Pollen: male reproductive phases in the life cycle of seed plants.

Polysaccharide: a polymer composed of repeating sugar units.

Porosity: the pore space per unit volume of a porous medium.

Postfire stand: a tree stand that develops after fire. The age structure of postfire stands is said to be even-aged.

Primary production: the biomass or energy incorporated into an ecosystem by the photosynthetic and chemosynthetic activity of plants, usually measured over a period of time (e.g., kg ha^{-1} year^{-1}).

Primary productivity: rate at which primary producers (photosynthetic organisms) capture and store energy from sunlight.

Prostrate form: with respect to the effect of snow, prostrate means the low growth form of a tree that develops because of exposure of the stem to extreme cold and abrasion by snow crystals at the snow–air interface. The tree remains prostrate because of the difficulty of growing above this interface.

Protein: a polymer composed of repeating amino acid units.

Psychrophile (psychrophilic): describes snow algae that grow optimally below 5°C, show abnormal growth at 10°C, and do not survive above 10°C (Hoham, 1975a, 1975b). The term also applies to bacteria that grow optimally below 16°C with an upper limit at 20°C (Morita, 1975).

Psychrotrophic: describes bacteria that can grow at 0°C but grow optimally at 20°C to 25°C (Morita, 1975).

Pukak: see *depth hoar*.

Pyrenoid: a protein structure associated with starch.

Radial growth: growth of a stem in diameter.

Radiation balance: balance between incoming and outgoing all-wave radiation.

Reaction wood: see *compression wood*.

Rebound: degree to which falling snow crystals bounce when they land.

Respiration (aerobic): the breakdown of sugars using an oxygen dependent pathway with ATP formation and the release of carbon dioxide.

Rhizopodium: a long thin cytoplasmic extension used for food capture by microorganisms.

Rhizosphere: the area of soil immediately surrounding plant roots, which is altered by their growth, respiration, and exchange of nutrients.

Rotifers: small microscopic bilateral animals with ciliated lobes at their head end and with a false body cavity.

Saltation: the skipping motion of blowing snow particles traveling just above the snowpack.

Saprophytic: describes a plant that is incapable of synthesising nutrients from inorganic matter and therefore must obtain food from dead and decaying organic matter.

Saturated hydraulic conductivity: the degree to which a saturated porous medium will conduct liquid flow when a vertical gradient of hydraulic head exists.

Scar: the reaction of a tree to physical damage of the cambium. A scar is made of central dead tissues covered by callous tissues.

Sea-level pressure: atmospheric pressure normalised to sea level, even when measured over land, in order to aid the identification of weather systems.

Seedling: the early stage of tree development following germination.

Sensible (turbulent) heat transfer: also called sensible convective transfer. It is the sensible energy flux caused by the movement of atmospheric heat via turbulent transfer in the atmosphere.

Setae: the sensory bristle-like hairs of invertebrates and other animals.

Shear stress: the force exerted by the wind on snow surfaces.

Shortwave radiation: visible light plus some of the near-infrared band. This radiation carries much of the solar heat received at the surface and can penetrate snow for a short distance, although most of it is reflected by snow.

Snow course: a series of snow sampling points across a snow cover or snow field usually spaced at equal intervals along a straight line.

Snow depth: the observed depth of snow on the ground.

Snow flush: pertaining to an area downslope from a snowbed, which receives meltwater during much of the summer.

Snow management: management of vegetation to produce distinctive snow accumulation patterns.

Snow storage capacity: the capacity of a canopy to further intercept snow (e.g., maximum canopy interception less the present amount intercepted).

Snowfall regime: a particular seasonal pattern (or climate) of snowfall.

Snowpack: the accumulation of snow at a given site.

Snowpatch: the snow remaining on the ground close to the end of snowmelt. Snow patches are the result of slowly melting snow due to shading or areas of high accumulation. Because of the short growing season of snowpatches, chionophilous flora develops. Chionophilous flora refers to plants that have the capability of completing their life cycle in a very short period of time and that support the cold and wet conditions due to the persistence of snowpatches.

Soil moisture tension: describes 1) the ability of soil to adhere to water and prevent percolation. 2) The molecular force that the surface areas of organic and inorganic soil particles exert on water. At field capacity the forces that hold the water in the soil are equal to the force of gravity, thereby stopping the downward percolation of water; at the wilting point the forces holding the water in the soil are so strong that the plants cannot extract it.

Solifluction: a downslope movement (flowing soil) of earth materials resulting from water-saturated soils. The process prevents development of typical soil profiles and influences development of plant cover.

Spectral reflectance: the physical property of all materials to reflect wavelengths of the light spectrum. Because of the unique spectral signatures of various materials, measurement of this property is used to characterise and identify aspects of surface images by remote sensing.

SPOT: Système Probatoire de l'Observation de la Terre; an Earth-resource satellite with high-resolution sensors launched by France in January 1986. The satellite possesses two sensing modes: i) panchromatic black and white with a resolution of 10 m over the range 0.51–0.73 μm and ii) multispectral with 20-m resolution in three channels (0.50–0.59 μm, 0.61–0.68 μm, and 0.79–0.89 μm).

Starch (true): complex carbohydrate composed of repeating units of the sugar α-glucose.

Strength: degree to which a material maintains its structural integrity when a force is applied.

365

Stubble field: an agricultural field with trimmed stems of the previous summer's crop; wheat stubble is often about 25 cm tall.

Sublimation: phase change from solid to vapour.

Subnivean: under the snow surface.

Sugar storage: accumulation of carbohydrates in plant tissue before the dormant period.

Suncup: a depression found in snow caused by more rapid snowmelt from sunlight due to the presence of particulate materials such as dust, pollen, or snow algae.

Supercooling: being able to cool invertebrate haemolymph to below its freezing point via an accumulation of cryoprotective compounds.

Supercooling point: a physiologically adapted minimum temperature below the freezing point of haemolymph and above which invertebrates do not freeze.

Supranivean: above the snow surface.

Suspension: the motion of blowing snow particles that are lifted by turbulence and carried by the wind.

SWE: snow water equivalent; the equivalent depth of water on the ground that would form if a snow cover were melted and if the meltwater did not run off or infiltrate. SWE is expressed as depth but has implied units of depth per unit area.

Symbiosis: a situation in which dissimilar organisms live together in close association. As originally defined, the term embraces all types of mutualistic and parasitic relationships.

Synoptic circulation: a pattern of winds, and associated weather, observed on a time scale of about a day and over a space scale of 1,000–2,000 km (e.g., a midlatitude depression or cyclone).

Synoptic-scale trough: an area, approximately 1,000- to 2,000-km horizontal scale, of low pressure, observable on a time scale of 1–3 days.

Systematics: the study of the patterns in life's diversity through taxonomy, phylogenetic reconstruction and classification.

Tardigrades: small microscopic unsegmented animals with four pairs of unjointed legs armed with claws and with two eyespots on their heads.

Taxonomy: the naming and identifying of species.

Temperature gradient metamorphism: metamorphism driven by temperature

366

gradients across the snowpack. Water vapour diffuses from warmer surfaces of crystals to colder ones, resulting in the sublimation and recrystallisation of ice and changes in crystal size and form.

Temperature index snowmelt models: semiempirical numerical schemes that use the average daily temperature in excess of the melting point to calculate daily snowmelt. These models work best where melt is dominated by long-wave or sensible heat fluxes.

Thermal conduction: heat flow without physical mixing of the material.

Thermal conductivity: the degree to which a material conducts heat along a temperature gradient.

Thermal hysteresis: the difference between the melting and freezing points of invertebrate haemolymph due to accumulated proteins.

Thermal preferendum (or thermal preference): the temperature range preferred by a species of animal.

Thermal resistivity: the degree to which a material, or mixture of materials, resists the flow of heat through it when a temperature gradient is applied.

Thermodynamic equilibrium: the balance between energy inputs and outputs, internal energy storage, internal phase change, and mass inputs and outputs.

Thermogenesis: metabolic generation of increased body temperature in endothermic animals.

Thermokarst: land with discontinuous permafrost. Some of the permafrost mounds may melt to create ponds that are similar to those made in karstic (calcareous or dolomitic) landscapes.

Thermoneutral zone: the range of temperature at which endothermic animals do not have to increase their food intake or metabolic rate to maintain their body temperature.

Thermophile (thermophilic): an organism that is tolerant of high temperatures, typical of thermal hot springs.

Threshold wind speed: the wind speed that is required to start or maintain blowing snow saltation. It is generally measured at a height of 10 m.

Tibiotarsae: the lower legs, feet, and claws of invertebrates.

Tree line: the extreme geographical position of erect trees. The northern tree line is south of the conifer species limit.

367

UV-A, UV-B, and UV-C: three designated types of ultraviolet light; A has a wavelength range of 0.32–0.40 μm and is harmless to plants; B has a wavelength range of 0.28–0.32 μm and can be harmful to the growth of plants; C has a wavelength range of 0.22–0.28 μm and can be destructive to DNA and harmful to plants and microbes.

VAM fungi: vesicular-arbuscular-mycorrhizal fungi; endomycorrhizal fungi that penetrate the wall but not the plasma membrane of cells in the cortex of plant roots. They are members of the zygomycete fungal group. About 70 percent of land plants are associated with this type of fungi. VAM fungi are now called arbuscular-mycorrhizal (AM) fungi.

Ventilation shafts: a more or less vertical shaft in the snow cover built by small mammals when snow density increases causing restricted airflow and increased subnivean concentrations of carbon dioxide.

Vitamins: organic molecules required by some algae for growth (auxotrophy), which include biotin, thiamine and cyanocobalamin.

Volvocales: an order of green algal flagellates, which is very common in snow.

Water deficit: the lack of water in plants. At the ecosystem level, the water deficit represents the difference between precipitation and runoff.

WDC: World Data Center for Glaciology; a centre that holds and distributes snow and ice data. There are four centres: WDC-A, University of Colorado, USA; WDC-B1, Russian Academy of Sciences, Moscow, Russia; WDC-C, Scott Polar Research Institute, Cambridge, UK; WDC-D, Chinese Academy of Sciences, Lanzhou, China.

Wet snow metamorphism: the process of change in grain shape in a snow and water mixture. Ice melts at the sharpest points of snow crystals and water refreezes along smooth concave surfaces because of the dependence of the freezing point on the curvature of the snow–water interface.

Winter active: animals active during winter months, often under the snow cover.

Xylem: the principal water-conducting tissue and support system of higher plants.

Yeast: single-celled fungi that belong primarily to the sac fungi, which reproduce asexually through budding.

Yellow–green algae: photosynthetic protists lacking brown pigments but have chlorophylls a and c and β-glucose food reserves called chrysolaminarin. Cells are usually surrounded by a carbohydrate cell wall.

Zoospore: an asexual cell with flagella that can develop into a vegetative cell that can further divide by mitosis to form more zoospores.

Zygospore: a $2N$ (diploid) zygote, usually with a thick wall, that goes into prolonged dormancy.

Zygote: a $2N$ (diploid) cell formed from the fusion of two gametes.

References

Hoham, R.W. 1975a. Optimum temperatures and temperature ranges for growth of snow algae. *Arctic Alpine Res.*, **7**, 13–24.

Hoham, R.W. 1975b. The life history and ecology of the snow alga *Chloromonas pichinchae* (Chlorophyta, Volvocales). *Phycologia*, **14**, 213–226.

Morita, R.Y. 1975. Psychrophilic bacteria. *Bacteriol., Rev.*, **39**, 144–167.

Index

371